Advances in Biochemistry in Health and Disease

Volume 19

Series Editor

Naranjan S. Dhalla, Institute of Cardiovascular Sciences, St. Boniface Hospital, Winnipeg, MB, Canada

Advances in Biochemistry in Health and Disease focus on the latest developments in biochemical research with implications for health and disease. This book series consists of original edited volumes and monographs, presented by leading experts in the field and provides an up to date and unique source of information for all those interested in the fundamental, biochemical processes of the latest and emerging topics and techniques.

Covering a wide variety of topics, this book series is a valuable source of information from those at the lab bench through to the Health Care workers.

More information about this series at http://www.springer.com/series/7064

Paramjit S. Tappia · Bram Ramjiawan ·
Naranjan S. Dhalla
Editors

Pathophysiology of Obesity-Induced Health Complications

Springer

Editors
Paramjit S. Tappia
Asper Clinical Research Institute
and Office of Clinical Research
St. Boniface Hospital
Winnipeg, MB, Canada

Bram Ramjiawan
Asper Clinical Research Institute
and Office of Clinical Research
St. Boniface Hospital
Winnipeg, MB, Canada

Naranjan S. Dhalla
Institute of Cardiovascular Sciences
St. Boniface Hospital Albrechtsen
Research Centre
Winnipeg, MB, Canada

Advances in Biochemistry in Health and Disease
ISBN 978-3-030-35360-5 ISBN 978-3-030-35358-2 (eBook)
https://doi.org/10.1007/978-3-030-35358-2

This Springer imprint is published by the registered company Springer Nature Switzerland AG
The registered company address is: Gewerbestrasse 11, 6330 Cham, Switzerland

Preface

Overweight and obesity are defined as abnormal or excessive fat accumulation that may impair health. Although obesity is preventable, according to the World Health Organization the global incidence of obesity has alarmingly nearly tripled since 1975. It is a global epidemic. Indeed, in 2016, more than 1.9 billion adults were overweight and 650 million of those were obese. The fundamental cause of obesity and overweight is an imbalance between caloric intake and expenditure. It is evident that the global trend has been moving towards an increased intake of energy-dense foods that are high in fat, coupled with a decrease of physical activity due to the increasingly sedentary lifestyles of the general population.

The risk of occurrence of obesity-induced health complications increases exponentially as body mass index increases. Accordingly, this book has compiled comprehensive information and new insights into the biochemistry and pathophysiology of the increase in the risk of obesity-associated non-communicable diseases—many of which increase the risk for premature death or can lead to serious chronic conditions that reduce the quality of life.

This book brings together an outline of the link between obesity and the onset of many adverse health conditions including diabetes (metabolic dysregulation), cardiovascular disease, mental health disorders, cancers, sleep disorders, liver and kidney disease and musculoskeletal disorders. The development of these diseases may be consequences of obesity-induced changes in cellular functioning in response to a pro-inflammatory environment, endocrine/metabolic dysfunction, activation of the autonomic nervous system, oxidative stress and impairment of several signal transduction pathways. Furthermore, we dedicate a section on some preventative strategies to combat obesity and related health complications.

While the major focus of this book is on the health consequences of obesity in adulthood, this book will also highlight the increase in obesity prevalence among children. This is of a major public health concern as obesity-related diseases not previously seen in children and adolescents are now of common occurrence.

The contributors to this book are international leading experts on obesity and associated health complications and highly knowledgeable within their respective areas of investigation. This book is also uniquely positioned to provide comprehensive perspectives on the biochemistry and pathophysiology of the effects of obesity. There are 19 chapters in 4 different parts in this book, comprising of Part I: General Aspects of Obesity and Health Complications (3 chapters); Part II: Metabolic Disturbances and Inflammation due to Obesity (5 chapters); Part III: Neurological and Visceral Complications due to Obesity (7 chapters) and Part IV; Strategies for the Prevention of Obesity-Induced Complications (4 chapters). Each section underscores the multifactorial nature of co-morbid conditions and complications.

A key intent of this volume is also to highlight the compendium of adverse consequences of obesity to human health and to provide current understanding of the cellular and biochemical mechanisms of obesity-induced health complications that will be of value not only to healthcare professionals, but will stimulate and motivate biomedical researchers and scientists to discover ways to prevent the epidemic of obesity and associated adverse health outcomes. Furthermore, this book will serve as a highly useful multidisciplinary resource for medical students, fellows, residents, and graduate students.

In summary, this monograph covers a broad range of topics related to the mechanisms of obesity-induced health complications. We hope that readers will understand and appreciate that obesity is significantly linked to increased risk and occurrence of lethal and non-lethal diseases and complications. Furthermore, the underlying message presented in the monograph is that the cause of obesity-related disorders is multifactorial and understanding the complexity and the full spectrum of the mechanisms involved may contribute to the development of novel interventions for the prevention and treatment of Obesity associated co-morbidities.

Winnipeg, Canada

Paramjit S. Tappia
Bram Ramjiawan
Naranjan S. Dhalla

Contents

Contributors

Sevginur Akdas Interdisciplinary Food, Metabolism and Clinical Nutrition Department, Institute of Health Sciences, Ankara University, Ankara, Turkey

Sreyoshi F. Alam College of Medicine, Alfaisal University, Riyadh, Saudi Arabia

Reena Badhwar Department of Pharmaceutics, Delhi Pharmaceutical Science and Research University, New Delhi, India

Satish Balan Department of Nephrology, Kerala Institute of Medical Sciences, Anayara, Trivandrum, India

Shalini Behl School of Forensic and Applied Sciences, University of Central Lancashire, Preston, Lancashire, UK

Lindsay Brown Functional Foods Research Group, University of Southern Queensland, Toowoomba, QLD, Australia;
School of Health and Wellbeing, University of Southern Queensland, Toowoomba, QLD, Australia

Harpal S. Buttar Department of Pathology and Laboratory Medicine, Faculty of Medicine, University of Ottawa, Ottawa, ON, Canada

Benjamin Bydak Faculty of Kinesiology and Recreation Management, University of Manitoba, Winnipeg, MB, Canada;
Children's Hospital Research Institute of Manitoba (CHRIM), Winnipeg, MB, Canada

Pascale Chavatte-Palmer UMR BDR, INRA, ENVA, Université Paris Saclay, Jouy en Josas, France

Dinesh Kumar Chellappan Department of Life Sciences, School of Pharmacy, International Medical University, Bukit Jalil, Kuala Lumpur, Malaysia

Trudi Collet Inovative Medicines Group, Institute of Health & Biomedical Innovation, Queensland University of Technology, Kelvin Grove, Brisbane, QLD, Australia

Kelsey H. Collins Department of Orthopaedic Surgery, Washington University at St. Louis, St. Louis, MO, USA

Shuchi H. Dave Institute of Pharmacy, Nirma University, Ahmedabad, India

Danielle Defries Department of Kinesiology and Applied Health, University of Winnipeg, Winnipeg, Manitoba, Canada

Kamal Dua Discipline of Pharmacy, Graduate School of Health, University of Technology Sydney, Ultimo, NSW, Australia;
Centenary Institute, Royal Prince Alfred Hospital, Camperdown, NSW, Australia;
School of Biomedical Sciences and Pharmacy, Priority Research Centre for Healthy Lungs, Hunter Medical Research Institute (HMRI), University of Newcastle, Callaghan, NSW, Australia

Harish Dureja Department of Pharmaceutical Sciences, Maharishi Dayanand University, Rohtak, Haryana, India

Yaqoot Fatima Centre for Rural and Remote Health (Mount Isa), James Cook University, Mount Isa, QLD, Australia

Vera Francisco SERGAS (Servizo Galego de Saude) and IDIS (Instituto de Investigación Sanitaria de Santiago), The NEIRID Group (Neuroendocrine Interactions in Rheumatology and Inflammatory Diseases), Santiago University Clinical Hospital, Santiago de Compostela, Spain

Paul K. Ganguly College of Medicine, Alfaisal University, Riyadh, Saudi Arabia

Rodolfo Gómez Musculoskeletal Pathology Group, SERGAS (Servizo Galego de Saude) and IDIS (Instituto de Investigación Sanitaria de Santiago), Research Laboratory 9, Santiago University Clinical Hospital, Santiago de Compostela, Spain

Miguel Angel Gonzalez-Gay Epidemiology, Genetics and Atherosclerosis Research Group on Systemic Inflammatory Diseases, Universidad de Cantabria and IDIVAL, Hospital Universitario Marqués de Valdecilla, Santander, Spain

Ramesh K. Goyal Delhi Pharmaceutical Sciences & Research University, Delhi, India

Oreste Gualillo SERGAS (Servizo Galego de Saude) and IDIS (Instituto de Investigación Sanitaria de Santiago), The NEIRID Group (Neuroendocrine Interactions in Rheumatology and Inflammatory Diseases), Santiago University Clinical Hospital, Santiago de Compostela, Spain

Monica Gulati School of Pharmaceutical Sciences, Lovely Professional University, Phagwara, Punjab, India

Gaurav Gupta School of Pharmaceutical Sciences, Jaipur National University, Jagatpura, Jaipur, India

David A. Hart Human Performance Laboratory, Faculty of Kinesiology, c/o McCaig Institute for Bone & Joint Health, University of Calgary, Calgary, AB, Canada;
Department of Surgery, McCaig Institute for Bone & Joint Health, University of Calgary, Calgary, AB, Canada;
Bone & Joint Health Strategic Clinical Network, Alberta Health Services, Edmonton, AB, Canada

Walter Herzog Human Performance Laboratory, Faculty of Kinesiology, c/o McCaig Institute for Bone & Joint Health, University of Calgary, Calgary, AB, Canada;
Department of Surgery, McCaig Institute for Bone & Joint Health, University of Calgary, Calgary, AB, Canada

V. S. Jayalekshmi Cardiovascular Diseases and Diabetes Biology, Rajiv Gandhi Centre for Biotechnology, Thiruvananthapuram, India

A. Kalaivani Department of Biochemistry, Centre for Biological Sciences, K. S. Rangasamy College of Arts and Science (Autonomous), Tiruchengode, Tamilnadu, India

Sreelekshmi Kamaladevan Division of Clinical Research, Kerala Institute of Medical Sciences, Anayara, Trivandrum, India

Chandrasekharan C. Kartha Society for Continuing Medical Education and Research, Kerala Institute of Medical Sciences, Anayara, Trivandrum, India

Ginpreet Kaur Department of Pharmacology, SPP School of Pharmacy & Technology Management, SVKM's NMIMS, Mumbai, Maharashtra, India

Santosh Kumar College of Medicine, Alfaisal University, Riyadh, Saudi Arabia

Francisca Lago Molecular and Cellular Cardiology Group, SERGAS (Servizo Galego de Saude) and IDIS (Instituto de Investigación Sanitaria de Santiago), Research Laboratory 7, Santiago University Clinical Hospital, Santiago de Compostela, Spain

Abdullah Al Mamun Institute for Social Science Research, University of Queensland, Brisbane, QLD, Australia

Meenu Mehta School of Pharmaceutical Sciences, Lovely Professional University, Phagwara, Punjab, India

Antonio Mera SERGAS (Servizo Galego de Saude), Division of Rheumatology, Santiago University Clinical Hospital, Santiago de Compostela, Spain

Delphine Mitanchez INSERM UMR _S 938, Saint Antoine Research Center, Sorbonne University, Paris, France;
Department of Neonatology, Bretonneau Hospital, CHRU, Tours, France

Praveen Murlidharan Department of Nephrology, Kerala Institute of Medical Sciences, Anayara, Trivandrum, India;
Division of Clinical Research, Kerala Institute of Medical Sciences, Anayara, Trivandrum, India

Long T. Nguyen Renal Medicine, Kolling Institute Level 9, Royal North Shore Hospital, The University of Sydney, St. Leonard, NSW, Australia

Patience Oluchukwu Obi Faculty of Kinesiology and Recreation Management, University of Manitoba, Winnipeg, MB, Canada;
Children's Hospital Research Institute of Manitoba (CHRIM), Winnipeg, MB, Canada

Kavita Pabreja Priority Research Centre for Healthy Lungs, The University Of Newcastle, Newcastle, NSW, Australia

Sunil K. Panchal Functional Foods Research Group, University of Southern Queensland, Toowoomba, QLD, Australia

Parijat Pandey Shri Baba Mastnath Institute of Pharmaceutical Sciences and Research, Baba Mastnath University, Rohtak, India

Bhoomika M. Patel Institute of Pharmacy, Nirma University, Ahmedabad, India

Jesús Pino SERGAS (Servizo Galego de Saude) and IDIS (Instituto de Investigación Sanitaria de Santiago), The NEIRID Group (Neuroendocrine Interactions in Rheumatology and Inflammatory Diseases), Santiago University Clinical Hospital, Santiago de Compostela, Spain

Carol A. Pollock Renal Medicine, Kolling Institute Level 9, Royal North Shore Hospital, The University of Sydney, St. Leonard, NSW, Australia

Harvinder Popli Department of Pharmaceutics, Delhi Pharmaceutical Science and Research University, New Delhi, India

Surya Ramachandran Cardiovascular Diseases and Diabetes Biology, Rajiv Gandhi Centre for Biotechnology, Thiruvananthapuram, India

Matthew Ramjiawan Asper Clinical Research Institute & Office of Clinical Research, St. Boniface Hospital, Winnipeg, MB, Canada

Raylene A. Reimer Human Performance Laboratory, Faculty of Kinesiology, c/o McCaig Institute for Bone & Joint Health, University of Calgary, Calgary, AB, Canada;
Department of Biochemistry & Molecular Biology, University of Calgary, Calgary, AB, Canada

Jaqueline L. Rios Human Performance Laboratory, Faculty of Kinesiology, c/o McCaig Institute for Bone & Joint Health, University of Calgary, Calgary, AB, Canada

Clara Ruiz-Fernández SERGAS (Servizo Galego de Saude) and IDIS (Instituto de Investigación Sanitaria de Santiago), The NEIRID Group (Neuroendocrine Interactions in Rheumatology and Inflammatory Diseases), Santiago University Clinical Hospital, Santiago de Compostela, Spain

Sonia Saad Renal Medicine, Kolling Institute Level 9, Royal North Shore Hospital, The University of Sydney, St. Leonard, NSW, Australia

Adeel Safdar Hamilton, ON, Canada

Saurabh Sajita School of Pharmaceutical Sciences, Lovely Professional University, Phagwara, Punjab, India

Ayesha Saleem Faculty of Kinesiology and Recreation Management, University of Manitoba, Winnipeg, MB, Canada;
Diabetes Research Envisioned and Accomplished in Manitoba (DREAM) Theme, Winnipeg, MB, Canada;
Developmental Origins of Chronic Diseases in Children Network (DEVOTION), Winnipeg, MB, Canada;
Biology of Breathing Research Theme, Winnipeg, MB, Canada;
Children's Hospital Research Institute of Manitoba (CHRIM), Winnipeg, MB, Canada

G. Saravanan Department of Biochemistry, Centre for Biological Sciences, K. S. Rangasamy College of Arts and Science (Autonomous), Tiruchengode, Tamilnadu, India

V. V. Sathibabu Uddandrao Department of Biochemistry, Centre for Biological Sciences, K. S. Rangasamy College of Arts and Science (Autonomous), Tiruchengode, Tamilnadu, India

Jaipaul Singh School of Forensic and Applied Sciences, University of Central Lancashire, Preston, Lancashire, UK

Avinash Kumar Singh Department of Pharmacology, School of Pharmaceutical Education and Research, Jamia Hamdard, New Delhi, India

Sachin Kumar Singh School of Pharmaceutical Sciences, Lovely Professional University, Phagwara, Punjab, India

Timothy Skinner Department of Psychology, University of Copenhagen, Copenhagen, Denmark

Isabella So Department of Internal Medicine-Molecular Medicine, Wake Forest School of Medicine, Winston-Salem, NC, USA

Paramjit S. Tappia Asper Clinical Research Institute & Office of Clinical Research, St. Boniface Hospital, Winnipeg, MB, Canada

S. Vadivukkarasi Department of Biochemistry, Centre for Biological Sciences, K. S. Rangasamy College of Arts and Science (Autonomous), Tiruchengode, Tamilnadu, India

Ridhima Wadhwa Faculty of Life Sciences and Biotechnology, South Asian University, Chanakyapuri, New Delhi, India

Deepika Yadav Department of Pharmaceutics, Delhi Pharmaceutical Science and Research University, New Delhi, India

Hariom Yadav Department of Internal Medicine-Molecular Medicine, Wake Forest School of Medicine, Winston-Salem, NC, USA

Nuray Yazihan Interdisciplinary Food, Metabolism and Clinical Nutrition Department, Institute of Health Sciences, Ankara University, Ankara, Turkey;
Internal Medicine, Pathophysiology Department, Faculty of Medicine, Ankara University, Sihhiye, Ankara, Turkey

Part I
General Aspects of Obesity and Health Complications

Chapter 1
Prevalence, Consequences, Causes and Management of Obesity

Paramjit S. Tappia and Danielle Defries

Abstract Obesity is a growing global health problem and is well-recognized to be a major contributing factor for increased risk of several non-communicable diseases including cardiovascular disease, diabetes and cancer in both the developed and developing world. This development is multi-factorial, but an increasingly sedentary lifestyle coupled with unhealthy dietary practices are key risk factors. Effective interventions for weight management would therefore not only be seen to reduce the epidemic of obesity, but also to lessen the risk for obesity-related morbidities. This article will briefly describe some factors that can cause obesity. Since men and women are different in their fat mass and distribution profile, and that ethnic groups are disproportionately affected by obesity, it is conceivable that disparities also exist in the occurrence of obesity and the consequential development of non-communicable diseases. Although the major adverse health outcomes due to obesity are mentioned, the influence and the role of sex, specifically women's health, and ethnicity in the increased risk as well as development of obesity-induced health complications will also be discussed.

Keywords Obesity · Non-communicable diseases · Women's health · Ethnicity · Weight management

---

P. S. Tappia (✉)
Asper Clinical Research Institute, St. Boniface Hospital, CR3129-369 Tache Avenue, Winnipeg, Manitoba R2H 2A6, Canada
e-mail: ptappia@sbrc.ca

D. Defries
Department of Kinesiology and Applied Health, University of Winnipeg, Winnipeg, Manitoba R3B 2E9, Canada

P. S. Tappia et al. (eds.), *Pathophysiology of Obesity-Induced Health Complications*, Advances in Biochemistry in Health and Disease 19,
https://doi.org/10.1007/978-3-030-35358-2_1

Introduction

A healthy body weight is dependent on age and development phase of an individual that is both achievable and sustainable. A healthy body weight would also be associated with normal blood pressure, normal glycemia and circulating lipid levels. The genetic makeup, family history of body type as well as good eating habits and physical activity are additional factors that can influence healthy body weight. The body mass index (BMI) is a well-established method for assessing healthy body weight, however; this ratio does not measure body fat and it is simply a measure of weight to height. On the other hand, body composition is an optimal tool for determining health risk. In this regard, several techniques have been developed, including underweight weighing, the BODPOD, which is an Air Displacement Plethysmograph (ADP) that uses whole-body densitometry to determine body composition (fat vs. lean). Similar in principle to underwater weighing, the BODPOD measures body mass (weight) using a very precise scale, and volume by sitting inside the BODPOD; the bioelectrical impedance analysis (BIA), which is a commonly used method for estimating body composition, in particular body fat and muscle mass. In BIA, a weak electric current flow is sent through the body and the voltage is measured to calculate impedance of the body. Lean tissue, which is >70% water, is a good conductor of electrical current, whereas fatty tissue, which is hydrophobic and low in water, is not. Thus, an increase in body fat is associated with increased electrical resistance combined with a skinfold thickness measure for subcutaneous fat.

A healthy body fat level in men is considered to be between 12 and 18% (3% is essential for insulation and to protect vital organs). For males, a body fat level of >25% would be considered as obese. For women, a healthy body fat level falls between 20 and 25% (12% is considered essential for normal reproductive function), whereas a body fat level between 30 and 35% in women is categorized as obese. Men and women are different in their fat mass and distribution pattern (Fig. 1.1). It should be noted that body fat distribution is determined by genetics and that the level of subcutaneous fat is positively correlated to increased risk of obesity, which is more common in women than men, whereas visceral fat is positively correlated to risk of obesity in men than women. In addition, a waist circumference of >102 cm in men and >88 cm in women is linked to an increase risk of obesity. In this regard, a CT scan would be used to determine levels of visceral fat, but it can also be readily estimated by measuring the waist circumference.

According to the World Health Organization (WHO) [1], worldwide obesity has nearly tripled since 1975. In 2016, more than 1.9 billion adults, 18 years and older, were overweight. Of these over 650 million were obese. Nearly 39% of adults aged 18 years and over were classified as overweight in 2016, and 13% were obese. It is noteworthy that most of the world's population live in countries where being

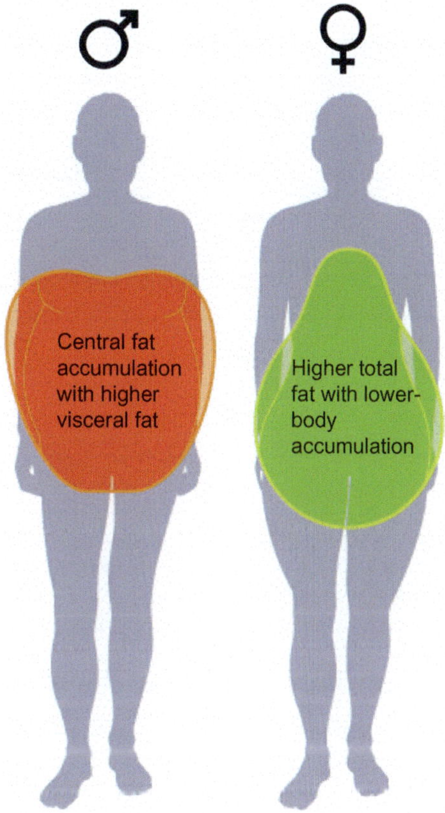

Fig. 1.1 Fat mass and distribution pattern in men and women. Women, compared to men, have higher percent body fat and deposit it in a different pattern, with relatively more adipose tissue in the hips and thighs in women and central obesity typical of men

overweight and obese has a higher mortality rate than malnutrition. It was reported that 41 million children under the age of 5 were overweight or obese in 2016. Furthermore, over 340 million children and adolescents aged 5-19 were overweight or obese in 2016. Most importantly, obesity is preventable. Being overweight and obese is a global epidemic that has been associated with a number of chronic diseases including hypertension, heart disease, diabetes and cancer [1–5]. Figure 1.2 shows the extent of the problem across the globe. It can be seen that obesity is a global concern affecting both developing and developed nations. Indeed, obesity is the most prevalent non-communicable disease in the 21st century. This chapter will briefly describe some factors that can cause obesity and the disparities that exist in its occurrence and consequent development of non-communicable diseases. In addition, the impact of obesity in women and the role of ethnicity in the increased risk as well as development of obesity-induced health complications will also be discussed.

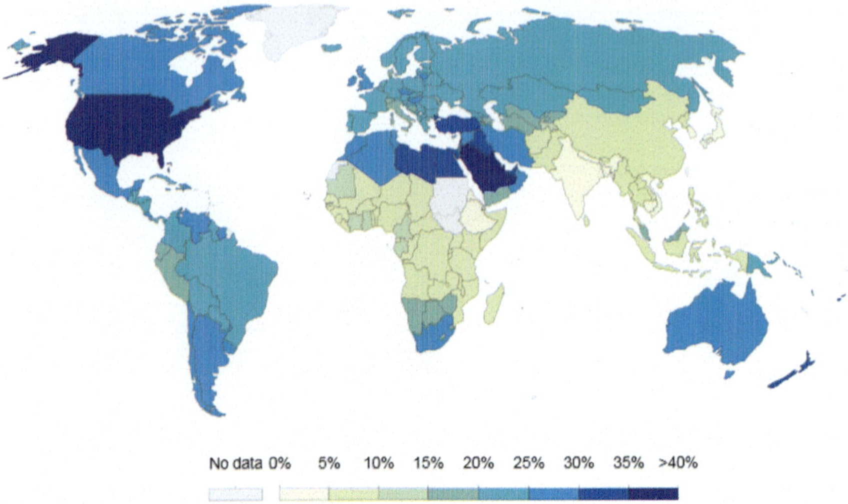

Fig. 1.2 Global distribution of obesity in 2016. Percentage of adults aged 18+ years of age who are defined as obese based on their BMI. A BMI ≥ 30 is defined as obese. body mass. *Source* WHO, Global Health Observatory

Energy Balance and Development of Overweight or Obesity

Energy expenditure is fuelled by the conversion of caloric energy from macronutrients to metabolic energy in the form of ATP. Energy balance occurs when caloric intake matches energy expenditure. When energy is consumed in excess of energy requirements, the extra is stored for later use. However, there is a limited capacity to store excess dietary carbohydrate as glycogen, and there is no storage form for excess protein. When glycogen stores are full, remaining carbohydrate must either be oxidized or metabolized to triglycerides and stored in adipose tissue. Similarly, excess protein remaining after synthesis of body proteins and non-protein nitrogenous compounds must either be oxidized, converted to glucose, or converted to triglycerides. In both situations, excess carbohydrate and protein is converted to triglyceride if immediate energy needs are satisfied. Adipose tissue has an unlimited capacity for triglyceride storage, and can increase to accommodate excess energy by increased sized of adipocytes or synthesis of new adipocytes from adipose progenitor cells [6]. Thus, increases in body weight during times of excess energy intake occur predominantly through expansion of white adipose tissue [7].

Sustained positive energy balance, where energy intake is consistently higher than expenditure, increases the risk of becoming overweight or obese. Several factors can contribute to this imbalance including family history, physical inactivity, and adverse social and behavioral patterns. However, it should be mentioned that such imbalance is necessary in pregnancy, infancy and childhood as well as time of growth. On the other hand, a negative energy balance, where energy intake is less than energy expenditure results in weight loss. Such a situation would be desirable under conditions of being overweight or obese, but would be undesirable, for example, during illness and anorexia.

Regulation of Energy Balance

Energy balance is controlled by the central nervous system and involves several intricate neural pathways, systems, and regions of the brain. Of these regions, the hypothalamus is considered principal in the homeostatic regulation of energy balance in response to peripheral energy stores and energy availability. The melanocortin system, consisting of agouti-related protein (AgRP)/neuropeptide Y (NPY)-expressing neurons and proopiomelanocortin (POMC)/cocaine and amphetamine regulated transcript (CART)-expressing neurons located in the arcuate nucleus, receives information regarding energy status via circulating hormones and relays this information to effector neurons in other areas of the hypothalamus that control energy intake and expenditure [8–10]. Although several hormones interact with this system, the two with the most influence are leptin and ghrelin. Leptin, a hormone produced by adipocytes in proportion to their size, promotes satiety and energy expenditure by activating POMC/CART neurons and inhibiting AgRP/NPY neurons [11]. Ghrelin, produced by endocrine cells in the stomach, triggers hunger and food intake by stimulating AgPR/NPY neurons [12]. Leptin signaling in the hypothalamus also increases energy expenditure through downstream effects on the sympathetic nervous system, heart rate, blood pressure, and brown adipose tissue thermogenesis [13]. Altered production and signaling of both leptin and ghrelin have been implicated in the development and maintenance of obesity [14]. Other hormones involved in regulation satiety and food intake include insulin, peptide YY, cholecystokinin, and glucagon-like peptide-1.

In addition to homeostatic regulation, hedonic mechanisms influence food intake, and can be powerful enough to override homeostatic regulation and promote energy intake in the absence of physiological hunger. Food cues, including the sight and smell of food, food advertisements, or triggered memories associated with pleasurable food experiences can elicit psychological and physiological responses, such as cravings, salivation, and hormone secretion, which enhance the desire to consume food and promote eating [15]. In the brain, the endocannabinoid and opioid systems within the nucleus accumbens reinforce the hedonic effects of food by stimulating dopamine release and triggering the reward system in response to enjoyable food [16]. Over-activity of the endocannabinoid system is thought to contribute to overeating and weight gain. However, recent research points to divergent effects of this system, with endocannabinoids also having a potential inhibitory effect on feeding behaviors [17].

Components of Energy Expenditure

Total energy expenditure (TEE) consists of basal energy expenditure (BEE), activity-induced energy expenditure (AEE), and diet-induced energy expenditure (DEE). BEE is the minimum energy required to sustain fundamental metabolic processes, including breathing, ion transport, and cardiovascular function, and constitutes anywhere from 60 to 75% of TEE [18]. Due to stringent measurement methods needed to accurately determine BEE, resting energy expenditure (REE) is used to estimate BEE, with REE being approximately 5–10% higher than BEE. REE is often expressed as resting metabolic rate (RMR), a measure of energy expenditure per minute. The two major factors accounting for variability in RMR are body size and body composition [18], with larger body size and higher lean body mass (brain, muscle, liver, blood, body fluids) proportional to RMR. Because of this, males generally have a higher RMR compared to females. Aging reduces RMR due to age-related loss of skeletal muscle mass, and can account for a 2–5% reduction in RMR for every decade after age 30 years [19]. Caloric restriction reduces RMR, with very low energy intake reducing RMR by as much as 10–20% [20]. Conversely, pregnancy, elevated body temperature caffeine, nicotine and thyroid hormone can increase RMR.

AEE is the most variable component of energy expenditure, with estimates ranging from 15–30% of BEE. However, given that AEE is a reflection of lifestyle, occupation, exercise duration and intensity, an accurate estimate is difficult to make. Energy expended for unintentional exercise is called non-exercise activity thermogenesis (NEAT). Generally, NEAT accounts for the majority of the energy expended for physical activity and varies enormously depending on an individual's daily activities [21]. Interestingly, high levels of spontaneous physical activity and therefore high NEAT appears to be protective against developing obesity [21].

Although food is consumed to yield energy, approximately 10% of TEE is required to digest, absorb, metabolize, and store nutrients derived from food [22]. This energy requirement, DEE, is also known as diet-induced thermogenesis due to the slight increase in body temperature observed for several hours after eating. DEE increases with meal size and is dependent on macronutrient composition, with dietary carbohydrate producing the highest DEE and dietary fat producing the lowest [23].

Genetic and Lifestyle Factors Affect Body Weight

It is estimated that genetic factors account for 40–75% of variation in BMI [24], with genome-wide association studies identifying more than 300 genes linked to BMI, waist-to-hip ratio, and adiposity [25]. Patterns and traits related to time spent in physical activity and sedentary behaviour are also influenced by genetic factors [26]. Several theoretical concepts have been developed to explain the genetics of obesity. Set point theory suggests that body weight is genetically determined, and internal genetically-determined mechanisms defend against changes in weight [27]. The thrifty gene theory suggests that genetic variants cause some individuals to store energy from food more efficiently and expend less energy at rest and during physical activity than people who do not express this gene [24]. While such energy economy was evolutionarily advantageous during times of energy scarcity, energy thriftiness has become a disadvantage in the current environment where energy is readily available. Other proposed theories include the alternative thrifty hypothesis, the drifty genotype hypothesis, the climate adaptation hypothesis, and the aggression control hypothesis [28–31]. Recent studies confirm that genetically predisposed individuals are indeed at greater risk for higher BMI, and that genetic predisposition combined with environmental factors has contributed to increasing incidence of obesity. However, BMI has increased for both genetically predisposed and non-predisposed individuals, implying that environmental factors may have a larger influence on the development of obesity than genetic factors [32].

An obesogenic environment is one where the availability, affordability, accessibility, and marketing of food, combined with lack of opportunities for physical activity and social norms surrounding food and physical activity, promotes excess energy intake, sedentary behavior, and weight gain [33]. Most environmental factors that contribute to the increased incidence of obesity have developed in response to a changing occupational landscape, societal norms, technological advances, and altered family dynamics. Historically, urban design has favored motorized transportation at the expense of active transportation, such as walking or cycling [34], and travel from suburban areas into the city for work often does not make active transportation feasible. Modern occupations are less physically demanding than in the past, with job-related energy expenditure now 100 kcal/day lower than in the past 50 years [35].

Changes in the amount and type of foods consumed also contribute to the rising incidence of obesity. Increasingly large portion sizes (both at home and restaurants) and value-size pricing have modified perceptions of normal and acceptable amounts to consume at an eating occasion [36–38]. Available portion size is directly related to the amount of food consumed at an eating occasion, and over-eating at one meal does not lead to compensatory reduction in food intake at future meals [36]. Interestingly, propensity to overeat in response to large portions does not appear to be affected by BMI, sex, dietary restraint, or socioeconomic status [39].

Convenience and fast foods, while time-saving and less expensive than unprocessed foods, are more energy dense, processed, and lacking in micronutrients [40]. Ultra-processed foods, defined as commercial foods produced from minimal or no whole food ingredients, have become increasingly prevalent and popular due to their ease of preparation, appealing taste, and extensive marketing. These foods tend to be more energy dense and higher in added sugars, salt, and artificial flavourings, colors, and preservatives compared to minimally processed foods [41]. Sales of ultra-processed foods have significantly increased over the past 30 years while those of unprocessed or minimally processed foods have gradually declined [42, 43], with energy intake from ultra-processed foods currently comprising up to

60% of total energy intake in U.S. adults [44]. Strong observational data supports the hypothesis that consuming ultra-processed foods has contributed to the high rates of excess weight and abdominal obesity in countries such as the U.S.A, Canada, the UK, Brazil, France, and Spain [45–49]. In a recent in-patient feeding study, participants consumed more calories when given ad libitum access to ultra-processed diet compared to ad libitum access to an diet composed of unprocessed foods, despite the two diets being matched for daily presented calories, sugar, fat, and macronutrients [50]. Additionally, participants on the ultra-processed diet gained weight, while those on the unprocessed diet lost weight, suggesting a direct role for consumption of ultra-processed foods on weight status.

Health Complications Due to Obesity

While 30 years ago the global focus was to combat childhood malnutrition and how to feed an increasing global population, today there is an additional challenge of managing obesity and concomitant non-communicable health complications [51]. There are several health complications that can develop as a consequence of being overweight or obese (summarized in Fig. 1.3). Since being obese increases the risk for developing hypertension and hypercholesterolemia, the risk for heart disease and occurrence of a stroke is elevated. Most individuals with type 2 diabetes are overweight or obese. Although diabetes and high blood pressure are the most common causes of chronic kidney disease, recent studies suggest that even in the absence of these risks, obesity itself may promote chronic kidney disease. Cancers of the colon, breast (post-menopausal women), endometrium (the lining of the

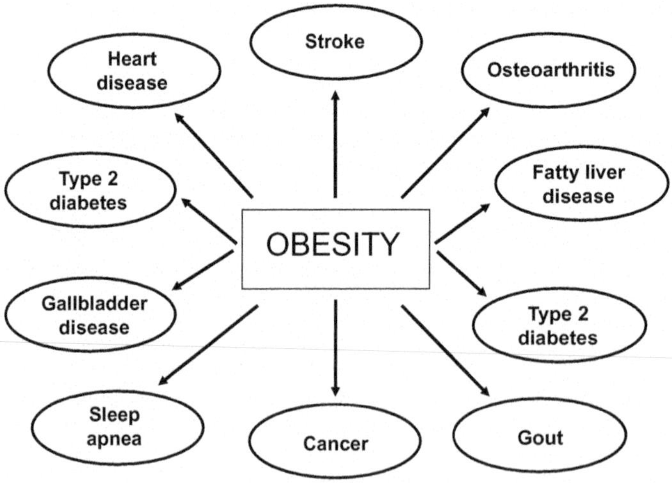

Fig. 1.3 Adverse health complications associated with obesity

uterus), kidney and esophagus have also been reported to be linked to obesity. Furthermore, some studies have reported an association between obesity and cancers of the gallbladder, ovaries, and pancreas. The incidence of gallbladder disease and gallstones are known to be more common in overweight and obese individuals. Osteoarthritis is a common joint condition that most often affects the knee, hip, or back.

With an excessive bodyweight, there is added pressure on joints, which degenerates the cartilage at a faster rate. On the other hand, gout, which is a condition that also affects the joints is caused by deposition of excessive uric acid crystals and is prevalent in overweight individuals. Sleep apnea is a breathing condition associated with being overweight that can lead to a brief interruption in normal breathing during sleep. In fact, sleep apnea can increase the risk for heart disease and stroke. Non-alcoholic fatty liver disease (NAFLD) due to fat accumulation in the liver causes liver injury. Fatty liver disease may lead to severe liver damage, cirrhosis, or even liver failure. Taken together, it is evident that understanding the cause of excessive weight gain as well as implementing measures that can prevent or treat it can result in a sustained weight reduction and normalization that would subsequently reduce the risk for obesity induced health complications.

Since the prevalence of being overweight and obesity among men and women varies greatly, more women are obese than men. Indeed, more than 2 of 3 women in the US are overweight or obese. Women, compared to men, have higher percent body fat and deposit it in a different pattern, with relatively more adipose tissue in the hips and thighs in women and central obesity typical of men [52]. Severe obesity is more prevalent in women than men worldwide, and obesity

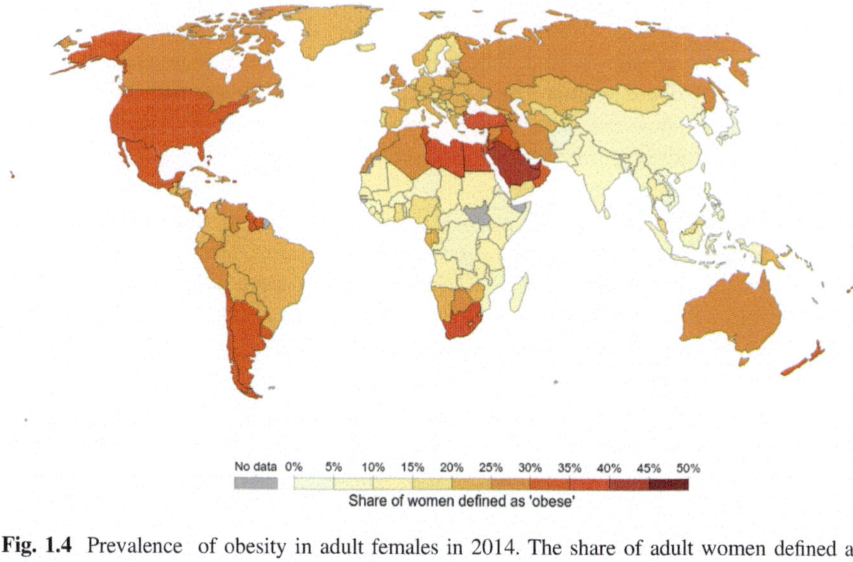

Fig. 1.4 Prevalence of obesity in adult females in 2014. The share of adult women defined as obese, measured as the percentage of women with a BMI value >30. *Source* Prevalence of weight categories in females, NCD Risk Factor Collaboration

pathophysiology and the resultant obesity-related disease risks differ in women and men. Although the underlying mechanisms are largely unknown, pre-clinical and human studies indicate that ovarian hormones may play a major role [53]. The underlying mechanisms are largely unknown. Pre-clinical and clinical research indicate that ovarian hormones may play a major role. Figure 1.4 shows the share of the global prevalence of obesity by women.

Consequences of Obesity in Women

Bodyweight is a major issue for women, however ethnicity and race as well as socioeconomic status can also have an effect on weight levels—which further affect the health status of the woman [54]. Central obesity seems to have a stronger impact in African-American women than general adiposity as measured by BMI [55]. There has been a global increase of obesity in women that are of reproductive age that has resulted in infertility/reduced fertility as well as an increase in the time taken to conceive [56, 57]. In addition, the development of obesity associated co-morbidities (i.e. type 2 diabetes and hypertension) increase the risk of adverse outcomes for both mother and child. Indeed, children of obese gravida are at a greater risk for the development of cardiometabolic disease in childhood and throughout adulthood [58].

Obese pregnant women are at a greater risk of premature pregnancy loss, increased risk of congenital fetal malformations, delivery of large for gestational age infants, spontaneous and premature birth, and stillbirth [59, 60]. During late stage of pregnancy the risk for gestational diabetes and pre-eclampsia are increased. Women with obesity can also experience difficulties during labor and delivery, and are more at risk of post-partum hemorrhage. With respect to long-term health complications in obese women, weight retention after delivery, and difficulties in subsequent pregnancy can occur [56]. It should be mentioned that aside from the physical complications of obesity, obesity has negative psychological consequences particularly in women including impaired body image, low self-esteem, eating disorders, stress, depression and poor quality of life [61]. In addition, it is interesting to note that emotional problems such as depression, anxiety, and stress are associated with an increase in BMI during pregnancy have also been reported to occur [62]. It is evident that women with obesity need support to lose weight before they conceive, and to minimize their weight gain in pregnancy to reduce the risk of complications for both mother and offspring [56]. It should also be noted that polycystic ovary syndrome is a common endocrine disorder that results in polycystic ovaries and is often seen concomitantly with obesity [63]. This condition also represents an increase in the risk in the development of cardiovascular, metabolic syndrome and diabetes.

It is known that there are large variations in obesity and breast cancer rates worldwide and across racial/ethnic groups, however; most studies evaluating the impact of obesity on breast cancer risk and survival have been conducted in non-Hispanic white women in the US or Europe [55]. Since there are differences in tumor hormone receptor subtype distribution, obesity prevalence, and risk factor profiles, among women of different racial/ethnic groups, it would be expected that differences also exist in breast cancer risk. Indeed, obesity and a sedentary lifestyle may be two important modifiable risk factors for breast carcinoma and thus may have a significant public health impact in women from various racial and ethnic backgrounds [64]. Despite the paucity of data, current evidence suggests a stronger adverse effect of obesity on breast cancer risk and survival in women of Asian ancestry. For African Americans and Hispanics, the strength of the associations appears to be more comparable to that of non-Hispanic whites, particularly when accounting for subtype and menopausal status [55]. In the US, African American women are more likely than non-Hispanic European women to be obese and to be diagnosed with triple-negative breast cancer [65]. With respect to other specific women's health issues during obesity, despite extensive research examining adiposity (BMI), a weak positive correlation has been observed between the risk of ovarian cancer, the most fatal gynecological cancer, and adiposity [66].

The prevalence of obesity is rapidly increasing in the US, particularly among women. Approximately 60–70% of hypertension in adults may be directly attributed to obesity. In addition, maternal obesity is a major risk factor for hypertensive disorders during pregnancy [67]. The underlying mechanisms for the association between obesity and cardiovascular disease (CVD) risk are multifactorial, but activation of the sympathetic nervous system is one significant contributing factor. Sex may influence the association between hypertension and sympathetic overactivity in obese people. Chronic hyperinsulinemia due to insulin resistance, high plasma levels of leptin, and obstructive sleep apnea may be responsible for sympathetic overactivity in obesity-related hypertension [67]. It is pointed out that weight gain in women in mid-life is related to an increase in central fat distribution as a consequence of diminishing levels of estrogen [68]. Central obesity results in dysglycemia, dyslipidemia, hypertension and CVD. Since CVD is the leading cause of death in postmenopausal women, the importance of weight management cannot be overstated.

Ethnic Differences in Obesity-Related Disease

Strategies for the treatment and prevention of obesity-related health complications may need to understand and address ethnic related differences in the occurrence of co-morbidities due to obesity as well as lifestyle factors that predispose ethnic groups to obesity [69]. Indeed, ethnic minorities are disproportionately affected by overweight and obesity that increases the risk for adverse health outcomes including CVD and diabetes [70].

An association between vitamin D status and obesity and obesity-induced co-morbidities has been proposed [71]. In this regard, ethnic minorities have higher rates of vitamin D insufficiency, which is correlated to obesity-related chronic diseases i.e. type 2 diabetes, CVD and metabolic syndrome. There is a high prevalence of obesity in American Indians of all ages and in both men and women [72] that has been linked to high rates of complications including type 2 diabetes, hypertension, dyslipidemia and respiratory problems. Such observations have been attributed to a high-fat, high-calorie diet coupled with a sedentary lifestyle. NAFLD has been reported to exist in approximately 30% of the world's population [73]. Epidemiological studies have concluded that ethnicity plays a role in complications and treatment response. The highest NAFLD prevalence is observed in Hispanic populations, exhibiting a worse disease progression. Interestingly, it has been reported that the Hispanic American population is at higher risk for obesity as well as diabetes and end-stage renal disease [74]. In contrast, African-Caribbean exhibit the lowest risk, with less severe steatosis and inflammation, lower levels of triglycerides, and less metabolic derangement, but conversely higher prevalence of insulin resistance. The prevalence of NAFLD in Asian cohorts is considered to be of epidemic proportions in these populations [73].

As already mentioned in the previous section, obese women experience higher rates of infertility and other pregnancy complications. Obese women have a lower chance of pregnancy following in vitro fertilization (IVF), which also appear to be related to racial/ethnic background. In this regard, compared with normal-weight women, failure to achieve a clinical intrauterine gestation is significantly more likely among obese women overall, normal-weight and obese Asian women, normal-weight Hispanic women, and overweight and obese Black women. Among women who do conceive, compared with normal-weight women, failure to achieve a live birth is significantly more likely among overweight and obese women overall, and among overweight and obese Asian women, overweight and obese Hispanic women, and normal-weight and obese Black women. Although weight loss should theoretically be the first line of therapy for obese women, other lifestyle factors, such as regular physical exercise, elimination of tobacco use and alcohol consumption, and stress management, may be of more immediate benefit in achieving conception [60].

South Asians are at higher risk than white Caucasians for the development of obesity and obesity-related non-communicable diseases, including insulin resistance, the metabolic syndrome, type 2 diabetes and coronary heart disease (CHD) [75]. Rapid nutrition and lifestyle transitions have contributed to acceleration of obesity-related non-communicable diseases in South Asians. Differences in determinants and associated factors for obesity-related non-communicable diseases between South Asians and White Caucasians include body phenotype (high body fat, high truncal, subcutaneous and intra-abdominal fat, and low muscle mass), biochemical parameters (hyperinsulinemia, hyperglycemia, dyslipidemia, hyperleptinemia, low levels of adiponectin and high levels of C-reactive protein), procoagulant state and endothelial dysfunction. Higher prevalence, earlier onset and increased complications of type 2 diabetes and CHD are often seen at lower levels of BMI and waist circumference in South Asians than white Caucasians.

Imbalanced nutrition, physical inactivity, perinatal adverse events and genetic differences are also important contributory factors. Other differences between South

Asians and white Caucasians include lower disease awareness and health-seeking behavior, delayed diagnosis due to atypical presentation and language barriers, and religious and sociocultural factors. All these factors result in poorer prevention, less aggressive therapy, poorer response to medical and surgical interventions, and higher morbidity and mortality in South Asians. During 2011 and 2012, more than a third of the US population was obese [76]. Significant racial and ethnic variations exist in the prevalence of obesity and diabetes. Generally, non-Hispanic blacks and Mexican Americans appear to be at higher risk for developing obesity as well as diabetes as compared to the non-Hispanic white population for both adults and children [76].

Although the prevalence of chronic kidney disease (CKD) is similar or slightly less in Hispanics than non-Hispanic whites, the prevalence of end-stage renal disease is almost 50% higher in Hispanics compared to non-Hispanic whites [77] that may be related to the greater prevalence of obesity in the US Hispanic population. It should be mentioned that since blood pressure is strongly related to body weight, the control of obesity is a key component in the prevention and control of hypertension [78]. Given the high prevalence of obesity in the African American population, especially among women [79], interventions for weight reduction would be highly beneficial as even a modest weight loss can not only prevent or reverse blood pressure elevations, but would also be seen to reduce the risk of obesity induced CVD, diabetes and hyperlipidemia [78].

Approaches for Obesity Management and Weight Reduction

Management of obesity as a chronic disease focuses on improving physical health, mental health, and overall wellbeing [80]. While preventing further weight gain may be the first goal of obesity management, weight reduction is indicated to improve obesity-related conditions such as hypertension, dyslipidemia, and type 2 diabetes. The first approach to obesity management focuses on lifestyle changes aimed at reducing energy intake and increasing physical activity to achieve fat loss [81]. Although seemingly straightforward, success of such lifestyle interventions relies heavily on behavioural change, which can be hampered by factors such as lack of time for physical activity or meal preparation, societal, social and family pressures, emotional state, physical limitations or injuries, socioeconomic status, limited knowledge on nutrition and physical activity, and lack of motivation [82]. Lifestyle interventions are more likely to be effective if individualized, with a patient's personal barriers considered before making specific dietary and exercise recommendations. Factors such as successful initial weight loss, lower starting BMI, better mental outlook at the start of weight loss, being male, and older age have been linked to higher adherence to and success with lifestyle interventions [83]. Maintaining initial weight loss can be challenging and often difficult for patients, with many experiencing cycles of weight loss and weight re-gain. Therefore, attention should be given to ensuring weight loss goals, and the methods used to achieve them, are realistic and sustainable. Specific lifestyle factors associated with sustained weight loss include one hour of physical activity per day, consuming a low calorie, low fat diet, eating breakfast regularly, consistency in weekly eating patterns, and self-monitoring of body weight [84, 85]. If weight loss can be maintained for 2 to 5 years, the likelihood of longer-term success, even after 10 years, significantly increases [84].

Very low energy diets (VLED) are the most aggressive form of caloric restriction used in obesity management, limiting caloric intake to 450–800 kcal per day. Because the volume of food associated with such a low caloric value is small, commercially available products fortified with vitamins and minerals are often used to prevent micronutrient deficiencies with this dietary approach. VLED should be undertaken with medical supervision to monitor micronutrient status, and also to adjust medications used for treating co-morbidities as they improve with weight loss. Initially weight loss may be significant with VLED. However, adherence to such a low energy diet can be challenging for patients and their usefulness for long-term sustained weight loss has been questioned. A recent systematic review and meta-analysis suggests that, if coupled with adequate support and behavioural programs, VLED can be well-tolerated and used more widely than is currently practiced [86]. Additional concerns regarding severe caloric restriction include loss of fat free mass, metabolic adaptations, and reduction in energy expenditure accompanying rapid and drastic weight loss that can oppose sustained weight loss [87]. However, fat free mass can be maintained with VLED with adequate protein intake and resistance exercise [88].

When lifestyle strategies alone are insufficient to elicit weight loss, complementary therapies may be initiated. Pharmacological agents for obesity management are generally only prescribed for patients with a BMI of ≥ 30 kg/m^2, or ≥ 27 kg/m^2 in the presence of one or more weight related co-morbidities, such as hypertension, dyslipidemia, or type 2 diabetes. Weight loss medications that were commonly prescribed for weight loss in the past, such as fen-phen and sibutramine, have been withdrawn from the market due to increased risk of serious cardiovascular complications [89, 90]. Five anti-obesity medications are currently approved by the American Food and Drug Administration, each with differing mechanisms of action: Orlistat (Xenical, Ali), Lorcaserin (Belviq), Phentermine/topiramate (Qsymia), Naltrexone/bupropion (Contrave), and Liraglutide (Saxenda). All of these medications induce weight loss (average of 5–10 kg lost) and are associated with improved cardiometabolic risk factors [91].

Orlistat, the oldest of these approved drugs, is available in a non-prescription version known as Ali, and both versions work by inhibiting pancreatic lipase and digestion of dietary triglycerides (TG). Undigested TG move through the intestinal tract and cause the side effects commonly reported with this medication, including abdominal pain, oily discharge, inability to hold or sudden urge to have a bowel movement, increased number of bowel movements, and malabsorption of fat soluble vitamins. The remaining four approved anti-obesity medications act on the central nervous system and/or neuroendocrine pathways to regulate energy intake and expenditure. Prescribing these newer medications in combination at lower doses can maximize effectiveness while minimizing side effects of each [82, 92]. These medications have favourable safety profiles, but have been associated with side effects including dry mouth, paresthesia, constipation, insomnia, dizziness, and dysgeusia, nausea, headache, and increased lipase enzymes [93].

The most effective medical intervention for sustained weight loss is bariatric surgery. Although specific eligibility criteria may vary, generally patients with a BMI >40 kg/m^2 or >35 kg/m^2 with co-morbidities such as type 2 diabetes, dyslipidemia, hypertension, sleep apnea or renal disease are eligible for this intervention [94]. The ultimate outcome of bariatric surgery is to reduce the size of the stomach, with different surgical procedures achieving this via different methods. The most commonly performed bariatric procedures are gastric banding, sleeve gastrectomy, and gastric bypass. Gastric banding involves surgically placing an adjustable and removable band around the upper stomach to create a small stomach pouch, while sleeve gastrectomy creates a small stomach pouch by surgically removing a large portion of the stomach. In Roux-en-y gastric bypass, the stomach is permanently divided to make a small pouch, which is then attached directly to the jejunum, and the remaining portion of the stomach and duodenum are bypassed. Each of these surgeries reduces stomach capacity and the amount of food that can be eaten, thereby reducing caloric intake. However, changes to circulating gastrointestinal hormones and the gut microbiome may also contribute to restoration of energy homeostasis, weight loss, and improvements in cardiometabolic risk factors observed after these surgical procedures [95, 96].

Bariatric surgeries carry risks common to any surgery. However, these surgeries have additional complications including abdominal hernias, stretched stomach pouches, gallstones, and dumping syndrome, where premature gastric emptying of hyperosmolar stomach contents into the small intestine leads to nausea, weakness, sweating, faintness, and diarrhea [97]. Due to the malabsorptive nature of these surgeries, nutrient deficiencies can be common, and supplementation is vital for these patients [98]. Deficiencies in calcium and vitamin D, combined with changes in gastrointestinal hormones, circulating adipokines, and decreased mechanical load from weight loss can compromise bone health [99]. Anemias are also a common complication of bariatric surgeries due to an increased susceptibility to vitamin B_{12}, folate, and iron deficiencies. In addition, bioavailability of oral medications may be impacted by bariatric surgery and medication doses must be closely monitored in the immediate post-surgery period [100]. As a result of the magnitude and speed of weight loss post-surgery, REE and the REE to fat free mass ratio may decrease, hindering sustained weight loss [101], which may account for the plateau or increase in weight that is often observed after the initial weight loss post-surgery [102, 103]. As such, lifestyle interventions are an important adjunct to surgery if weight loss is to be sustained.

Conclusions

It is evident that obesity and its related co-morbidities have become major threats to world health. Greater risks for adverse health complications occur during obesity are associated with ethnic backgrounds and sex differences. Effective strategies are required for obesity prevention and treatment that will also reduce the burden of obesity associated non-communicable diseases. Indeed, to be effective, educational and environmental interventions are required that are culture and gender specific. Although weight loss is the first line of therapy, there are of course, other lifestyle factors including regular physical activity and stress management that can reduce obesity and associated health complications and the global epidemic of obesity.

Acknowledgements Infrastructural support was provided by the St. Boniface Hospital Research Foundation and the University of Winnipeg.

References

1. World Health Organization (2019) https://www.who.int/news-room/fact-sheets/detail/obesity-and-overweight. Accesed 11 July 2019
2. Wagner KH, Brath H (2012) A global view on the development of non communicable diseases. Prev Med 54(Suppl):S38–S41
3. Taherogorabi Z, Khazaei M, Moodi M, Chamani E (2016) From obesity to cancer: a review on proposed mechanisms. Cell Biochem Funct 34:533–545
4. Dugani S, Gaziano TA (2016) 25 by 25: achieving global reduction in cardiovascular mortality. Curr Cardiol Rep 18:10
5. Jagannathan R, Patel SA, Ali MK, Narayan KMV (2019) Global updates on cardiovascular disease mortality trends and attribution of traditional risk factors. Curr Diab Rep 19:44
6. Berry DC, Jiang Y, Graff JM (2016) Emerging roles of adipose progenitor cells in tissue development, homeostasis, expansion and thermogenesis. Trends Endocrinol Metab 27:574–585
7. Cuthbertson DJ, Steele T, Wilding JP, Halford JC, Harrold JA, Hamer M, Karpe F (2017) What have human experimental overfeeding studies taught us about adipose tissue expansion and susceptibility to obesity and metabolic complications? Int J Obes 41:853–865
8. Aponte Y, Atasoy D, Sternson SM (2011) AGRP neurons are sufficient to orchestrate feeding behavior rapidly and without training. Nat Neurosci 14:351–355
9. Fan W, Boston BA, Kesterson RA et al (1997) Role of melanocortinergic neurons in feeding and the agouti obesity syndrome. Nature 385:165–168
10. Krashes MJ, Koda S, Ye C et al (2011) Rapid, reversible activation of AgRP neurons drives feeding behavior in mice. J Clin Invest 121:1424–1428
11. Oswal A, Yeo GS (2007) The leptin melanocortin pathway and the control of body weight: lessons from human and murine genetics. Obes Rev 8:293–306
12. Shaw AM, Irani BG, Moore MC et al (2005) Ghrelin-induced food intake and growth hormone secretion are altered in melanocortin 3 and 4 receptor knockout mice. Peptides 26:1720–1727
13. Pandit R, Beerens S, Adan RAH (2017) Role of leptin in energy expenditure: the hypothalamic perspective. Am J Physiol Regul Integr Comp Physiol 312:R938–R947
14. Cui H, López M, Rahmouni K (2017) The cellular and molecular bases of leptin and ghrelin resistance in obesity. Nat Rev Endocrinol 13:338–351

15. Belfort-DeAguiar R, Seo D (2018) Food cues and obesity: overpowering hormones and energy balance regulation. Curr Obes Rep 7:122–129
16. Lutter M, Nestler EJ (2009) Homeostatic and hedonic signals interact in the regulation of food intake. J Nutr 39:629–632
17. Lau BK, Cota D, Cristino L, Borgland SL (2017) Endocannabinoid modulation of homeostatic and non-homeostatic feeding circuits. Neuropharmacology 124:38–51
18. Westerterp KR (2017) Control of energy expenditure in humans. Eur J Clin Nutr 71:340–344
19. St-Onge MP, Gallagher D (2010) Body composition changes with aging: the cause or the result of alterations in metabolic rate and macronutrient oxidation? Nutrition 26:152–155
20. Keys A, Brozek J, Henschel A et al (1950) The biology of human starvation. University of Minnesota Press, Minneapolis, USA
21. Teske JA, Billington CJ, Kotz CM (2008) Neuropeptidergic mediators of spontaneous physical activity and non-exercise activity thermogenesis. Neuroendocrinology 87:71–90
22. van Baak MA (2008) Meal-induced activation of the sympathetic nervous system and its cardiovascular and thermogenic effects in man. Physiol Behav 94:178–186
23. Tappy L (1996) Thermic effect of food and sympathetic nervous system activity in humans. Reprod Nutr Dev 36:391–397
24. Reddon H, Patel Y, Turcotte M et al (2018) Revisiting the evolutionary origins of obesity: lazy versus peppy-thrifty genotype hypothesis. Obes Rev 19:1525–1543
25. Goodarzi MO (2018) Genetics of obesity: what genetic association studies have taught us about the biology of obesity and its complications. Lancet Diabetes Endocrinol 6:223–236
26. den Hoed M, Brage S, Zhao JH et al (2013) Heritability of objectively assessed daily physical activity and sedentary behavior. Am J Clin Nutr 98:1317–1325
27. Farias MM, Cuevas AM, Rodriguez F (2011) Set-point theory and obesity. Metab Syndr Relat Disord 9:85–89
28. Stöger R (2008) The thrifty epigenotype: an acquired and heritable predisposition for obesity and diabetes? BioEssays 30:156–166
29. Speakman JR (2007) A nonadaptive scenario explaining the genetic predisposition to obesity: the "predation release" hypothesis. Cell Metab 6:5–12
30. Sellayah D, Cagampang FR, Cox RD (2014) On the evolutionary origins of obesity: a new hypothesis. Endocrinology 155:1573–1588
31. Belsare PV, Watve MG, Ghaskadbi SS, Bhat DS, Yajnik CS, Jog M (2010) Metabolic syndrome: aggression control mechanisms gone out of control. Med Hypotheses 74:578–589
32. Brandkvist M, Bjørngaard JH, Ødegård RA et al (2019) Quantifying the impact of genes on body mass index during the obesity epidemic: longitudinal findings from the HUNT Study. BMJ 366:l4067
33. World Health Organization (2016) Report of the commission on ending childhood obesity. http://apps.who.int.uml.idm.oclc.org/iris/bitstream/10665/204176/1/9789241510066_eng.pdf?ua=1 Version current 1 2016. Accessed 12 July 2019
34. Garfinkel-Castro A, Kim K, Hamidi S, Ewing R (2017) Obesity and the built environment at different urban scales: examining the literature. Nutr Rev 75:51–61
35. Church TS, Thomas DM, Tudor-Locke C et al (2011) Trends over 5 decades in U.S. occupation-related physical activity and their associations with obesity. PLoS ONE 6:e19657
36. Livingstone MB, Pourshahidi LK (2014) Portion size and obesity. Adv Nutr 5:829–834
37. Nielsen SJ, Popkin BM (2003) Patterns and trends in food portion sizes, 1977–1998. JAMA 289:450–453
38. Zheng M, Rangan A, Meertens B, Wu JHY (2017) Changes in typical portion sizes of commonly consumed discretionary foods among Australian adults from 1995 to 2011–2012. Nutrients 9. pii: E577
39. Ello-Martin JA, Ledikwe JH, Rolls BJ (2005) The influence of food portion size and energy density on energy intake: implications for weight management. Am J Clin Nutr 82:236S–241S

40. Wellard L, Havill M, Hughes C et al (2015) Energy-dense fast food products cost less: an observational study of the energy density and energy cost of Australian fast foods. Aust N Z J Public Health 39:544–545
41. Moubarac JC, Parra DC, Cannon G, Monteiro CA (2014) Food classification systems based on food processing: significance and implications for policies and actions: a systematic literature review and assessment. Curr Obes Rep 3:256–272
42. Marrón-Ponce JA, Tolentino-Mayo L, Hernández-F M, Batis C (2018) Trends in ultra-processed food purchases from 1984 to 2016 in Mexican households. Nutrients 11:2018
43. Solberg SL, Terragni L, Granheim SI (2016) Ultra-processed food purchases in Norway: a quantitative study on a representative sample of food retailers. Ultra-processed food purchases in Norway: a quantitative study on a representative sample of food retailers. Public Health Nutr 19:1990–2001
44. Martínez Steele E, Baraldi LG, Louzada ML et al (2016) Ultra-processed foods and added sugars in the US diet: evidence from a nationally representative cross-sectional study. BMJ Open 6:e009892
45. Adams J, White M (2015) Characterization of UK diets according to degree of food processing and associations with socio-demographics and obesity: cross-sectional analysis of UK National Diet and Nutrition Survey (2008–12). Int J Behav Nutr Phys Act 12:160
46. Julia C, Martinez L, Allès B et al (2018) Contribution of ultra-processed foods in the diet of adults from the French NutriNet-Santé study. Public Health Nutr 21:27–37
47. Louzada ML, Baraldi LG, Steele EM et al (2015) Consumption of ultra-processed foods and obesity in Brazilian adolescents and adults. Prev Med 81:9–15
48. Mendonça RD, Pimenta AM, Gea A, de la Fuente-Arrillaga C et al (2016) Ultra-processed food consumption and risk of overweight and obesity: the University of Navarra Follow-Up (SUN) cohort study. Am J Clin Nutr 104:1433–1440
49. Canella DS, Levy RB, Martins AP et al (2014) Ultra-processed food products and obesity in Brazilian households (2008–2009). PLoS ONE 9:e92752
50. Hall KD, Ayuketah A, Brychta R et al (2019) Ultra-processed diets cause excess calorie intake and weight gain: an inpatient randomized controlled trial of ad libitum food intake. Cell Metab 30:67–77.e3
51. Prentice AM (2006) The emerging epidemic of obesity in developing countries. J Epidemiol 35:93–99
52. Lee MJ, Fried SK (2017) Sex-dependent depot differences in adipose tissue development and function; role of sex steroids. J Obes Metab Syndr 26:172–180
53. Leeners B, Geary N, Tobler PN, Asarian L (2017) Ovarian hormones and obesity. Hum Reprod Update 23:300–321
54. Bowen DJ, Tomoyasu N, Cauce AM (1991) The triple threat: a discussion of gender, class, and race differences in weight. Women Health 17:123–143
55. Bandera EV, Maskarinec G, Romieu I, John EM (2015) Racial and ethnic disparities in the impact of obesity on breast cancer risk and survival: a global perspective. Adv Nutr 6:803–819
56. Poston L, Caleyachetty R, Cnattingius S et al (2016) Preconceptional and maternal obesity: epidemiology and health consequences. Lancet Diabetes Endocrinol 4:1025–1036
57. Kelly-Weeder S, Cox CL (2006) The impact of lifestyle risk factors on female infertility. Women Health 44:1–23
58. Chandrasekaran S, Neal-Perry G (2017) Long-term consequences of obesity on female fertility and the health of offspring. Curr Opin Obstet Gynecol 29:180–187
59. Marchi J, Berg M, Dencker A et al (2015) Risks associated with obesity in pregnancy, for the mother and baby: a systematic review of reviews. Obes Res 16:621–638
60. Luke B (2017) Adverse effects of female obesity and interaction with race on reproductive potential. Fertil Steril 107:868–877
61. Chu DT, Minh Nguyet NT, Nga VT et al (2019) An update on obesity: mental consequences and psychological interventions. Diabetes Metab Syndr 13:155–160

62. Faria-Schützer DB, Surita FG, Nascimento SL et al (2017) Psychological issues facing obese pregnant women: a systematic review. J Matern Fetal Neonatal Med 30:88–95
63. Meier RK (2018) Polycystic ovary syndrome. Nurs Clin North Am 53:407–420
64. McTiernan A (2000) Associations between energy balance and body mass index and risk of breast carcinoma in women from diverse racial and ethnic backgrounds in the U.S. Cancer 88:1248–1255
65. Dietze EC, Chavez TA, Seewaldt VL (2018) Obesity and triple-negative breast cancer: disparities, controversies, and biology. Am J Pathol 188:280–290
66. Tworoger SS, Huang T (2016) Obesity and ovarian cancer. 208:155–176
67. Fu Q (2019) Sex differences in sympathetic activity in obesity and its related hypertension. Ann NY Acad Sci 1454:31–41
68. Kapoor E, Collazo-Clavell ML, Faubion SS (2017) Weight gain in women in midlife: a concise review of the pathophysiology and strategies for management. Mayo Clin Proc 92:1552–1558
69. Abate N, Chandalia M (2003) The impact of ethnicity on type 2 diabetes. J Diabetes Complic 17:39–58
70. Nesbitt SD, Ashaye MO, Stettler N et al (2004) Overweight as a risk factor in children: a focus on ethnicity. Ethn Dis 14:94–110
71. Renzaho AM, Halliday JA, Nowson C (2011) Vitamin D, obesity, and obesity-related chronic disease among ethnic minorities: a systematic review. Nutrition 27:868–879
72. Story M, Stevens J, Himes J et al (2003) Obesity in American-Indian children: prevelance, consequences, and prevention. Prev Med 37:S3–S12
73. Szanto KB, Li J, Cordero P, Oben JA (2019) Ethnic differences and heterogeneity in genetic and metabolic makeup contributing to nonalcoholic fatty liver disease. Diabetes Metab Syndr Obes 12:357–367
74. Yracheta JM, Alfonso J, Lanaspa MA et al (2015) Postgrad Med 127:503–510
75. Misra A, Khurana L (2011) Obesity-related non-communicable diseases: South Asians vs white Caucasians. Int J Obes 35:167–187
76. Bhupathiraju SN, Hu FB (2016) Epidemiology of obesity and diabetes and their cardiovascular complications. Circ Res 118:1723–1735
77. Desai N, Lora CM, Lash JP, Ricardo AC (2019) CKD and ESRD in US Hispanics. Am J Kidney Dis 73:102–111
78. Kumanyika SK (1997) The impact of obesity on hypertension management in African Americans. J Health Care Poor Underserved 8:352–364
79. Kumanyika S (1987) Obesity in black women. Epidemiol Rev 9:31–50
80. Obesity Canada (2019) https://obesitycanada.ca/managing-obesity/. Accessed 10 July 2019
81. Semlitsch T, Stigler FL, Jeitler K et al (2019) Management of overweight and obesity in primary care-A systematic overview of international evidence-based guidelines. Obes Rev 20:1218–1230
82. Sweeting AN, Caterson ID (2017) Approaches to obesity management. Intern Med J 47:734–739
83. Burgess E, Hassmén P, Pumpa KL (2017) Determinants of adherence to lifestyle intervention in adults with obesity: a systematic review. Clin Obes 7:123–135
84. Wing RR, Phelan S (2005) Long-term weight loss maintenance. Am J Clin Nutr 82:222S–225S
85. Thomas JG, Bond DS, Phelan S et al (2014) Weight-loss maintenance for 10 years in the National Weight Control Registry. Am J Prev Med 46:17–23
86. Parretti HM, Jebb SA, Johns DJ et al (2016) Clinical effectiveness of very-low-energy diets in the management of weight loss: a systematic review and meta-analysis of randomized controlled trials. Obes Rev 17:225–234
87. Fothergill E, Guo J, Howard L et al (2016) Persistent metabolic adaptation 6 years after "The Biggest Loser" competition. Obesity (Silver Spring). 24:1612–1619

88. Hansen TT, Hjorth MF, Sandby K et al (2019) Predictors of successful weight loss with relative maintenance of fat-free mass in individuals with overweight and obesity on an 8-week low-energy diet. Br J Nutr 27:1–12
89. Connolly HM, Crary JL, McGoon MD et al (1997) Valvular heart disease associated with fenfluramine-phentermine. N Engl J Med 337:581–588
90. James WP, Caterson ID, Coutinho W et al (2010) Effect of sibutramine on cardiovascular outcomes in overweight and obese subjects. N Engl J Med 363:905–917
91. Daneschvar HL, Aronson MD, Smetana GW (2016) FDA-approved anti-obesity drugs in the United States. Am J Med 129(879):e1–e6
92. Sweeting AN, Hocking SL, Markovic TP (2015) Pharmacotherapy for the treatment of obesity. Mol Cell Endocrinol 418:173–183
93. Krentz AJ, Fujioka K, Hompesch M (2016) Evolution of pharmacological obesity treatments: focus on adverse side-effect profiles. Diabetes Obes Metab 18:558–570
94. International Federation for the Surgery of Obesity and Metabolic Disorders (2019) Are you a candidate. https://www.ifso.com/are-you-a-candidate/. Accessed 11 July 2019
95. Madsbad S, Dirksen C, Holst JJ (2014) Mechanisms of changes in glucose metabolism and bodyweight after bariatric surgery. Lancet Diabetes Endocrinol 2:152–164
96. Tremaroli V, Karlsson F, Werling M et al (2015) Gastroplasty induce long-term changes on the human gut microbiome contributing to fat mass regulation. Cell Metab 22:228–238
97. Ahmad A, Kornrich DB, Krasner H et al (2019) Prevalence of dumping syndrome after laparoscopic sleeve gastrectomy and comparison with Laparoscopic Roux-en-Y gastric bypass. Obes Surg 29:1506–1513
98. Stroh C, Manger T, Benedix F (2017) Metabolic surgery and nutritional deficiencies. Minerva Chir 72:432–441
99. Gregory NS (2017) The effects of bariatric surgery on bone metabolism. Endocrinol Metab Clin North Am 46:105–116
100. Angeles PC, Robertsen I, Seeberg LT et al (2019) The influence of bariatric surgery on oral drug bioavailability in patients with obesity: a systematic review. Obes Rev 20:1299–1311
101. Lamarca F, Melendez-Araújo MS, Porto de Toledo I et al (2019) Relative energy expenditure decreases during the first year after bariatric surgery: a systematic review and meta-analysis. Obes Surg 29:2648–2659
102. Chang SH, Stoll CR, Song J et al (2014) The effectiveness and risks of bariatric surgery: an updated systematic review and meta-analysis, 2003-2012. JAMA Surg 149:275–287
103. Gloy VL, Briel M, Bhatt DL et al (2013) Bariatric surgery versus non-surgical treatment for obesity: a systematic review and meta-analysis of randomised controlled trials. BMJ 347: f5934

Chapter 2
Adipocytes Under Environmental Assault: Targets for Obesity?

Shalini Behl and Jaipaul Singh

Abstract In the recent years, there has been a tremendous concern over the possible health threat posed by endocrine-disrupting chemicals (EDCs). These are mostly synthetic chemicals found in various materials such as organo-chlorinated pesticides, industrial chemicals, plastics and plasticizers, fuels, heavy metals, additives or contaminants in food, and personal care products. These chemicals are present in the environment and are with widespread use. Human exposure to EDCs occurs via ingestion of food, dust and water, via inhalation of gases and particles in the air, and through the skin. Data from several animal models, human clinical observations, and epidemiological studies converge to implicate their association with altered reproductive function in males and females, increased incidence of breast cancer, abnormal growth patterns and neuro-developmental delays in children, disruption of adipocyte function, as well as changes in immune function. The EDCs exert their insulting effects by interfering with hormone biosynthesis, metabolism, or action resulting in a deviation from normal homeostatic control or reproduction. The mechanisms of EDCs involve divergent pathways including (but not limited to) estrogenic, anti-androgenic, thyroid, peroxisome proliferator-activated receptor γ, retinoid, and actions through other nuclear receptors; steroidogenic enzymes; neurotransmitter receptors and systems; and many other pathways that are highly conserved in wildlife and humans. Emerging data from in vitro as well as in vivo models suggest new targets (i.e. adipocyte differentiation and mechanisms involved in weight homeostasis) of abnormal programming by EDCs, and provide strong evidence to support the scientific term 'obesogen'. The emerging idea of a link between EDCs and obesity expands the focus on obesity from intervention and treatment to include prevention and avoidance of these chemical modifiers. Because expansion of the adipocyte pool is critical for safely storing excess lipid, an understanding how these signaling axes

S. Behl (✉) · J. Singh
School of Forensic and Applied Sciences, University of Central Lancashire, Preston, Lancashire PR1 2HE, UK
e-mail: SBehl@uclan.ac.uk; behl.shalini@gmail.com

J. Singh
e-mail: jsingh3@uclan.ac.uk

P. S. Tappia et al. (eds.), *Pathophysiology of Obesity-Induced Health Complications*, Advances in Biochemistry in Health and Disease 19,
https://doi.org/10.1007/978-3-030-35358-2_2

can be altered by EDCs is critical in appreciating how environmental contaminants might contribute to the development of metabolic diseases.

Keywords Adipose · Obesity · Metabolic syndrome · Endocrine disruptors

Introduction

The human body is regarded as one of the prodigies that is very well equipped with self defense mechanisms, thereby prevailing the existence of the Homo sapiens through evolution. However, with the increasing amounts of toxins that are being released in the environment, even human bodies go through some dramatic side effects.

Obesity is a complex, multifactorial disease and the worldwide prevalence of obesity had more than doubled since 1980. Being overweight and obese in humans are increasing in both developed and developing countries and they are rapidly becoming a worldwide public health concern [1]. In 2010, the World Health Organization estimated that >700 million people worldwide will have obesity and nearly 2 billion people will be overweight by 2020 [2]. The current figure is 700 million with 39% men and 40% women who are obese, which represents around 13% of the world's population. What is more worrying is that the prevalence of childhood obesity is also increasing, and it is a strong risk factor for adult obesity. Obesity, being a multi-factorial disorder, can lead to a number of co morbidities, including type 2 diabetes mellitus, cancer, gall bladder disease, sleep apnea, high blood pressure, insulin resistance, inflammation, breathlessness, the metabolic syndrome, nonalcoholic fatty liver disease and gestational diabetes mellitus [3]. Although, this increase in obesity is generally attributed to changes in diet, sedentary life style, low physical activity and genetic predisposition [4], complex endocrine and other chemical signaling mechanisms exist which change resting and non-resting energy expenditure to promote constant weight in a person [5, 6].

In spite of several research studies that have been carried out to make sure that the understanding of the pathogenesis and therapy of diseases can be improved, it appears that only 10% of the diseases are caused by human genes and the rest are the repercussions of the environmental exposures. With increase in human demands over time, the amount of industrial pollutants released in atmosphere has also been increased, thereby exposing humans to more contaminations now than ever. Although the human body is capable enough to excrete water-soluble pollutants, the same is not true for lipid soluble pollutants. Besides, there are numerous fat-soluble compounds which tend to accommodate in the body's adipose tissue, where they may continue to exist for an indefinite duration of time [7].

Any foreign chemical that enters the living human body tends to trigger a reaction. This interaction between xenobiotics (the foreign chemical bodies) and the living systems proceeds in several phases. The first phase, also known as the exposure phase, is initiated at the time when a living organism is exposed to foreign substances. This may later be taken up or rejected by the living organism. The next phase is where the distribution of this foreign chemical substance throughout the organism takes place. Following the delivery, metabolism takes place as facilitated by enzymes and is subject to chemical changes which may or may not nullify the exogenous substance. The former mentioned phases are categorized as "toxicokinetics" whereas the following phase is "toxicodynamic". In this phase, the interaction happens between the chemical, its constituents and the constituents of the living organism. Correspondingly, an additional phase is subject to taking place where pathological or functional changes may occur. In addition, the metabolism of glucose and cholesterol is also affected by the environmental toxins. Moreover, it also interferes with adipokine secretion and hence induces insulin resistance and cell multiplication. Thus, this implicates in the pathogenesis of cardio-metabolic and malignant diseases [8].

Adipose Tissue and Adipogenesis

Adipose tissue is an areolar connective tissue that regulates and maintains body temperature, attaches the skin, and shields internal organs. It is further sub-divided into two types namely brown adipose tissue (BAT) and white adipose tissue (WAT). While WAT serves as the reservoir of energy, where triglycerides are stored in large lipid droplets, BAT mainly serves in the regulation of body temperature and energy expenditure [9]. It is now a well-accepted fact that mesenchymal stem cells (MSCs) act as precursor cells to proliferate and undergo differentiation to become mature adipocytes. The process of differentiation of the fibroblast like pre-adipocytes into mature lipid-laden, insulin-responsive adipocytes is known as adipogenesis [10].

The adipocyte life cycle starts with differentiation of adipocytes from either committed embryonic stem cells or mesenchymal stem cells. This includes a growth phase followed by growth arrest, clonal expansion, and a complex sequence of changes in gene expression leading to storage of lipid and finally death [11].

Once triggered to mature, pre-adipocytes proliferate and undergo growth arrest followed by a round of cell division known as clonal expansion. These cells then commit to differentiate into mature adipocytes. This process is accompanied by a dramatic increase in expression of adipocyte-specific genes. Several adipokines, hormones, enzymes, nutritional and environmental factors influence adipocyte life cycle contributing to overall adipose tissue growth and development.

Differentiation of pre-adipocytes into adipocytes involves a comprehensive network including a number of transcription factors responsible for the expression of key proteins that induce mature adipocyte formation. This highly regulated transcriptional cascade begins with re-entry of growth arrested pre-adipocytes into the cell cycle where they undergo several rounds of mitosis. This initial phase is known as mitotic clonal expansion (MCE) and is accompanied by the transient expression of C/EBP-β and C/EBP-δ. These transcription factors subsequently stimulate transcription of PPAR, which in turn can activate C/EBP-α. PPAR-γ and C/EBP-α exist in a positive feedback loop to propagate differentiation and induction of late adipogenic genes including aP2 and FAS in the terminal differentiation phase [12].

Modulation of Adipocyte Metabolism by Various Endocrine Disruptors

Endocrine-disrupting chemicals (EDCs) are exogenous chemicals or mixtures of chemicals that interfere with any aspect of endogenous hormonal signaling, affecting not only production, release, and transport of hormones, but also their cellular metabolism, binding action, and elimination [13, 14]. They have been shown to significantly alter the function (gene expression, hormone secretion) of white adipose tissue as well as adipose tissue mass (adipocyte number and/or volume). Figure 2.1 shows the effect of endocrine disrupting chemicals on the adipose cells. Adipose cells may either increase in number or size.

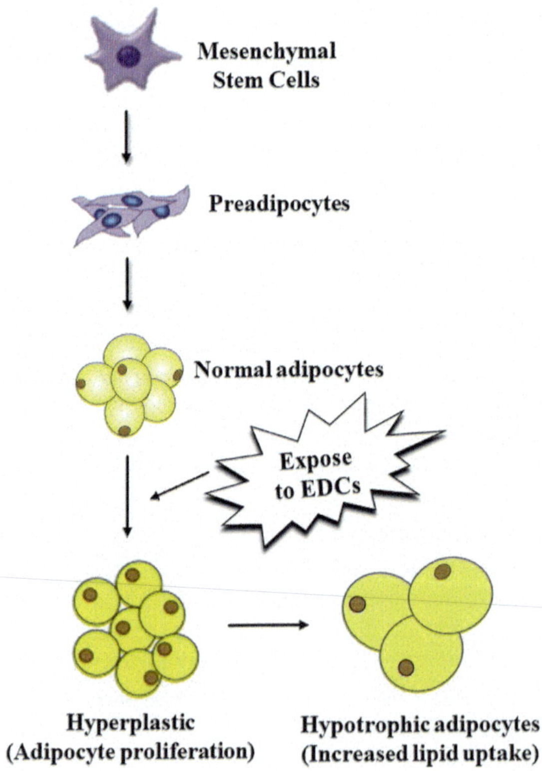

Fig. 2.1 Effect of endocrine disrupting chemicals on the adipocytes

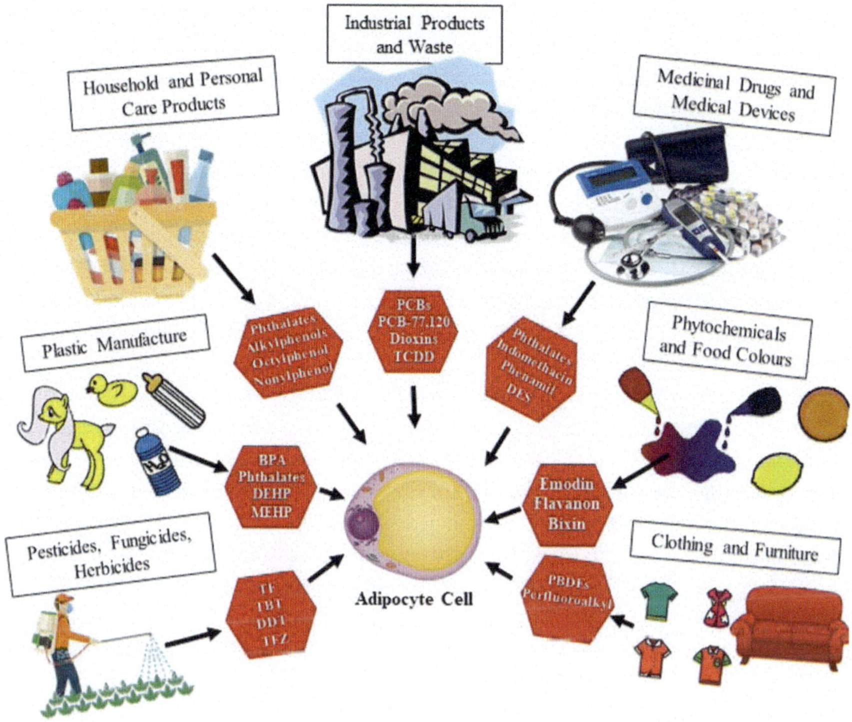

Fig. 2.2 Diagram showing various sources that may release/act as endocrine disrupting chemicals

EDCs are heterogeneous, rapidly growing group of natural, or man-made chemical compounds, including synthetic chemicals used as industrial solvents, plastics, plasticizers, fungicides, pesticides, heavy metals, and pharmaceutical agents [13, 15]. Figure 2.2 shows the ubiquitous presence of endocrine disrupting chemicals.

Bisphenol A

Bisphenol A (BPA) was first synthesized in 1891 and was discovered to be estrogenic in 1936 [16]. It is used in a wide array of manufacturing, food packaging, toys, and plastic manufacturing industry for various applications, such as lining of plastic containers [17], and aluminum cans [18], receipt paper [19–21], and some dental sealants [22]. BPA exposure has been linked to a wide range of negative health effects, including an increased risk of miscarriages, obesity, and cancer [23]. Furthermore, BPA is known to accumulate in fat, with 50% of breast adipose tissue from women containing BPA [24]. It has been known to rapidly metabolize to non-bioactive forms and has a short half-life of approximately 4–5 h in adult humans, with lower metabolic rates in the fetus and infants [25, 26]. Measurements of bioactive or free BPA in human serum is controversial at present, with some documenting nanograms per milliliter quantities in samples using contamination-free conditions [21, 25, 27], whereas others report that ordinary exposures result in picograms per milliliter levels or lower [28]. Although relevant internal exposure remains a critical issue that is still unresolved, it is noteworthy that industrial exposures, vulnerable populations, and individual variations in metabolism and susceptibility must be taken into consideration. Considerable and compelling evidence has been presented in both animals [29–34] as well as humans [35, 36], showing an association between increase in weight gain and body fat after developmental exposure with BPA.

Perinatal exposure of mice to BPA has been shown to increase adipose tissue mass, raises serum cholesterol level and may predispose to the development of metabolic syndrome [31, 37, 38]. A series of studies conducted in animal models highlighted that exposure to BPA in either adult animals [39, 40] or in utero [37], resulted in the development of insulin resistance. In vitro studies conducted on cultured pre-adipocyte cell lines, treatment with BPA ranging between 0–100 μg/ml for two days early in differentiation was reported to accelerate differentiation [41, 42] via a PI3 K-dependent pathway. Enhanced adipogenesis was observed in cells treated with either BPA alone or in combination with 5 μg/ml of insulin.

Dioxins

2,3,7,8-Tetrachlorodibenzo-*p*-dioxins (TCDDs) are released into the environment primarily as an unwanted by-product of manufacturing processes, such as incineration and burning of fossil fuels. Volcanic eruptions and forest fires also contribute to the overall environmental burden of dioxins. They may be formed during the chlorine bleaching process used by pulp and paper mills, and as a by-product from the manufacture of certain chlorinated organic chemicals, such as chlorinated phenols [43]. Ingestion of contaminated food is the main source of dioxin exposure in human and animal populations [44–47]. These compounds are highly resistant to both chemical as well as biological degradation and thereby, pose a significant degree of bioaccumulation and environmental persistence [48]. Reduction in pollution levels and stricter emission standards and regulations in developed countries has nevertheless led to a significant decline in TCDD exposure over the past decades. The current TCDD body burden of citizens in these countries is estimated to be approximately two parts per trillion-lipid adjusted (ppt), down from an estimated 20 ppt in the early 1970s [49].

TCDDs are widespread environmental contaminants and potent endocrine disruptor [50, 51]. These organic pollutants are hypothesized to have a either direct or indirect role in the pathogenesis of metabolic disorders, including obesity, metabolic syndrome, and diabetes [52–54]. A number of experimental studies support a link between TCDD exposure and metabolic alterations. Its exposure has been associated with wasting syndrome in rodents fed a normal diet [55], and with accelerated weight gain in mice fed a high-fat diet [56]. Both in vivo as well as in vitro studies have shown effects on glucose homeostasis following TCDD exposure. These include reduced glucose uptake in adipose tissue, liver, and pancreas; altered glucose tolerance; and impaired insulin secretion [57–60]. Studies on mice models have shown exposure to TCDDs is associated with increased serum triglycerides, cholesterol, and blood pressure and with earlier onset and greater severity of atherosclerosis [61].

Several epidemiologic studies have been conducted to establish a link between exposure to TCDDs and other dioxin-like compounds with metabolic syndrome. An increased risk for diabetes has been reported in Vietnam veterans [62, 63], phenoxyherbicide production workers [64] and Yucheng accident cohort members [65], but not in Great Lakes sport fish consumers [66], Finnish fishermen and their wives [67], young U.S. adults [68], or elderly Swedish adults [69]. Higher risks among women than men were reported in the Yucheng cohort [65], but few studies have had sufficient power to examine the risk of diabetes in women separately. Recent cross-sectional studies also suggest a positive association of dioxin-like compounds with metabolic syndrome and its individual components [70–72].

Phytoestrogens

Phyto-oestrogens are naturally occurring plant-based compounds that are structurally and/or functionally similar to mammalian oestrogens and their active metabolites. Among the several phyto-oestrogen classes, the most widely studied and hormonally active are the isoflavones and coumestans. Isoflavones are abundantly present in soya beans, legumes, berries, wine, grains, nuts and soya-fortified foods [73]. Naturally, Isoflavones are found as biologically inactive glycoside conjugates containing either glucose or carbohydrate moieties. The unconjugated from (aglycone) is the bioactive form. Among the soy-based foods, fermented soya, such as tem-peh or miso, typically contains higher aglycone levels. Once the isoflavones are consumed, they are rapidly metabolized and absorbed in the body. They enter the systemic circulation predominantly as conjugates with limited bioavailability and bioactivity, leaving only a tiny fraction of the 'free' bioactive form in systemic circulation. Typically, metabolites are less bioactive than the parent compounds but equol, a metabolite of daidzein, is a notable exception [74]. It appears that only 30–50% of individuals are capable of bio-converting daidzein to its more oestrogenic metabolite equol with vegetarians and individuals of Asian descent being most likely [75, 76]. Some of the factors that can purportedly contribute to inter-individual differences with regard to the production and absorption of isoflavones are age, genetics, gut physiology, diet and health status, particularly the use of antibiotics.

One of the increasing concerns with regard to the use of phytoestrogens is its possibility to interfere with the organizational role of estrogen in the developing brain and reproductive system. Experiments on various animal models with manipulation of estrogen during specific critical windows of development throughout gestation and early infancy, have been shown to lead to a myriad of adverse health outcomes including malformations in the ovary, uterus, mammary gland and prostate, early puberty, reduced fertility, disrupted brain organization, and reproductive tract cancers [77–81].

Notably, soya has been shown to impact reproductive healthy in women. The suppression of circulating steroid hormone levels and attenuation of the pre-ovulatory gonadotropin surge have been repeatedly observed. In one meta-analysis, it was concluded that isoflavone intake moderately increases cycle length and suppresses luteinizing hormone and follicle-stimulating hormone levels [82]. A clinical case-report conducted in 2008, reported three women (aged 35–56 years) experiencing a suite of symptoms related to excessive soya intake (estimated to exceed 40 g/d), including abnormal uterine bleeding, endometrial pathology and dysmenorrhea, all of which resolved when soya intake was discontinued or reduced [83].

Intake of adult soya has shown to impact mood and anxiety-related behaviors [84]. In human subjects, nearly all studies along these lines have focused on post-menopausal women and evidence for improvement of mood is minimal and sporadic [15]. Results across animal studies are mixed and sex-dependent with females generally shows decreased anxiety compared to males who show heightened effects [85, 86]. This pattern tends to abrogate or reverse expected sex differences in assessments of anxiety-related behaviors [87] and appears to involve the neuropeptides and vasopressin.

Pesticides

Since the discovery of DDT in 1939, numerous pesticides (organochlorides, organophosphates, carbamates) have been developed and used extensively worldwide with few guidelines or restrictions to kill unwanted organisms in crops, public areas, homes and gardens, and parasites in medicine. Humans are exposed to pesticides due to their occupations or through dietary and environmental exposure (water, soil, air). For several years, there have been concerns about the impact of environmental factors on the occurrence of human pathologies. The World Health Organization (WHO) has reported that roughly three million pesticide poisonings occur annually, resulting in 220,000 deaths worldwide [88]. In some cases, it has been suggested that diseases such as cancer, allergies, neurological disorders and reproductive disorders may be connected to pesticide exposure.

Pesticides may interfere with the pro-synthesis, transport, metabolism and elimination of hormones, thereby decreasing the concentration of natural hormones.

For instance, the thyroid hormone synthesis can be inhibited by several pesticides including amitrole, cyhalothrin, fipronil, ioxynil, maneb, mancozeb, pentachloronitro-benzene, prodiamine, pyrimethanil, thiazopyr, ziram, zineb [89–91].

The pesticides have also been shown to disrupt reproductive and sexual development. These effects seem to depend on several factors, including gender, age, diet, and occupation with age in particular being one of the most sensitive. Human foetuses, infants and children show greater susceptibility compared to healthy adult subjects [92–94]. Much of the damage occurs during the stages of gametogenesis and the early development of the foetus [94–96]. The role of pesticides on fertility has also been well- documented [97].

Several previous studies have reported that the exposure to pesticides may affect spermatogenesis, leading to poor semen quality and reduced male fertility. Furthermore, an increasing number of epidemiological studies tend to link environmental exposure to pesticides and hormone-dependent cancer risks. Increased levels of pesticides have been found in fat samples from women with breast cancer [98]. This increases the risk of breast cancer by four folds [99]. One of the major studies conducted in Spain between 1999 and 2009 shows that among a total of 2,661 cases of breast cancer reported in the female population, approximately 2,173 (81%) were observed in areas of high pesticide contamination [100].

A link between pesticide exposure and prostate cancer has also been observed. Several epidemiological studies have consistently demonstrated a higher risk in agricultural populations than in the general population [13, 101, 102]. For instance, in a multi-site case-control study conducted in five rural areas of Italy, a significantly higher rate of prostate cancer was found in farmers who are routinely exposed to pesticides. These observations are consistent with other studies in the USA and Sweden where the farmers and commercial pesticide applicators had slightly and/or significantly higher rate of prostate cancer than the general population [103, 104].

Apart from humans, at environmental level, wildlife remains vulnerable to the endocrine disrupting effects of pesticides. The relation between pesticide exposure and endocrine disruption have been largely noted in invertebrates [105–109], reptiles [81, 110–113], fish [114–116], birds [114, 117–120] and mammals [121–124]. Most of these studies show that pesticides, in particular organochlorine pesticides, affect the reproductive function. A study on *Daphnia magna* has shown that endosulfan sulphate disrupts the ecdysteroidal system (regulating processes such as molting and embryonic development) and juvenile hormone activity (regulating the sex ratio) of crustaceans [125, 126]. Another example is the influence of linuron on reproductive hormone production, testosterone production in rats being significantly reduced after in utero exposure to linuron, whereas progesterone production was not affected [127].

Polychlorinated Biphenyls

These are persistent organic pollutants (POPs) that are consumed because of their bioaccumulation through the food chain. They are man-made synthetic chemical mixtures, which were widely used in industry and their exposure to humans remains ubiquitous because of improper disposal and bioaccumulation in the environment. POPs can accumulate at high levels in adipose tissue and might be a contributing factor to the onset of cardiovascular, endocrine, and metabolic diseases [128, 129]. Data supporting the association between polychlorinated biphenyls and metabolic disease continue to be reported, and a survey conducted by NHANES reported an association between waist circumference and BMI in individuals with detectable levels of persistent organic pollutants [130].

Phthalates

Phthalates are esters of phthalic acid and are a class of chemicals that are mainly used as plasticizers to impart flexibility, transparency and durability in plastic materials. These compounds are found in a variety of household and consumer products such as adhesives, paints, packaging, children's toys, electronics, flooring, medical devices, personal care products, air fresheners, dietary supplements, pharmaceuticals and textiles [131, 132]. Exposure to humans is via oral ingestion, inhalation, and dermal contact. Lately, a number of phthalates have been documented to possess endocrine-disrupting properties based on their ability to interfere with normal reproductive function and hormone signalling. Several epidemiological studies have been conducted to investigate the relationship between phthalates exposure in human and health outcomes, including hormonal regulation of steroid and thyroid hormones, reproductive effects, pregnancy, precocious puberty, obesity and infertility [133, 134].

Either diethylhexylphthalate, or its metabolite monohexylphthalate has been reported to be linked with obesity in both animal as well as human models [135–137]. Prenatal and neonatal exposure to these compounds in pregnant mice was reported increase body weight, number and size of adipocytes, and activation of PPAR-γ in male offspring, suggesting its sexually dimorphic effect. In another study, exposure to diethylhexylphthalate in utero and during lactation showed increased weight gain, which persisted into adulthood [138]. Both of these studies were linked to behavioural changes in offspring through four generations suggesting that early-life exposure to at least some obesogens might exert permanent and transgenerational effects. A probable reasoning for these changes may be the epigenetic modification of imprinted genes, thereby, implicating the potential for trans-generational epigenetic inheritance.

Cigarette Smoke and Nicotine

Several epidemiological studies, conducted to examine the effects of maternal smoking during pregnancy on body weight of offspring during childhood or adulthood, have shown a positive and probable causal association between maternal smoking and increased risk of obesity or overweight in offspring. This conclusion is based on the very consistent pattern of overweight and/or obesity observed in children whose mothers smoked during pregnancy, along with findings of obese offspring from laboratory animals exposed to nicotine during pregnancy [128]. This consistent association between maternal smoking during pregnancy, low birth weight, and increased risk of overweight and/or obesity in offspring has been widely reported in the literature. Moreover, evaluation of literature in several meta-analyses indicated a 50–64% increase in obesity due to smoking during pregnancy [139–142].

Concluding Remarks

A number of man-made synthetic chemicals such as pesticides, plasticizers, antimicrobials, and flame retardants currently used in a variety of industrial and agricultural applications are leading to widespread contamination of the environment. Even though their intended uses are beneficial, effects on human health are a global concern. These so-called endocrine-disrupting chemicals (EDCs) are substances of exogenous origin that exert various endocrine functions in specific doses, including hormonal synthesis/transportation and adverse health effects in an organism and their descendants. They can disrupt hormonal balance and result in developmental and reproductive abnormalities. While the unbalance between calorie intake and expenditure is thought to be the most common cause of obesity, emerging in vitro, in vivo, and epidemiological studies link human EDCs exposure with obesity, metabolic syndrome, and type 2 diabetes. A number of studies are available showing a large variety of compounds to be epidemiologically correlated with the occurrence of obesity and type 2 diabetes. However, detailed mechanistic information on how these organic pollutants interfere with adipocyte metabolism is lacking. Presently available data is sparse and fragmented. However, more research efforts on the interaction of compounds with beta-cell function and/or mass in animal models at human relevant concentrations are needed to further evaluate the hypothesis that environmental pollutants can be additional risk factors for obesity and diabetes development. Targeting adipocyte biology may help to elucidate the underlying pathways that regulate beta cell function and reveal key regulator and signaling genes and molecules targeted by and susceptible for xenobiotic chemicals. Multigenerational studies in animal and human models with long-term exposures at low doses will definitely ameliorate risk assessment of such endocrine disrupting chemicals. Simultaneously, more targeted epidemiological research and awareness should be undertaken and broadened towards the general population. In-depth information on different exposure scenarios and effect levels of these chemicals in the popular media can help individuals to act on a precautionary principle to reduce exposure of themselves and their children ahead of any regulatory actions and therefore have some degree of self-determination in their own exposure to such toxins.

References

1. Hebert JR et al (2013) Scientific decision making, policy decisions, and the obesity pandemic. Mayo Clin Proc 88(6):593–604. https://doi.org/10.1016/j.mayocp.2013.04.005
2. WHO (2010) Global status report on non-communicable diseases. http://apps.who.int/iris/bitstream/10665/44579/1/9789240686458_eng.pdf
3. Sturm R (2002) The effects of obesity, smoking, and drinking on medical problems and costs. Health Aff Health Aff 21(2):245–253. https://doi.org/10.1377/hlthaff.21.2.245
4. Gennuso K et al (2013) Sedentary behavior physical activity, and markers of health in older adults. Med Sci Sports Exerc. https://doi.org/10.1249/MSS.0b013e318288a1e5
5. Hamilton M et al (2007) 'Role of low energy expenditure and sitting in obesity, metabolic syndrome, type 2 diabetes, and cardiovascular disease 56(November):2655–2667. https://doi.org/10.2337/db07–0882.cvd
6. Goodman JE et al (2009) Weight-of-evidence evaluation of reproductive and developmental effects of low doses of Bisphenol A. Crit Rev Toxicol 39(1):1–75. Taylor & Francis. https://doi.org/10.1080/10408440802157839
7. Yanev S, Chaldakov G (2012) Adipose tissue: a master in toxicology. Adipobiology. https://doi.org/10.14748/adipo.v4.281
8. Yanev S, Chaldakov GN (2012) Adipotoxicology of obesity and related diseases. Biomed Rev 23:53–60
9. Girard J, Lafontan M (2008) Impact of visceral adipose tissue on liver metabolism and insulin resistance. Part II: Visceral adipose tissue production and liver metabolism. Diab Metab. https://doi.org/10.1016/j.diabet.2008.04.002
10. Lefterova MI, Lazar MA (2009) New developments in adipogenesis. Trends Endocrinol Metab 20(3):107–114. https://doi.org/10.1016/j.tem.2008.11.005
11. Gregoire F (2001) Adipocyte differentiation: from fibroblast to endocrine cell. Exp Biol Med. https://doi.org/10.1177/153537020122601106
12. Rosen ED et al (2000) Transcriptional regulation of adipogenesis, pp 1293–1307
13. Diamanti-Kandarakis E et al (2009) Endocrine-disrupting chemicals: an Endocrine Society scientific statement. Endocr Rev 30(4):293–342. The Endocrine Society. https://doi.org/10.1210/er.2009-0002
14. Gore AC et al (2015) EDC-2: The Endocrine Society's second scientific statement on endocrine-disrupting chemicals. Endocr Rev. Endocr Soc 36(6):E1–E150. https://doi.org/10.1210/er.2015-1010
15. Zoeller RT et al (2012) Endocrine-disrupting chemicals and public health protection: a statement of principles from The Endocrine Society. Endocrinology 153(9):4097–4110. Endocrine Society. https://doi.org/10.1210/en.2012-1422
16. Dodds EC, Lawson W (1936) Synthetic strogenic agents without the phenanthrene nucleus. Nature 137(3476):996. https://doi.org/10.1038/137996a0
17. Le HH et al (2008) Bisphenol A is released from polycarbonate drinking bottles and mimics the neurotoxic actions of estrogen in developing cerebellar neurons. Toxicol Lett 176(2):149–156. https://doi.org/10.1016/j.toxlet.2007.11.001

18. Hugo ER et al (2008) Bisphenol A at environmentally relevant doses inhibits adiponectin release from human adipose tissue explants and adipocytes. Environ Health Perspect 116 (12):1642–1647. National Institute of Environmental Health Sciences. https://doi.org/10.1289/ehp.11537
19. Biedermann S, Tschudin P, Grob K (2010) Transfer of Bisphenol A from thermal printer paper to the skin. Anal Bioanal Chem 398(1):571–576. https://doi.org/10.1007/s00216-010-3936-9
20. Liao C, Kannan K (2011) Widespread occurrence of Bisphenol A in paper and paper products: implications for human exposure. Environ Sci Technol 45(21):9372–9379. American Chemical Society. https://doi.org/10.1021/es202507f
21. Liao C, Liu F, Kannan K (2012) Bisphenol S, a new Bisphenol analogue, in paper products and currency bills and its association with Bisphenol A residues. Environ Sci Technol 46 (12):6515–6522. American Chemical Society. https://doi.org/10.1021/es300876n
22. Maserejian NN et al (2014) Dental sealants and composite restorations and longitudinal changes in immune function markers in children. Int J Paediatr Dent 24(3):215–225. https://doi.org/10.1111/ipd.12064
23. Rochester JR (2013) Bisphenol A and human health: a review of the literature. Reprod Toxicol 42:132–155. https://doi.org/10.1016/j.reprotox.2013.08.008
24. Fernandez MF et al (2007) Bisphenol-A and chlorinated derivatives in adipose tissue of women. Reprod Toxicol 24(2):259–264. https://doi.org/10.1016/j.reprotox.2007.06.007
25. Gerona RR et al (2013) 'Bisphenol-A (BPA), BPA glucuronide, and BPA sulfate in midgestation umbilical cord serum in a northern and central California population. Environ Sci Technol 47(21):12477–12485. https://doi.org/10.1021/es402764d
26. Patterson TA et al (2013) Concurrent determination of Bisphenol A pharmacokinetics in maternal and fetal rhesus monkeys. Toxicol Appl Pharmacol 267(1):41–48. https://doi.org/10.1016/j.taap.2012.12.006
27. Veiga-Lopez A et al (2015) Impact of gestational Bisphenol A on oxidative stress and free fatty acids: human association and interspecies animal testing studies. Endocrinology 156 (3):911–922. Endocrine Society. https://doi.org/10.1210/en.2014-1863
28. Teeguarden J et al (2013) Are typical human serum BPA concentrations measurable and sufficient to be estrogenic in the general population? Food Chem Toxicol 62:949–963. https://doi.org/10.1016/j.fct.2013.08.001
29. Rubin BS et al (2001) Perinatal exposure to low doses of Bisphenol A affects body weight, patterns of estrous cyclicity, and plasma LH levels. Environ Health Perspect 109(7):675–680. https://doi.org/10.1289/ehp.01109675
30. Miyawaki J et al (2007) Perinatal and postnatal exposure to Bisphenol A increases adipose tissue mass and serum cholesterol level in mice. J Atheroscler Thromb 14(5):245–252. https://doi.org/10.5551/jat.E486
31. Ryan KK et al (2010) Perinatal exposure to Bisphenol-A and the development of metabolic syndrome in CD-1 mice. Endocrinology 151(6):2603–2612. The Endocrine Society. https://doi.org/10.1210/en.2009-1218
32. Anderson OS et al (2013) Perinatal Bisphenol A exposure promotes hyperactivity, lean body composition, and hormonal responses across the murine life course', FASEB journal: official publication of the Federation of American Societies for Experimental Biology. Fed Am Soc Exp Biol 27(4):1784–1792. https://doi.org/10.1096/fj.12-223545
33. Hatch EE et al (2015) Prenatal diethylstilbestrol exposure and risk of obesity in adult women. J Dev Orig Health Dis. 6(3)201–207. Cambridge University Press. https://doi.org/10.1017/s2040174415000033
34. Yang M et al (2016) Bisphenol A promotes adiposity and inflammation in a nonmonotonic dose-response way in 5-week-old male and female C57BL/6J mice fed a low-calorie diet. Endocrinology 157(6):2333–2345. https://doi.org/10.1210/en.2015-1926
35. Lang IA et al (2008) Association of urinary Bisphenol A concentration with medical disorders and laboratory abnormalities in adults. JAMA 300(11):1303–1310. https://doi.org/10.1001/jama.300.11.1303

36. Braun JM et al (2014) 'Early-life Bisphenol A exposure and child body mass index: a prospective cohort study. Environ Health Perspect. 122(11):1239–1245. NLM-Export. https://doi.org/10.1289/ehp.1408258
37. Alonso-Magdalena P et al (2010) Bisphenol A exposure during pregnancy disrupts glucose homeostasis in mothers and adult male offspring. Environm Health Perspect 118(9):1243–1250. National Institute of Environmental Health Sciences. https://doi.org/10.1289/ehp.1001993
38. Wei J et al (2011) Perinatal exposure to Bisphenol A at reference dose predisposes offspring to metabolic syndrome in adult rats on a high-fat diet. Endocrinology 152(8):3049–3061. https://doi.org/10.1210/en.2011-0045
39. Alonso-Magdalena P et al (2006) The estrogenic effect of Bisphenol A disrupts pancreatic beta-cell function in vivo and induces insulin resistance. Environ Health Perspect. 114 (1):106–112. National Institute of Environmental Health Sciences. https://doi.org/10.1289/ehp.8451
40. Batista TM et al (2012) Short-term treatment with Bisphenol-A leads to metabolic abnormalities in adult male mice. PloS One 7(3):e33814–e33814. Public Library of Science. https://doi.org/10.1371/journal.pone.0033814
41. Masuno H et al (2002) Bisphenol A in combination with insulin can accelerate the conversion of 3T3-L1 fibroblasts to adipocytes. J Lipid Res 43:676–684
42. Masuno H et al (2005) Bisphenol A accelerates terminal differentiation of 3T3-L1 cells into adipocytes through the phosphatidylinositol 3-kinase pathway. Toxicol Sci 84(2):319–327. https://doi.org/10.1093/toxsci/kfi088
43. U.S.Deaprtment of Public Health and Services (1998) Toxicological profile for chlorinated dibenzo-p-dioxins
44. Schecter A et al (2002) Characterization of dioxin exposure in firefighters, residents, and chemical workers in the Irkutsk Region of Russian Siberia. Chemosphere 47(2):147–156. https://doi.org/10.1016/S0045-6535(01)00197-7
45. Harrad S et al (2003) Human dietary intake and excretion of dioxin-like compounds. J Environ Monit: JEM. https://doi.org/10.1039/b211406b
46. Pompa G, Caloni F, Fracchiolla ML (2003) Dioxin and PCB contamination of fish and shellfish: assessment of human exposure. Rev Int Situat Vet Res Commun. https://doi.org/10.1023/B:VERC.0000014134.23782.10
47. Malisch R, Kotz A (2014) Dioxins and PCBs in feed and food—review from European perspective. Sci Total Environ 491–492:2–10. https://doi.org/10.1016/j.scitotenv.2014.03.022
48. Birnbaum LS (1994) Endocrine effects of prenatal exposure to PCBs, dioxins, and other xenobiotics: implications for policy and future research. Environ Health Perspect 102 (8):676–679. https://doi.org/10.1289/ehp.94102676
49. Aylward LL, Hays SM (2002) Temporal trends in human TCDD body burden: decreases over three decades and implications for exposure levels. J Eposure Sci Environ Epidemiol 12 (5):319–328. https://doi.org/10.1038/sj.jea.7500233
50. Zook DR, Rappe C (1994) Environmental sources, distribution, and fate of polychlorinated dibenzodioxins, dibenzofurans, and related organochlorines. In: Schecter A (ed) BT-dioxins and health. Springer, Boston, MA US, pp 79–113. https://doi.org/10.1007/978-1-4899-1462-0_3
51. Birnbaum L, Tuomisto J (2000) Non-carcinogenic effects of TCDD in animals. Food Addit Contam. https://doi.org/10.1080/026520300283351
52. Swedenborg E et al (2009) Endocrine disruptive chemicals : mechanisms of action and involvement in metabolic disorders. https://doi.org/10.1677/jme-08-0132
53. Casals-Casas C, Desvergne B (2011) Endocrine disruptors: from endocrine to metabolic disruption. Annu Rev Physiol Annu Rev 73(1):135–162. https://doi.org/10.1146/annurev-physiol-012110-142200

54. Hectors TLM et al (2011) Environmental pollutants and type 2 diabetes: a review of mechanisms that can disrupt beta cell function. Diabetologia 54(6):1273–1290. https://doi.org/10.1007/s00125-011-2109-5
55. Seefeld MD, Keesey RE, Peterson RE (1984) Body weight regulation in rats treated with 2,3,7,8-tetrachlorodibenzo-p-dioxin. Toxicol Appl Pharmacol 76(3):526–536. https://doi.org/10.1016/0041-008X(84)90357-0
56. Zhu BT et al (2008) Effect of 2,3,7,8-tetrachlorodibenzo-p-dioxin administration and high-fat diet on the body weight and hepatic estrogen metabolism in female C3H/HeN mice. Toxicol Appl Pharmacol 226(2):107–118. https://doi.org/10.1016/j.taap.2007.08.018
57. Enan E, Liu PCC, Matsumura F (1992) 2,3,7,8-tetrachlorodibenzo-p-dioxin, pp 19785–19791
58. Enan E, Matsumura F (1993) 2,3,7,8-tetrachlorodibenzo-p-dioxin induced alterations in protein phosphorylation in guinea pig adipose tissue. J Biochem Toxicol. https://doi.org/10.1002/jbt.2570080206
59. Ishida T et al (2005) 2,3,7,8-tetrachlorodibenzo-p-dioxin-induced change in intestinal function and pathology: evidence for the involvement of arylhydrocarbon receptor-mediated alteration of glucose transportation. Toxicol Appl Pharmacol 205(1):89–97. https://doi.org/10.1016/j.taap.2004.09.014
60. Kurita H et al (2009) Aryl hydrocarbon receptor-mediated effects of 2,3,7,8-tetrachlorodibenzo-p-dioxin on glucose-stimulated insulin secretion in mice. J Appl Toxicol 29(8):689–694. John Wiley & Sons, Ltd. https://doi.org/10.1002/jat.1459
61. Dalton T et al (2001) Dioxin exposure is an environmental risk factor for ischemic heart disease. Cardiovasc Toxicol. https://doi.org/10.1385/CT:1:4:285
62. Michalek JE, Pavuk M (1980) Diabetes and cancer in veterans of operation ranch hand after adjustment for calendar period, days of spraying, and time spent in Southeast Asia, pp 330–340. https://doi.org/10.1097/jom.0b013e31815f889b
63. Kang HK et al (2006) Health status of army chemical corps Vietnam veterans who sprayed defoliant in Vietnam. Am. J. Ind Med. Wiley 49(11):875–884. https://doi.org/10.1002/ajim.20385
64. Vena J et al (1998) Exposure to dioxin and nonneoplastic mortality in the expanded IARC international cohort study of phenoxy herbicide and chlorophenol production workers and sprayers. Environ Health Perspect 106 Suppl 2(Suppl 2):645–653. https://doi.org/10.1289/ehp.98106645
65. Wang SL et al (2008) Increased risk of diabetes and polychlorinated biphenyls and dioxins: a 24-year follow-up study of the Yucheng cohort. Diabetes Care 31(8):1574–1579. American Diabetes Association. https://doi.org/10.2337/dc07-2449
66. Turyk M et al (2009) Organochlorine exposure and incidence of diabetes in a cohort of Great Lakes sport fish consumers. Environ Health Perspect 117(7):1076–1082. National Institute of Environmental Health Sciences. https://doi.org/10.1289/ehp.0800281
67. Turunen AW et al (2008) Mortality in a cohort with high fish consumption. Int J Epidemiol 37(5):1008–1017. https://doi.org/10.1093/ije/dyn117
68. Lee DH et al (2010) Low dose of some persistent organic pollutants predicts type 2 diabetes: a nested case-control study. Environ Health Perspect 118(9):1235–1242. National Institute of Environmental Health Sciences. https://doi.org/10.1289/ehp.0901480
69. Lee DH et al (2011) Polychlorinated biphenyls and organochlorine pesticides in plasma predict development of type 2 diabetes in the elderly. Diab Care 34(8):1778 LP–1784. https://doi.org/10.2337/dc10-2116
70. Lee DH et al (2007) Relationship between serum concentrations of persistent organic pollutants and the prevalence of metabolic syndrome among non-diabetic adults: results from the National Health and Nutrition Examination Survey 1999–2002. Diabetologia 50 (9):1841–1851. https://doi.org/10.1007/s00125-007-0755-4
71. Uemura H et al (2009) Prevalence of metabolic syndrome associated with body burden levels of dioxin and related compounds among Japan's general population. Environ Health Perspect 117(4):568–573. National Institute of Environmental Health Sciences. https://doi.org/10.1289/ehp.0800012

72. Chang J et al (2010) Dioxin exposure and insulin resistance in taiwanese living near a highly contaminated area 21(1). https://doi.org/10.1097/ede.0b013e3181c2fc6e
73. Kurzer MS, Xu X (1997) DIETARY PHYTOESTROGENS. Annu Rev Nutr Annu Rev 17 (1):353–381. https://doi.org/10.1146/annurev.nutr.17.1.353
74. Setchell KDR, Brown NM, Lydeking-Olsen E (2002) The clinical importance of the metabolite equol—a clue to the effectiveness of soy and its isoflavones. J Nutr 132 (12):3577–3584. https://doi.org/10.1093/jn/132.12.3577
75. Lampe JW et al (1998) Urinary equol excretion with a soy challenge: influence of habitual diet. Proc Soc Exp Biol Med 217(3):335–339. SAGE Publications. https://doi.org/10.3181/00379727-217-44241
76. Setchell KDR et al (2003) Bioavailability, disposition, and dose-response effects of soy isoflavones when consumed by healthy women at physiologically typical dietary intakes. J Nutr 133(4):1027–1035. https://doi.org/10.1093/jn/133.4.1027
77. Gorski RA (1963) Modification of ovulatory mechanisms by postnatal administration of estrogen to the rat. Am J Physiol-Legacy Content 205(5):842–844. American Physiological Society. https://doi.org/10.1152/ajplegacy.1963.205.5.842
78. Lindzey J, Korach KS (1997) Developmental and physiological effects of estrogen receptor gene disruption in mice. Trends Endocrinol Metab 8(4):137–145. https://doi.org/10.1016/S1043-2760(97)00007-6
79. Simerly RB (2002) Wired for reproduction: organization and development of sexually dimorphic circuits in the mammalian forebrain. Annu Rev Neurosci Annu Rev 25(1):507–536. https://doi.org/10.1146/annurev.neuro.25.112701.142745
80. Crain DA et al (2008) Female reproductive disorders: the roles of endocrine-disrupting compounds and developmental timing. Fertil Steril 90(4):911–940. https://doi.org/10.1016/j.fertnstert.2008.08.067
81. Newbold RR (2008) Prenatal exposure to diethylstilbestrol (DES). Fertil Steril. Elsevier 89 (2):e55–e56. https://doi.org/10.1016/j.fertnstert.2008.01.062
82. Hooper L et al (2009) Effects of soy protein and isoflavones on circulating hormone concentrations in pre- and post-menopausal women: a systematic review and meta-analysis, Hum Reprod Update 15(4):423–440. Oxford University Press. https://doi.org/10.1093/humupd/dmp010
83. Chandrareddy A et al (2008) Adverse effects of phytoestrogens on reproductive health: a report of three cases. Complement Ther Clin Pract 14(2):132–135. https://doi.org/10.1016/j.ctcp.2008.01.002
84. Lephart ED et al (2004) Behavioral effects of endocrine-disrupting substances: phytoestrogens. ILAR J 45(4):443–454. https://doi.org/10.1093/ilar.45.4.443
85. Patisaul H et al (2005) Dietary soy supplements produce opposite effects on anxiety in intact male and female rats in the elevated plus-maze. Behav Neurosci. https://doi.org/10.1037/0735-7044.119.2.587
86. .Patisaul HB (2005) Phytoestrogen action in the adult and developing brain. J Neuroendocrinol 17(1):57–64. John Wiley & Sons, Ltd. https://doi.org/10.1111/j.1365-2826.2005.01268.x
87. Patisaul HB et al (2012) Anxiogenic effects of developmental Bisphenol A exposure are associated with gene expression changes in the juvenile rat amygdala and mitigated by soy 7 (9):e43890–e43890. PloS One. Public Library of Science. https://doi.org/10.1371/journal.pone.0043890
88. WHO (1992) Our planet, our health. WHO; Geneva, Switzerland: 1992. Report of the WHO Commission on Health and Environment
89. Akhtar N (1996) Insecticide-induced changes in secretory activity of the thyroid gland in rats. J Appl Toxicol 16(5):397–400. Wiley Ltd. https://doi.org/10.1002/(sici)1099-1263(199609)16:5%3c397::aid-jat362%3e3.0.co;2-y

90. Cocco P (2002) On the rumors about the silent spring: review of the scientific evidence linking occupational and environmental pesticide exposure to endocrine disruption health effects. Cadernos de Saúde Pública. scielo, pp 379–402
91. Leghait J et al (2009) Fipronil-induced disruption of thyroid function in rats is mediated by increased total and free thyroxine clearances concomitantly to increased activity of hepatic enzymes. Toxicology 255(1):38–44. https://doi.org/10.1016/j.tox.2008.09.026
92. Birnbaum LS, Fenton SE (2003) Cancer and developmental exposure to endocrine disruptors. Environ Health Perspect 111(4):389–394. https://doi.org/10.1289/ehp.5686
93. Goldman L et al (2004) Environmental pediatrics and its impact on government health policy. Pediatrics 113(Supplement 3):1146 LP–1157. http://pediatrics.aappublications.org/content/113/Supplement_3/1146.abstract
94. Sharpe RM (2006) Pathways of endocrine disruption during male sexual differentiation and masculinisation. Best Pract Res Clin Endocrinol Metab 20(1):91–110. https://doi.org/10.1016/j.beem.2005.09.005
95. Sultan C et al (2001) Environmental xenoestrogens, antiandrogens and disorders of male sexual differentiation. Mol Cell Endocrinol 178(1):99–105. https://doi.org/10.1016/S0303-7207(01)00430-0
96. Skakkebæk NE (2002) Endocrine disrupters and testicular dysgenesis syndrome. Hormon Res Paediatr 57(suppl 2):43. https://doi.org/10.1159/000058100
97. Roeleveld N, Bretveld R (2008) The impact of pesticides on male fertility. Curr Opin Obstet Gynecol. https://doi.org/10.1097/GCO.0b013e3282fcc334
98. Falck Jr F et al (1992) Pesticides and PCB residues in human breast lipids and their relation to breast cancer. Arch Environ Health
99. Davis DL et al (1993) Medical hypothesis: xenoestrogens as preventable causes of breast cancer. Environ Health Perspect 101(5):372–377. https://doi.org/10.1289/ehp.93101372
100. Parron Carreño T et al (2010) Increased breast cancer risk in women with environmental exposure to pesticides. Toxicol Lett. https://doi.org/10.1016/j.toxlet.2010.03.614
101. Alavanja MCR et al (2005) Cancer incidence in the agricultural health study. Scand J Work Environ Health (1):39–45. http://www.sjweh.fi/show_abstract.php?abstract_id=895
102. Prins GS (2008) Endocrine disruptors and prostate cancer risk. Endocr Relat Cancer 15(3):649–656. https://doi.org/10.1677/erc-08-0043
103. Dich J, Wiklund K (1998) Prostate cancer in pesticide applicators in Swedish agriculture. Prostate 34(2):100–112. Wiley Ltd. https://doi.org/10.1002/(sici)1097-0045(19980201)34:2%3c100::aid-pros4%3e3.0.co;2-o
104. Alavanja MCR et al (2003) Use of agricultural pesticides and prostate cancer risk in the agricultural health study cohort. Am J Epidemiol 157(9):800–814. https://doi.org/10.1093/aje/kwg040
105. Bryan GW et al (1986) The Decline of the gastropod Nucella lapillus around South-West England: evidence for the effect of tributyltin from antifouling paints. J Mar Biol Assoc UK 66(3):611–640. Cambridge University Press. https://doi.org/10.1017/s0025315400042247
106. Short J et al (1989) Occurrence of tri-n-butyltin-caused imposex in the North Pacific marine snail Nucella lima in Auke Bay, Alaska. Mar Biol. https://doi.org/10.1007/BF00428480
107. Ellis DV, Agan Pattisina L (1990) Widespread neogastropod imposex: a biological indicator of global TBT contamination? Mar Pollut Bull 21(5):248–253. https://doi.org/10.1016/0025-326X(90)90344-8
108. Heidrich DD, Steckelbroeck S, Klingmuller D (2001) Inhibition of human cytochrome P450 aromatase activity by butyltins. Steroids 66(10):763–769. https://doi.org/10.1016/S0039-128X(01)00108-8
109. Gooding MP et al (2003) The biocide tributyltin reduces the accumulation of testosterone as fatty acid esters in the mud snail (Ilyanassa obsoleta). Environ Health Perspect 111(4):426–430. https://doi.org/10.1289/ehp.5779

110. Bishop CA et al (1995) Chlorinated hydrocarbons in early life stages of the common snapping turtle (Chelydra serpentina serpentina) from a coastal wetland on lake Ontario, Canada. Environ Toxicol Chem 14(3):421–426. Wiley Ltd. https://doi.org/10.1002/etc.5620140311
111. Guillette LJ Jr et al (1994) Developmental abnormalities of the gonad and abnormal sex hormone concentrations in juvenile alligators from contaminated and control lakes in Florida. Environ Health Perspect 102(8):680–688. https://doi.org/10.1289/ehp.94102680
112. Guillette LJ Jr et al (1996) Reduction in penis size and plasma testosterone concentrations in juvenile alligators living in a contaminated environment. Gen Comp Endocrinol 101(1):32–42. https://doi.org/10.1006/gcen.1996.0005
113. Guillette L et al (1999) Serum concentrations of various environmental contaminants and their relationship to sex steroid concentrations and phallus size in juvenile American alligators. Arch Environ Contam Toxicol. https://doi.org/10.1007/PL00006617
114. Fry DM, Toone CK (1981) DDT-induced feminization of gull embryos. Science 213 (4510):922 LP–924. https://doi.org/10.1126/science.7256288
115. Munkittrick KR et al (1991) Impact of bleached kraft mill effluent on population characteristics, liver MFO activity, and serum steroid levels of a Lake Superior white sucker (Catostomus commersoni) population. Can J Fish Aquat Sci 48(8):1371–1380. NRC Research Press. https://doi.org/10.1139/f91-164
116. Purdom CE et al (1994) Estrogenic effects of effluents from sewage treatment works. Chem Ecol 8(4):275–285. Taylor & Francis. https://doi.org/10.1080/02757549408038554
117. Hand JL, Southern E (1985) Ecology and behavior of gulls. In: Proceedings of an international symposium of the colonial waterbird group and the pacific seabird group, San Francisco, California. Studies in Avian Biology No. 10
118. Crisp TM et al (1998) Environmental endocrine disruption: an effects assessment and analysis. Environ Health Perspect 106(Suppl 1):11–56. https://doi.org/10.1289/ehp.98106s111
119. Tyler CR, Jobling S, Sumpter JP (1998) Endocrine disruption in wildlife: a critical review of the evidence. Crit Rev Toxicol 28(4):319–361. Taylor & Francis. https://doi.org/10.1080/10408449891344236
120. Vos JG et al (2000) Health effects of endocrine-disrupting chemicals on wildlife, with special reference to the european situation. Crit Rev Toxicol 30(1):71–133. Taylor & Francis. https://doi.org/10.1080/10408440091159176
121. Reijnders PJH (1986) Reproductive failure in common seals feeding on fish from polluted coastal waters. Nature 324(6096):456–457. https://doi.org/10.1038/324456a0
122. Norstrom RJ, Muir DCG (1994) Chlorinated hydrocarbon contaminants in arctic marine mammals. Sci Total Environ 154(2):107–128. https://doi.org/10.1016/0048-9697(94)90082-5
123. Facemire CF, Gross TS Guillette Jr LJ (1995) Reproductive impairment in the Florida panther: nature or nurture?. Environ Health Perspect 103(Suppl 4):79–86. https://doi.org/10.1289/ehp.103-1519283
124. Oskam I et al (2003) Organochlorines affect the major androgenic hormone, testosterone, in male polar bears (Ursus maritimus) at Svalbard. J Toxicol Environ Health. Part A. https://doi.org/10.1080/15287390390211342
125. Palma P et al (2009) Assessment of the pesticides atrazine, endosulfan sulphate and chlorpyrifos for juvenoid-related endocrine activity using Daphnia magna. Chemosphere 76 (3):335–340. https://doi.org/10.1016/j.chemosphere.2009.03.059
126. Palma P et al (2009) Effects of atrazine and endosulfan sulphate on the ecdysteroid system of Daphnia magna. Chemosphere 74(5):676–681. https://doi.org/10.1016/j.chemosphere.2008.10.021
127. Wilson VS et al (2009) The herbicide linuron reduces testosterone production from the fetal rat testis during both in utero and in vitro exposures. Toxicol Lett 186(2):73–77. https://doi.org/10.1016/j.toxlet.2008.12.017

128. Thayer KA et al (2012) Role of environmental chemicals in diabetes and obesity: a National Toxicology Program workshop review. Environ Health Perspect 120(6):779–789. National Institute of Environmental Health Sciences. https://doi.org/10.1289/ehp.1104597
129. Bourez S et al (2013) The dynamics of accumulation of PCBs in cultured adipocytes vary with the cell lipid content and the lipophilicity of the congener. Toxicol Lett 216(1):40–46. https://doi.org/10.1016/j.toxlet.2012.09.027
130. Elobeid MA et al (2010) Endocrine disruptors and obesity: an examination of selected persistent organic pollutants in the NHANES 1999-2002 data. Int J Environ Res Public Health 7(7):2988–3005. Molecular Diversity Preservation International (MDPI). https://doi.org/10.3390/ijerph7072988
131. Heudorf U, Mersch-Sundermann V, Angerer J (2007) Phthalates: toxicology and exposure. Int J Hyg Environ Health 210(5):623–634. https://doi.org/10.1016/j.ijheh.2007.07.011
132. Hannon PR, Flaws JA (2015) The effects of phthalates on the ovary. Front Endocrinol 6:8. Frontiers Media S.A. https://doi.org/10.3389/fendo.2015.00008
133. Huang P et al (2012) Phthalates exposure and endocrinal effects : an epidemiological review phthalates exposure and endocrinal effects : an epidemiological review. https://doi.org/10.6227/jfda.2012200401
134. Kamrin MA (2014) Phthalate risks, phthalate regulation, and public health : a review. https://doi.org/10.1080/10937400902729226
135. Hao C et al (2012) The endocrine disruptor mono-(2-ethylhexyl) phthalate promotes adipocyte differentiation and induces obesity in mice. Biosci Rep 32(6):619–629. Portland Press Ltd. https://doi.org/10.1042/bsr20120042
136. Yaghjyan L et al (2016) Maternal exposure to di-2-ethylhexylphthalate and adverse delivery outcomes: A systematic review. Reprod Toxicol 65:76–86. https://doi.org/10.1016/j.reprotox.2016.07.002
137. Wassenaar PNH, Legler J (2017) Systematic review and meta-analysis of early life exposure to di(2-ethylhexyl) phthalate and obesity related outcomes in rodents. Chemosphere 188:174–181. https://doi.org/10.1016/j.chemosphere.2017.08.165
138. Schmidt JS et al (2012) Effects of di(2-ethylhexyl) phthalate (DEHP) on female fertility and adipogenesis in C3H/N mice. Environ Health Perspect 120(8):1123–1129. National Institute of Environmental Health Sciences. https://doi.org/10.1289/ehp.1104016
139. Oken E, Levitan EB, Gillman MW (2008) Maternal smoking during pregnancy and child overweight: systematic review and meta-analysis. Int J Obes (2005) 32(2):201–210. https://doi.org/10.1038/sj.ijo.0803760
140. Ino T (2010) Maternal smoking during pregnancy and offspring obesity: meta-analysis. Pediatr Int 52(1):94–99. Wiley Ltd. (10.1111). https://doi.org/10.1111/j.1442-200x.2009.02883.x
141. Riedel C et al (2014) Parental smoking and childhood obesity: higher effect estimates for maternal smoking in pregnancy compared with paternal smoking—a meta-analysis. Int J Epidemiol 43(5):1593–1606. https://doi.org/10.1093/ije/dyu150
142. Rayfield S, Plugge E (2017) Systematic review and meta-analysis of the association between maternal smoking in pregnancy and childhood overweight and obesity. J Epidemiol Community Health 71(2):162 LP–173. https://doi.org/10.1136/jech-2016-207376

Chapter 3
Obesity and Its Complications Pathogenesis

Isabella So and Hariom Yadav

Abstract Obesity became an epidemic public health problem and continuously increasing every year. For the longest time, the pathologic framework for obesity had been simple: caloric intake is greater than the energy output, which by far still true, however, the underlying reasons for increased energy intake and reduced energy expenditure remain largely unknown. Emerging data indicate that obesity can be oftentimes secondary to an underlying pathology, which can be due to genetics, inflammation, microbiome dysbiosis, metabolism, drugs and several other reasons. In this chapter, we described the pathology of obesity and underlying factors are linked or contributing in its progression.

Keyword Obesity · Diet · Nutrition · Microbiome · Genes · Life style · Inflammation · Comorbidities · Diabetes · Cancer · Arthritis · Drugs

Introduction

In the present climate of health and nutrition, global trends indicate that by 2030, almost 50% of the world's adult population will be considered overweight or obese [1]. The rising costs of obesity to the healthcare systems are compounded by accompanying comorbidities such as Type 2 Diabetes, cardiovascular disease, dementia, cancer, and chronic pain. These comorbidities put the population at risk for reversing the improvements in life expectancy in the last 50 years, especially when examining the model of the United States. The US's rising BMI levels, which are already the highest of any other country, have led to slowed rates of improvement in mortality by 23%, compared to an average 2.21% slowed rate seen in other countries [2].

I. So · H. Yadav (✉)
Department of Internal Medicine-Molecular Medicine, Wake Forest School of Medicine, Winston-Salem, NC 27101, USA
e-mail: hyadav@wakehealth.edu

P. S. Tappia et al. (eds.), *Pathophysiology of Obesity-Induced Health Complications*, Advances in Biochemistry in Health and Disease 19,
https://doi.org/10.1007/978-3-030-35358-2_3

Perhaps what makes obesity the deadly condition that it is are its accompanying comorbidities that further decrease mortality in patients. Obesity has a known association with several diseases that affect different organs and systems. Type 2 Diabetes (T2D) has long been established as comorbidity of obesity, due to the mechanism of insulin resistance development [3]. While the association is clear, the physiological link between cardiovascular disease (CVD) and obesity is not as explicit, however may be due to higher concentrations of visceral fat leading to higher rates of cardiac inflammation. Cancer has a known association with obesity, especially cancers of gallbladder, esophagus, thyroid, kidney, uterus, colon, and breast. The underlying mechanisms of all cancer due to obesity are not always clear, although may be largely attributed to chronic inflammation. For uterus and breast cancers, it is hypothesized to be due to elevated estrogen levels synthesized from fat tissue in obese women [3]. Chronic pain's relationship to obesity is hypothesized to be due to many differing metabolic components, such as C-reactive protein, an inflammatory mediator, and leptin, an energy-regulating hormone dispensed by adipocytes. Osteoarthritis is a non-inflammatory arthritis that mostly affects weight-bearing joints such as the spine, hip, knee, and ankle [4]. Obesity's role in osteoarthritis is a direct one in which increased weight leads to increased mechanical force on these joints, becoming a leading cause of disability. Dementia and other neuropsychiatric processes' causal connection from obesity has been documented by many, but the mechanism remains largely unclear.

The classic understanding of obesity is an increased caloric intake relative to energy expenditure. Consequently, when the Body Mass Index (BMI) reaches 30.0 kg/m^2 or higher it is considered obesity, which is characterized by the accumulation of excess fat in adipose tissues. However, obesity can also occur secondary to medications, disease, and other dysregulation of the body. For instance, lifestyle and diet contribute to and control the balance between caloric intake and output. Genetics and epigenetics can predispose individuals to different tendencies in storing adipose tissue or driving appetite. The microbiome of the gastrointestinal system can alter the immunologic and absorptive ability of the gut. And, while many illnesses can cause large shifts in metabolism leading to higher adiposity, many pharmacological drugs can enact those similar effects. Figure 3.1 summarizes several factors that can contribute to the development of obesity and conversely several conditions that obesity can lead to.

This chapter will break down and further discuss the commonly known mechanisms in obesity, examine the comorbidities, and analyze strides made in present day diagnoses and treatments. It is important to acknowledge that different etiologies for obesity exist from person-to-person, and in this way treatments must be considered on an individual level.

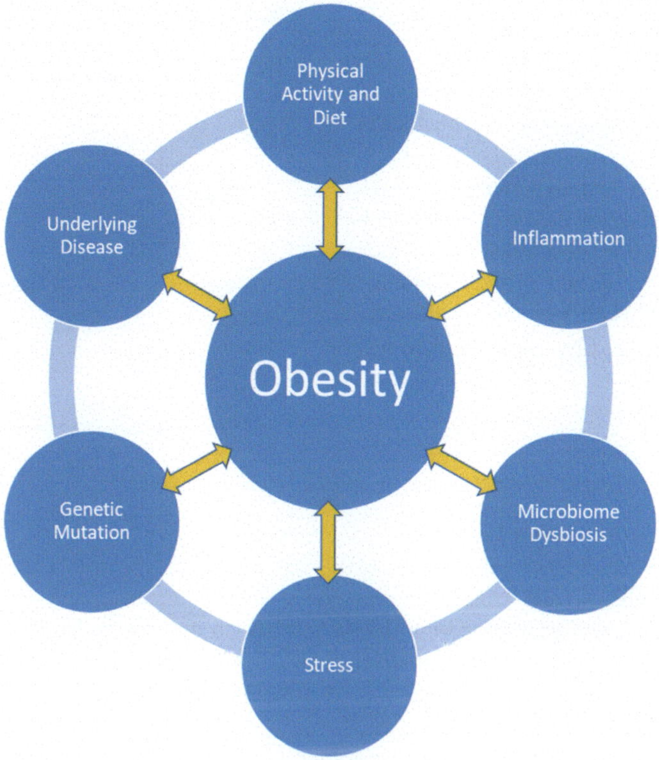

Fig. 3.1 Obesity has a two way relationship with several factors, all of which also must be in balance with each other for prevention

Common Mechanisms Involved in Obesity Pathology

There are many mechanisms involved in the pathology of obesity, and many that contribute to complications of obesity. Of these, the most significant are lifestyle, inflammation, microbiome dysbiosis, and genetic mutations.

Physical Activity and Diet

Examining a patient's lifestyle as a cause for obesity mainly accounts for physical activity and diet. In general, physical activity and diet are the first behavioral factors that are modified in the initial treatment of obesity or pre-obesity. A sedentary lifestyle leads to a lower probability of utilizing physical activity or exercise to lose or maintain weight. This occurs by lowering energy expenditure and expediting age-related loss of lean mass, coactively resulting on fat weight gain. In a 6-year

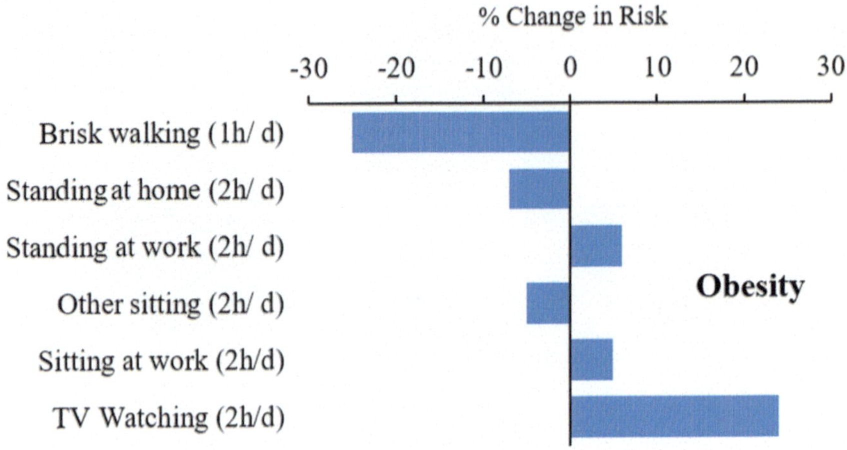

Fig. 3.2 The variable changes in obesity risk based on daily activities [5]

study following 50,277 women with a BMI <30 free from diagnosed cardiovascular disease, diabetes, and cancer, it was found that 2 hr increments per day in TV-watching was associated with a 23% increase in obesity, and 2 hr increments per day sitting at work was associated with a 5% increase in obesity (Fig. 3.2). However, for every 1 hr daily spent brisk walking, there was a 24% reduction in obesity [5].

Furthermore, the total energy expenditure (TEE) is a sum of the basic components of the resting metabolic rate (RMR), thermic effect of feeding (TEF), and physical activity (PA). Physical activity is a particularly important component of energy balance, and its deficiency is associated with higher rates of chronic diseases and reduced life expectancy. In an analysis of the 57 million deaths in 2008, physical inactivity was determined to cause 9% (5.3 million cases) of all premature mortality, and found that physical inactivity was on par with the risk factors established with smoking [6].

Similarly, diet is an important determinant of an individual's tendency towards overweight or obesity. The international food supply since the 1970s has shifted the average diet towards increased caloric intake, increased carbohydrates and refined fats, increased glycemic index, and increased portion sizes at each meal. Specifically, between 1962 and 1990, vegetable oils caused a 10–13% global increase in fat availability, with an overall decrease availability of fruits, fiber, and other complex carbohydrates [7]. These high levels of dietary fats poorly affect metabolism, but furthermore change gut epithelial permeability, ultimately compromising macronutrient absorption, inflammation, and microbiome flora [8]. Similarly, the decrease in fiber means a decrease in bowel motility, intestinal absorption, colonic microbiota, and fermentation.

A randomized study evaluating 20 inpatient adults in 2 separate ad libitum diets ("ultra-processed" and "whole food") matching calories, energy density, macronutrients, sugar, sodium, and fiber for 14 days each demonstrated a 508 ± 106 kcal/day increase in energy intake and 0.9 ± 0.3 kg gain in weight for the ultra-processed diet [9]. These observations under the ultra-processed diet were accompanied by an increased consumption specifically of carbohydrates (280 ± 54 kcal/day) and fat (230 ± 53 kcal/day). The data suggest that removing ultra-processed foods, which are enhanced by their palatability and convenience, can effectively prevent obesity [9].

Stress/Sleep

Psychological stress plays many indirect and direct roles on the development of obesity over time. Stress impacts decision-making capacity and self-regulation. Obesity itself, while it might be the result of stress, can also trigger further a cycle of stress and weight gain. Stress also causes the release of glucocorticoids, which cause physiological changes in the hypothalamic-pituitary-adrenal axis, impacting reward-processing in the brain and can lead to "emotional eating" [10]. When stress becomes chronic or develop to the state of a mental illness such as depression or anxiety, these poor eating behaviors may become more consistent and result in constant caloric excess. The HUNT-2 study, which assessed 65,648 adults between 20 and 89 years of age for their waist-hip ratio (WHR), found that WHR, which is associated with abdominal obesity, was elevated in a dose-response pattern with both anxiety and depression, with the highest prevalence of anxiety and depression occurring in individuals with a WHR above the 95th percentile [11].

Similar to and derived from stress, sleep duration and quality impact the development of obesity and its complications. Sleep is an important component of neuroendocrine function, and consequently of glucose metabolism as well. Glucose control and release at night depend on the deeper stage of non-REM sleep, when brain glucose utilization decreases and parasympathetic activity increases [12]. Leptin, another hormone that is released by adipocytes and enterocytes, is known to signal satiety and energy balance to the brain. Ghrelin acts in an opposite way, promoting hunger. Both leptin and ghrelin interact with the arcuate nucleus to modulate food intake, and are both observed to be affected by sleep restriction. In a short-term view of the effect of sleep on hunger hormones, one study observed 12 young (mean age 22 ± 2 years) lean (BMI 23.6 ± 2.9 kg/m^2) men who slept four hours per night for two days after sleeping ten hours per night for two days [13]. After the two days with four hours of sleep, there was an 18% decrease in leptin and a 28% increase in ghrelin[13].

Inflammation

Obese people have higher low chronic low-grade inflammation. This inflammation is not only seen in the circulation levels but also often seen in the metabolically important organs such as liver, brain, pancreas, and adipose tissue. As a result, the body undergoes a dysregulated immune system, which is often seen as the connection between obesity and its metabolic and vascular complications. Additionally, this can cause the increased risk of cancer and infectious disease in obese patients. It is estimated that 3.6% of all new cancer diagnoses are due to excess adiposity [14].

Adipose tissue is the body's method of maintaining physiologic homeostasis through lipids as a long-term energy reservoir. During obesity, overnutrition occurs with excess caloric intake or reduced caloric expenditure, and adipose tissue enlarges and expands to adapt to higher demands of energy storage. The classic signs of adipose inflammation are the enlarged adipocytes which turn apoptotic, are then surrounded by macrophages, and appear as crown-like structures. These interactions between adipocytes and the immune cells lead to enhanced pro-inflammatory factors, affecting peripheral target tissues and thereby inducing insulin resistant and hyperinsulinemia, hyperglycemia, hyperlipidemia, and vascular injury [14]. However, the precise mechanism(s) and source of triggering this chronic inflammation are not known, and warranted the further research. Inflammation is a natural response for bacterial antigens, hence microbes living in and around our body (microbiome) can contribute in the induction of low grade inflammation of the obese people.

Gut bacteria can instigate obesity- and insulin-resistance (IR)-associated inflammatory state through lipopolysaccharide (LPS; a component of gram-negative bacterial cell-wall), which triggers inflammation by binding to toll-like receptor-4 (TLR-4) complex at the surface of innate immune cells. Deletion of TLR-4 prevents high-fat diet (HFD)-induced IR, suggesting that TLR-4 is implicated in metabolic diseases. HFD-induced obese mice demonstrate fewer gut bifidobacteria and eubacteria and increased circulating LPS levels. In humans, a similar-grade endotoxemia associated to IR, and a high-fat-high-carbohydrate meal induces significant elevations in postprandial plasma LPS, hinting that endotoxemia might play a pathological role in obesity-associated inflammatory state and that the food ingestion may affect plasma endotoxin levels [15].

Microbiome Dysbiosis

Emerging research suggest that the gut of obese people harbor an abnormal (dysbiosis) microbiome. The intestinal microbiome is made up of the numerous and diversified microbial species in the gastrointestinal tract. In the context of obesity-related complications, gut microbes are known to influence host metabolism by means of signaling pathways with effects on inflammation, fat deposition, and insulin resistance [16]. The major roles played by the intestinal microbiome include: (1) resistance to infection from pathogens through direct competition for nutrients and adhesion sites, and production of antimicrobial substances; (2) increased proliferation and differential of the epithelial layer for mucosal surface maintenance; (3) increased development of lymphoid tissue via dendritic cell maturation and lymphocytic differentiation of B and T cells; and (4) production of energy from non-digestible dietary starches [17].

The majority of microorganisms in the adult gut fall under two bacterial phyla, the gram-negative Bacteroides and the gram-positive Firmicutes. Under anaerobic conditions, species belonging to these geni will produce short-chain fatty acids (SCFAs). SCFAs such as acetate (2 carbons) and butyrate (4 carbons) can be utilized by colonocytes to induce secretion of glucagon-like peptide (GLP-1) and peptide YY (PYY). These hormones play a major role in gut homeostasis by increasing nutrient absorption from the intestinal lumen. However, the person-to-person composition differs greatly depending on genetics, age, diet, geographic location, and medication. Dysbiosis occurs when there is an imbalance in the ratio between beneficial and pathogenic bacterial species, and can be induced by these external factors. Gut dysbiosis is known to instigate obesity and T2D. For example, energy-rich diets and obesity involve increased intestinal *Firmicutes-Bacteroidetes* ratio in rodents and humans. Likewise, T2D patients have altered gut microbiota with reduced butyrate-producing genera (e.g., *Roseburia, Subdoligranulum, Clostridiales*) wherein metformin-treatment reduces/reverses changes in Firmicutes population [15].

In one of our studies, we used leptin-deficient ($Lep^{ob/ob}$) mice alongside C57BL/6J control mice fed on identical diets in order to perform comparative and correlative analyses of the gut microbiome composition, gut permeability, intestinal structural changes, tight junction-mucin formation, cellular turnover, and stemness genes. We discovered that associated with the obesity-induced mice were increased cell death, increased intestinal permeability, changes in villi/crypt length, and decreased expression of tight junctions and mucus synthesis, in conjunction with dysbiotic gut microbiome signature [18].

Genetic Mutation

While several environmental and social factors such as sedentary lifestyles, consumption of high-calorie foods, and urbanization have been most heavily emphasized as the causes for obesity, there has been considerable evidence based on twin, adoption, and family studies that indicates 40–70% of BMI variation occurs due to genetic factors [19]. In general, specific genetic syndromes such as Prader-Willi and Bardet-Biedl are recognized for dysmorphic characteristics and developmental dysfunction that result in obesity. However, more recently, monogenic defects have been pinpointed as important causes of obesity.

Monogenic defects typically occur from the disruption of the signaling pathways for leptin and melanocortin. They can result from genetic mutations in the locus of leptin, leptin receptor, proopiomelanocortin (POMC), or other downstream enzymes and neuropeptides of leptin [20]. It has been suggested that the genetic contribution in obese patients is due to multiple genes amplifying an effect, meaning that obesity is a "polygenic" disease state. In a study taking 1,924 T2D patients and 2,938 population controls for 490,032 autosomal single nucleotide polymorphisms (SNPs), the strongest association made with obesity and BMI was consistently found with variants of the FTO (fat mass and obesity) gene on chromosome 16 [20].

Additionally, "epigenetic theory" plays a role in the increased risk of diabetes and obesity. Unregulated elevated consumption of fat and sugar by pregnant women may spur epigenetic changes during embryo development in a child, increasing susceptibility to diabetes in the next generation [21]. Pollution is similarly believed to be involved during the pregnancy and development phase.

Secondary to Primary Disease

Polycystic Ovary Syndrome (*PCOS*) is identified by ovarian cysts, irregular menstrual periods, hirsutism, infertility, acne, and obesity due to dysregulation of androgen secretion resulting in functional ovarian hyperandrogenism [22]. The American College of Obstetrics and Gynecology reports that about 80% of women with PCOS are obese [23]. Hyperinsulinemia is associated with direct hypothalamic outcomes that can favor disordered gonadotropin secretion. In a study where obese mice have a selective knockout of the insulin receptor in the pituitary, the result was normal gonadotropin secretion and improved fertility, providing a possible bridge between the insulin receptor and PCOS [24].

Cushing's Syndrome (*CS*) is a primary adrenal gland disease caused by excess cortisol secretion, a rare disease affecting approximately 1 per million in the U.S. It can be identified clinically by hypercortisolism, diabetes, hypertension, sudden weight gain, and central obesity [25]. In a study examining the frequency of CS in 150 obese patients with no specific clinical symptoms of CS, 9.33% of patients were found to have CS, suggesting that patients with obesity should be screened for CS [26].

Hypothyroidism is characterized by an underactive thyroid gland unable to produce enough thyroid hormone, resulting in symptoms associated with a slowed metabolism. In a study assessing 27,097 individuals with a BMI $\geq$30 kg/m^2, hypothyroidism correlated with higher BMI and higher prevalence of obesity regardless of smoking status [27].

Diagnosis and Treatment of Obesity

The initial diagnosis of obesity begins with a calculation of the BMI to evaluate the degree of excess weight on a patient. The classification that BMI $\geq$ 30 kg/m^2 indicates obesity is a recommendation based on risk of cardiovascular disease (CVD) adopted by the National Institutes of Health (NIH) and the World Health Organization (WHO) [28]. Furthermore, within the diagnosis of obesity, there are three more sub-classifications, on a scale of I to III:

- Class I—30.0 to 34.9 kg/m^2
- Class II—35.0 to 39.9 kg/m^2
- Class III—$\geq$40 kg/m^2 (also referred to as severe, extreme, or massive obesity).

However, the calculated risks associated with these different BMIs are determined for Caucasian, Hispanic, and black populations. They are shifted in Asian and South Asian populations, where a lower threshold for BMI can be associated with the same risk, due to a higher tendency for body fat percentage at the same BMI compared to white individuals [29]. In conjunction with BMI, waist measurement aids in assessing obesity at the level of the abdomen. Increased cardiovascular and metabolic risk can be found at a waist circumference of ≥40 in (102 cm) in men and ≥35 in (88 cm) in women [30]. This measurement provides more information regarding risk that might not be apparent with BMI alone. Abdominal obesity, also known as central or visceral adiposity, is associated with an increased risk for heart disease, diabetes, hypertension, dyslipidemia, and non-alcoholic fatty liver disease [31]. Moving forward into management of obesity begins with evaluating the patient for the cause. Analyzing their history, physical examination, and measurements for fasting glucose, thyroid hormones, liver enzymes, and fasting lipids will aid in determining the appropriate intervention.

Treatment

BEHAVIORAL: Lifestyle management is a first-line treatment that modifies eating and physical activity behaviors with low risk of harm or financial incapacitation. In a study pairing pharmacotherapy intervention with high-intensity behavioral counseling, 60–65% of patients lost 5% or more of their initial weight in the span of one year [32]. Some cornerstones of behavioral therapy include self-documentation of food intake, physical activity, and weight, which can all be facilitated by smartphone applications and other electronic devices.

DIET: The known baseline mantra for diet modification for weight loss is that energy intake must be less than energy expenditure. It is important to monitor the chosen diet for nutritional deficiencies such as vitamin D and iron, which are common when choosing the route of calorie repletion [33]. In a study assessing 811 overweight patients assigned to one of four diets (low-fat, average protein; low-fat, high-protein; high-fat, average-protein; high-fat, high-protein), particular diets may have demonstrated differences in weight loss at 6 months, but by the end of the 24 month study, there were no significant differences between diets [33]. However, the main issue found with dietary methods of weight loss is the high likelihood of regain and/or binge eating. For instance, the regimen for very low calorie diets (VLCDs) is a 400–800 kcal per day often in the form of shakes and other liquids, and can induce 15–25% loss of the initial weight in the first 16 weeks. This weight loss is twice the amount of other typical low-calorie diets, however it is expensive to maintain and 35–50% of individuals will regain the weight in the 1–2 years following [34]. Ultimately, the overall success of a diet depends on adherence and financial availability of an individual, and not one specific diet's ratio, with the exception that it requires a caloric deficit, will dictate the patient's progress.

MEDICATIONS: The choice to begin drug therapy for obese patients varies among physicians, and it is only executed after failure to lose weight with the initial lifestyle and diet intervention.

Orlistat is the oldest of the drugs used for long-term obesity treatment, approved in 1999. It acts as a gastric and pancreatic lipase inhibitor that induces fat malabsorption, leading to weight loss. Orlistat causes the excretion of 30% of all ingested fat through the stool. Due to this consistent fat loss, there are increased risks of hyperoxaluria and oxalate nephrolithiasis as well as fat-soluble vitamin malabsorption [35]. A study analyzed 1,111 patients taking 120 mg TID vs. placebo and found that the mean change in body weight after one year was 7.9–9.4 kg, and that more patients avoided weight regain in the maintenance phase than those on placebo [36].

Phentermine-Extended Release Topiramate is a drug combination approved in 2012 by the FDA that acts as a carbonic anhydrase inhibitor. The mechanism of action causing weight loss is not known, but it is known that carbonic anhydrase is an enzyme present in mitochondria and involved in gluconeogenesis and lipogenesis [35]. Perhaps because this drug was first approved in 1996 as an antiepileptic, the adverse effects of phentermine-topiramate include insomnia, reversible memory loss, attention disturbance, anxiety, and headache, alongside other common effects such as paresthesia, dry mouth, and constipation.

Lorcaserin works as a selective serotonin 2C receptor agonist. It activates the pro-opiomelanocortin (POMC) neural system, increasing satiety and decreasing hunger. Lorcaserin's most common adverse effects are headache, nausea, constipation, dizziness, fatigue, xerostomia, and dry eyes. A 52-week study randomized subjects to lorcaserin or placebo with heavy behavioral modification, lowering caloric intake 600 kcal below estimated requirements and requiring physical activity 30 min per day. At least 10% change in weight was achieved in up to a 22.6% of patients receiving lorcaserin, compared to a 9.7% of those on placebo [37].

Naltrexone-bupropion is a combination drug utilizing the dopamine and norepinephrine reuptake inhibition of bupropion alongside the opioid receptor antagonism of naltrexone. The POMC neural system in the arcuate nucleus of the hypothalamus release alpha-melanocyte-stimulating hormone (MSH) and beta-endorphin. Alpha-MSH is responsible for the appetite-suppression, while beta-endorphin induces auto-inhibitory feedback by activating opioid receptors [35]. Adverse effects include symptoms such as constipation, headache, vomiting, dizziness, and insomnia. The CORI study tracked subjects over 56 weeks who were randomized to 32 mg and 16 mg doses in naltrexone-buproprion or placebo supplemented by a 500 kcal deficit meal plan, and found that up to 48% and 39% of subjects lost more than 5% of their body weight with naltrexone-bupropion 32 mg and 16 mg respectively, compared to a 16% of patients on placebo. Additionally, there were improvements in waist circumference, insulin resistance, and triglycerides in naltrexone-bupropion as compared to the placebo [35].

Liraglutide was initially a drug to manage T2D at a lower dose, but is now approved to treat chronic weight management. As an analog of GLP-1, liraglutide acts as an incretin hormone that is released from L-cells in the small intestine and colon to regulate appetite physiologically [35]. It is taken daily by subcutaneous injection, and are associated with the adverse effects of transient nausea and vomiting. During the 56-week SCALE Maintenance study, subjects who had already successfully lost over 5% of their body weight during an 1200–1400 kcal/day meal plan for 4–12 weeks were randomized to liraglutide 3 mg or placebo. The liraglutide group lost an additional 6.2% of weight compared to 0.2% for the placebo group, and 81.4% of the liraglutide subjects maintained weight loss compared to 48.9% of those on placebo [38].

SURGERY: There are three primary forms of bariatric surgery performed (Fig. 3.3). The Roux-en-Y gastric bypass ensures that 95% of the volume of the stomach and duodenum are reduced by anastomosing the Roux limb of the jejunum to the upper fundus, which leads to restricted food intake. Vertical-sleeve gastrectomy removes about 70% of the stomach to accelerate gastric emptying [32]. Laparoscopic adjustable banding is the least invasive, placing an inflatable silicone band around the fundus of the stomach to reduce the gastric volume to approximately 30 ml [32]. In a study assessing 205 patients (101 having undergone sleeve gastrectomy, 104 gastric bypass), they found a higher likelihood of gastric reflux worsening after sleeve gastrectomy (31.8%) as compared to gastric bypass (6.3%).

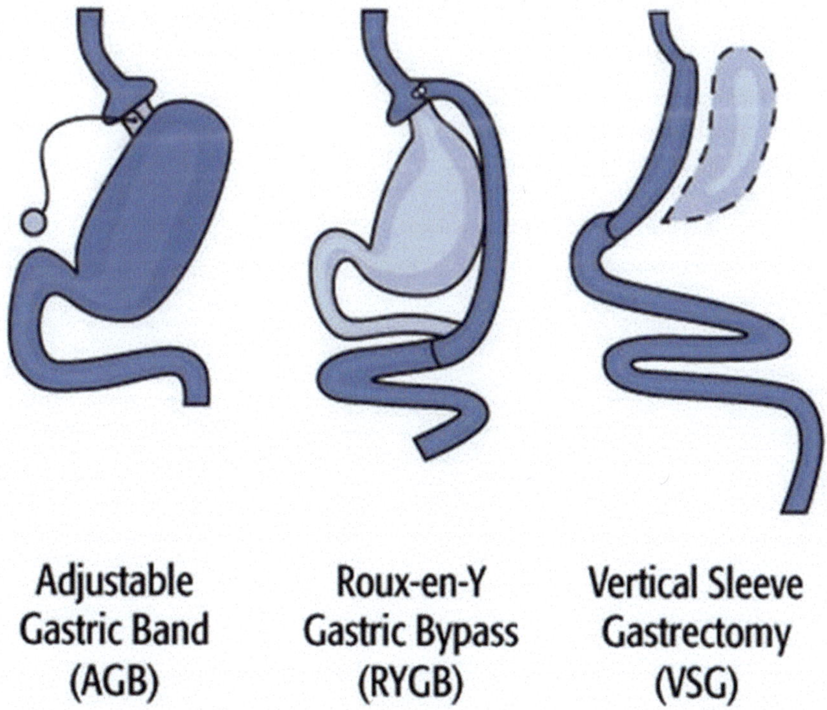

Fig. 3.3 Diagram of bariatric surgical options. Image adopted from—Walter Pories, M.D. FACS

However, with 61.1% of sleeve gastrectomy patients and 68.3% of Roux-en-Y gastric bypass patients losing excess BMI, there was no significant difference in loss of excess BMI between the two procedures at 5 years of follow-up after the surgery [39]. Another study with 245 patients in each of the primary three bariatric surgeries concluded with similar results, but also noted that the adjustable gastric band had a higher likelihood of complication, and that corrective or revisional surgery was required in 21% of adjustable gastric band patients as compared to 9% of sleeve gastrectomy patients [40]. Current limitations of these surgeries are the high initial costs, risks of complications, and weight regain.

Conclusions

If the prevalence of obesity continues to increase at the same rate as it has in the last four decades, then the poor health outcomes of individuals will only continue to worsen, with new diseases expected to follow as well. At this time, the best long-term action and prevention includes precision nutrition and exercise programs. While the exact mechanistic views of obesity are still not clear, we can observe many different pathophysiologies that result in overweight. With financial and health burdens rising, it is critical now for healthcare workers to identify and prevent prevalent signs and factors of obesity in their patients to improve outcomes for all.

References

1. Schuffham PA (2017) Economic burden of obesity: a systemic literature review. Int J Environ Res Public Health 14(4):435
2. Preston SH, Andrew Stokes YCV (2018) The role of obesity in exceptionally slow US mortality improvement. Proc Natl Acad Sci USA 115(5):957–961
3. Segula D (2014) Complications of obesity in adults: a short review of the literature. Malawi Med J 25(1):20–24
4. Pandey R, Kumar N, Paroha S, Prasad R, Yadav M, Jain S, Yadav H (2013) Impact of obesity and diabetes on arthritis: an update. Health (Irvine California) 5(1):143–156
5. Hu FB, Li TY, Colditz GA et al (2003) Television watching and other sedentary behaviors in relation to risk of obesity and type 2 diabetes mellitus in women. JAMA 289(14): 1785–1791
6. Lee IM, Shiroma EJ, Lobelo F, Puska P, Balire SN et al (2012) Effect of physical inactivity on major non-communicable diseases worldwide: an analysis of burden of disease and life expectancy. The Lancet 380(9838):219–229
7. Kac G, Pérez-Escamilla R (2013) Nutrition transition and obesity prevention through the life-course. Int J Obes Suppl 3:6–8
8. Yadav H, Rane SG (2015) Dietary fatty acids: friends or foes? obesity (Silver Spring) 23 (7):1329
9. Hall KD, Ayuketah A, Brychta R, Cai H (2019) Ultra-processed diets cause excess calorie intake and weight gain: an inpatient randomized controlled trial of ad libitum food intake. Cell Metab 30(1):66–77

10. Geiker NR, Astrup A, Hjorth MF, Sjödin A, Pijls L, Markus CR (2018) Does stress influence sleep patterns, food intake, weight gain, abdominal obesity and weight loss interventions and vice versa? Obes Rev 19(1):81–97
11. Rivenes AC, Harvey SB, Mykletun A (2009) The relationship between abdominal fat, obesity, and common mental disorders: results from the HUNT study. J Psychosom Res 66 (4):269–275
12. Morselli L, Leproult R, Balbo M, Spiegel K (2011) Role of sleep duration in the regulation of glucose metabolism and appetite. Best Pract Res Clin Endocrinol Metab 24(5):687–702
13. Spiegel K, Tasali E, Penev P, Van Cauter E (2004) Brief communication: sleep curtailment in healthy young men is associated with decreased leptin levels, elevated ghrelin levels, and increased hunger and appetite. Ann Intern Med 141(11):846–850
14. Deng T, Lyon CJ, Bergin S, Caligiuri MA, Hsueh WA (2016) Obesity, inflammation, and cancer. Annu Rev Pathol: Mech Dis 11:421–449
15. Whang A, Nagpal R, Yadav H (2019) Bi-directional drug-microbiome interactions of anti-diabetics. The Lancet 39:591–602
16. Ley R (2010) Obesity and the human microbiome. Curr Opin Gastroenterol 26(1):5–11
17. Cox AJ, West NP, Cripps AW (2015) Obesity, inflammation, and the gut microbiota. The Lancet 3(3):207–215
18. Nagpal R, Newman TM, Wang S, Jain S, Lovato JF, Yadav H (2018) Obesity-linked gut microbiome dysbiosis associated with derangements in gut permeability and intestinal cellular homeostasis independent of diet. J Diab Res
19. Helene Choquet DM (2011) Genetics of obesity: what have we learned? Curr Genomics 12 (3):169–179
20. Stavroula Paschou, H.M., Christos Mantzoros, *Obesity: Genetics, Pathogenesis, Therapy.* Principles of Diabetes Mellitus, 2015
21. Yadav H, Jain S (2016) Possible mystery behind higher susceptibility of type 2 diabetes in Asian Indians: is it diet, genetics or something else. J Nutr Health Food Eng 5
22. Rosenfield RL, Ehrmann DA (2016) The Pathogenesis of Polycystic Ovary Syndrome (PCOS): The hypothesis of PCOS as functional ovarian hyperandrogenism revisited. Endocr Rev 37(5):467–520
23. Legro R (2012) Obesity and PCOS: implications for diagnosis and treatment. Semin Reprod Med 30(6):496–506
24. Brothers KJ, Wu S, DiVall SA, Messmer MR, Kahn CR, Miller RS, Radovick S, Wondisford FE, Wolfe A (2010) Rescue of obesity-induced infertility in female mice due to a pituitary-specific knockout of the insulin receptor (IR). Cell Metab 12(3):295–305
25. Apovian C (2016) Obesity: definition, comorbidities, causes, and burden. Am J Manag Care 22(7):176–185
26. Tiryakioglu O, Ugurlu S, Yalin S, Yirmibescik S, Caglar E, Yetkin DO, Kadioglu P (2010) Screening for cushing's syndrome in obese patients. Clinics (Sao Paulo) 65(1):9–13
27. Åsvold BO, Bjøro T, Vatten LJ (2009) Association of serum TSH with high body mass differs between smokers and never-smokers. J Clin Endocrinol Metab 94(12):1
28. WHO (2000) Obesity: preventing and managing the global epidemic. Report of a WHO consultation. World Health Organ Tech Rep Ser 894:1–253
29. Deurenberg P, Yap M, Van Staveren WA (1998) Body mass index and percent body fat: a meta analysis among different ethnic groups. Int J Obes Relat Metab Disord 22(12): 1164–1171
30. al, M.J.e., 2013 AHA/ACC/TOS guideline for the management of overweight and obesity in adults: a report of the American College of Cardiology/American Heart Association Task Force on Practice Guidelines and The Obesity Society. Circulation, 2014. **129**(25): p. 102–138
31. Jacobs EJ, Newton CC, Wang Y, Patel AV et al (2010) Waist circumference and all-cause mortality in a large US cohort. Arch Intern Med 170(15):1293–1301
32. Heymsfield SB, Wadden TA (2017) Mechanisms, pathophysiology, and management of obesity. N Eng J Med 376:254–266

33. Dietz WH, Baur LA, Hall K, Puhl RM, Taveras EM, Uauy R, Kopelman P (2015) Management of obesity: improvement of health-care training and systems for prevention and care. The Lancet 385(9986):20–26
34. Wadden TA, Butryn ML, Wilson C (2007) Lifestyle modification for the management of obesity. Gastroenterology 133(1):371
35. Brett EM (2018) Pharmacotherapy for weight management. Bariatr Endocrinol 395–411
36. Hollander PA, Elbein SC, Hirsch IB, Kelley D (1998) Role of orlistat in the treatment of obese patients with type 2 diabetes: a 1-year randomized double-blind study. Diab Care 21 (8):1288–1294
37. Fidler MC, Sanchez M, Raether B, Weissman NJ, Smith SR, Shanahan WR, Anderson CM (2011) A one-year randomized trial of lorcaserin for weight loss in obese and overweight adults: the BLOSSOM trial. J Clin Endocrinol Metab 96(10):3066–3077
38. Wadden TA, Hollander P, Klein S, Niswender K, Woo V, Hale PM, Aronne L (2013) Weight maintenance and additional weight loss with liraglutide after low-calorie-diet-induced weight loss: the SCALE maintenance randomized study. Int J Obes 37:1443–1451
39. Peterli R, Wölnerhanssen BK, Peters T et al (2018) Effect of laparoscopic sleeve gastrectomy vs laparoscopic Roux-en-Y gastric bypass on weight loss in patients with morbid obesity. JAMA 319(3):255–265
40. Dogan K, Gadiot RP, Aarts EO, Betzel B, van Laarhoven CJ, Biter LU et al (2015) Effectiveness and safety of sleeve gastrectomy, gastric bypass, and adjustable gastric banding in morbidly obese patients: a multicenter, retrospective, matched cohort study. Obes Surg 25 (7):1110–1118

Part II
Metabolic Disturbances and Inflammation Due to Obesity

Chapter 4
Extracellular Vesicles and Circulating miRNAs—Exercise-Induced Mitigation of Obesity and Associated Metabolic Diseases

Patience Oluchukwu Obi, Benjamin Bydak, Adeel Safdar and Ayesha Saleem

Abstract Obesity is a progressive chronic disease that is defined by increased adiposity and dysregulated blood lipid and glucose profiles. This, coupled with insulin resistance and hypertension, leads to the development of the metabolic syndrome in obese patients. Furthermore, obesity correlates strongly with an elevated risk and progression of a number of different cancers. Endurance exercise is a gold standard method for rescuing obesity and the associated metabolic dysfunction. Physical activity stimulates fat loss, evokes metabolic adaptations and induces browning of white adipose tissue. The 'brown' fat depots are thermogenically active, thereby facilitating energy expenditure and weight loss. New evidence suggests that the systemic effects of exercise are mediated by extracellular vesicles (EVs). These are released from all cell types, and contain canonical myokines that are shed from skeletal muscle, as well as a plethora of other molecular cargo including miRNA, mRNA, DNA, metabolites and proteins. Pro-metabolic

P. O. Obi · B. Bydak · A. Saleem (✉)
Faculty of Kinesiology and Recreation Management, University of Manitoba, 120 Frank Kennedy Centre, Winnipeg, MB R3T 2N2, Canada
e-mail: ayesha.saleem@umanitoba.ca; asaleem@chrim.ca

A. Safdar
Hamilton, ON, Canada

A. Saleem
Diabetes Research Envisioned and Accomplished in Manitoba (DREAM) Theme, Winnipeg, MB, Canada

Developmental Origins of Chronic Diseases in Children Network (DEVOTION), Winnipeg, MB, Canada

Biology of Breathing Research Theme, Winnipeg, MB, Canada

P. O. Obi · B. Bydak · A. Saleem
Children's Hospital Research Institute of Manitoba (CHRIM), 600A - 715 McDermot Avenue, John Buhler Research Centre, Winnipeg, MB R3E 3P4, Canada

P. S. Tappia et al. (eds.), *Pathophysiology of Obesity-Induced Health Complications*, Advances in Biochemistry in Health and Disease 19,
https://doi.org/10.1007/978-3-030-35358-2_4

myokines include proteins, as well as miRNAs that are linked to rescuing obesity and associated metabolic syndrome. Circulating miRNAs have been shown to be useful biomarkers of pathological conditions including obesity, cancer and the metabolic syndrome. EVs and their enclosed molecular cargo offers a viable therapeutic target for future studies designed to mimic exercise and recapitulate the beneficial effects of exercise in obese subjects.

Keywords Exercise · Extracellular vesicles · Exosomes · Obesity · Metabolic syndrome · Diabetes · Cancer · Browning of fat · Myokines · miRNA

Introduction

Epidemiological data demonstrate that a sedentary lifestyle coupled with excessive caloric intake results in the emergence of chronic metabolic diseases and associated co-morbidities, such as type 2 diabetes mellitus (T2DM), obesity and cardiovascular diseases [1]. In 2015, the approximate prevalence of diabetes in Canada was 3.4 million, or roughly 9.3% of the population. This is predicted to rise to 5 million or 12.1% of the population by 2025, a 44% increase in 10 years [2].

Physical activity induces pro-metabolic health and rescues obesity and T2DM. Remarkably, the pro-metabolic beneficial adaptations of exercise are observed not only in the exercising muscle, but also in non-exercised tissues such as brain, liver, pancreas, and fat illustrating the fascinating systemic effects of physical activity. Indeed, exercise is a positive physiological regulator of metabolic health and the first line of therapy for metabolic diseases. Physical activity has been shown to reduce disability scores and mortality, improve glucose control and insulin sensitivity in obese and T2DM populations [3]. Consequently, the World Health Organization (WHO) has adopted a global strategy to increase physical activity in order to mitigate the prevalence and economic burden of chronic diseases such as obesity.

Obesity and the Metabolic Syndrome

Obesity

Obesity is a medical condition generally defined using the body-mass index (BMI), obtained by dividing weight in kilograms by height in meters squared, that provides an estimation of excess weight relative to height. BMI can be broken down into five different categories: underweight (BMI of <18.5), normal weight (BMI of 18.5–24.9), overweight (BMI of 25–29.9), and obese (BMI of ≥ 30). The obesity category can be further compartmentalized into three grades of severity: grade 1 (BMI of 30–34.9), grade 2 (BMI of 35–39.9), and grade 3 (BMI of >40) [4].

Obesity can be caused by many factors. Lifestyle-related factors include meal portion sizes, sedentary activity levels, as well as poor sleeping patterns [5]. Genetic factors also play a role in the development of obesity, with heritability accounting for 40–70% of variance in adiposity, although only 30% of observed variance in BMI is said to be due to genetic variance [5–7]. This range of genetic influence indicates that lifestyle factors play a large role in the obesity epidemic. Finally, maternal and developmental factors such as gestational obesity and perinatal environmental exposure can also influence the likelihood of developing obesity [5].

Maintaining mitochondrial health is integral to promoting optimal energy generation and local cell function. Mitochondrial dysfunction refers to a loss in mitochondrial function, content and biogenesis. Production of reactive oxygen species (ROS) due to inflammation associated with obesity contributes to the development of mitochondrial dysfunction that characterizes obesity [8]. The increase in oxidative stress promotes further inflammation creating a postive feedback loop which exacerbates the pathology. Mitochondrial dysfunction in obesity can also be due to excessive calorie and nutrient consumption, which promotes cellular apoptosis. Impaired mitochondrial function in adipocytes subsequently promotes insulin insensitivity, which over time can lead to the development of T2DM. Cardiomyocyte mitochondrial deterioration, normally associated with aging, has been reported in obese patients under 55 years of age [9]. This suggests that obesity may induce premature aging in cardiomyocytes and could potentiate impaired cardiovascular function in obese individuals. Obesity is associated with a wide range of systemic effects, and has been illustrated to be a risk factor for several diseases such as T2DM, cardiovascular disease, renal insufficiency, respiratory disorders, musculoskeletal disorders, and cancer. Furthermore, obesity is characterized by insulin resistance, hypertension, dyslipidemia, that leads to the development of the metabolic syndrome [8, 10, 11].

Metabolic Syndrome

Metabolic syndrome is classified by abdominal obesity, dyslipidemia (high triglyceride and cholesterol levels with low high-density lipoprotein (HDL) content), hyperglycemia, and hypertension (Engin, 2017; Han and Lean, 2016). The National Cholesterol Education Program (NCEP) has mandated specific criteria for the diagnosis of metabolic syndrome. A waist circumference $\geq$ 102 cm (40 in) in men or >88 cm (35 in) in women; triglyceride levels $\geq$ 1.7 mmol/L (150 mg/dl); HDL cholesterol levels <1.03 mmol/L (40 mg/dl) in men or <1.29 mmol/L (50 mg/dl) in women; blood pressure $\geq$ 130/85 mmHg; and a fasting plasma glucose level $\geq$ 6.1 mmol/L (110 mg/dl) [7].

There are several factors involved in the onset of metabolic syndrome. Like in obesity, lifestyle factors such as a high calorie and cholesterol-rich diet in combination with a sedentary lifestyle, especially during childhood, play a substantial role in the development of the metabolic syndrome. Additionally, smoking and alcohol consumption, and adult weight-gain are also contributing risk factors. Interestingly, genetic variance in fat distribution and insulin resistance can affect how an individual responds to lifestyle factors that predispose them to developing metabolic syndrome [7].

Existing evidence supports a chronic inflammatory state in metabolic syndrome is associated with increased levels of pro-inflammatory cytokines and inflammation markers [5]. Adipose tissue is comprised of a heterogenous population of cells including pre-adipocytes, adipocytes, stromal cells, eosinophils, and macrophages. Both adipocytes and stromal cells induce inflammatory cytokine release, such as the tumour necrosis factor alpha (TNF-α), from adipose tissue [10, 12]. The dysregulation of adipokines (cytokines released from adipose tissue) such as TNF-α and interleukin 6 (IL-6) is associated with systemic metabolic dysfunction [11]. These pro-inflammatory cytokines also promote the development of hepatic insulin resistance [12]. Additionally, individuals with high abdominal adiposity with normal weight exhibit higher amounts of TNF-α and a greater overall pro-atherogenic profile than individuals with a higher BMI and lower abdominal obesity [13]. This indicates that abdominal obesity and waist circumference are primary contributors to the overall inflammatory state associated with metabolic syndrome, and that individuals who may be slightly overweight with relatively high abdominal obesity are still at risk of negative systemic effects.

Metabolic syndrome is known to increase the risk of developing cardiovascular disease, T2DM, cancer and all-cause mortality [7]. Chronic lipolysis due to increased adipose tissue mass potentiates higher levels of plasma free fatty acids travelling to the liver. This, in combination with inflammatory cytokines, can trigger insulin resistance and eventually prediabetes if left untreated [10]. TNF-α is associated with cancer and insulin resistance, and IL-6 has been linked with breast and lung tumours, as well as cancers of the liver, prostate, and cervix [10]. TNF-α antagonists have been shown to decrease fasting glucose levels in obese patients [14], indicating that the pro-inflammatory state associated with metabolic syndrome can further alter glucose homeostasis. Similarly to the metabolic syndrome, obesity is also linked to other metabolism-related pathologies such as cancer.

Obesity and Its Link to Cancer

Almost 1 in 2 Canadians can expect to be diagnosed with cancer at some point, with an overall mortality rate from cancer of 1 in 4 [15]. Although an estimated 206,200 new cancer cases were diagnosed in 2017, overall mortality rates of cancer have been declining since 1988. Cancer prevalence in Canada is reflected in its overall costs to society. From 2005 to 2012, total public expenditures on cancer care in Canada rose from $2.9 billion to approximately $7.5 billion, respectively [16]. These costs, which more than doubled over a seven year period, are primarily due to hospital care expenditures and physician care [16].

Unfortunately, obesity is linked to increased cancer risk and cancer development. Specific cancers associated with obesity include breast cancer in post-menopausal women, colon cancer in men, esophegeal, endometrial, gall bladder, renal cancer and adenocarcinomas to name a few [17]. Given the obesity epidemic, and that cancer is predicted to be the leading cause of death in the United States by 2030 [18], it is important to delineate the underlying mechanistic link between the two.

While the relationship between obesity and cancer is strong, the underlying cellular pathways that connect the two have yet to be fully established. A calorie rich diet is purported to be a driving cause for both obesity and colorectal cancer [5, 17]. Additionally, dysregulated hormonal signalling such as high levels of circulating estrogen can induce cancer cell growth [19]. Obesity is characterized with higher levels of circulating insulin, which potentiaties the release of growth factors such as the insulin like growth factor-1 (IGF-1) from the liver, and also mediates insulin resistance. Both insulin and IGF-1 are correlated with an elevated risk of cancer and cancer-induced mortality [20, 21]. The increased adipose tissue mass in obesity promotes the release of proinflammatory adipokines, leading to sustained chronic inflammation. Higher levels of adipokines such as leptin and IL-6 are associated with an elevated risk of several types of cancer, with leptin being linked to colon and breast cancer in particular [17, 22]. Leptin also suppresses the expression of a beneficial adipokine called adiponectin that inhibits cancer cell growth, proliferation, and its expression is inversely linked to cancer risk [23]. Elevated plasma IL-6 and TNF-α levels are associated with a higher risk of colorectal adenomas, which can become cancerous [17]. Higher ratios of collagen: adipose tissue in breast tissue are also associated with hightened rates of breast cancer [17]. Moreover, research shows that obesity can also facilitate development of cancerous malignancies from pre-cancerous conditions. Chang et al. [24] reported that overweight and obese individuals with monoclonal gammopathy of undetermined significance (MGUS) were more likely to progress from MGUS to multiple myeloma than their non-overweight or obese counterparts.

Concomitantly, with rising rates of obesity, recent studies demonstrate that the incidence of obesity-related cancers has increased in 20–49 year olds. Obesity-related cancers, such as breast, colon and rectal, kidney, endometrial, thyroid, gastric cardia, meningioma, and ovarian cancer all have peak incidence rates in populations over 50 years old [25]. However, all of the aforementioned cancers have shown increases of at least 5% in 20–54 year olds, with thyroid cancer and meningioma showing 23.9% and 16.8% new cases, respectively [25]. These increases in cancer rates in young people indicate that overweight and obesity may be contributing to increased rates of cancers in young people that are generally only seen in older populations.

The economic and health burden associated with obesity and related metabolism-related disorders such as the metabolic syndrome and cancer warrants urgent therapeutic intervention. A vast array of pharmacological drugs and interventions for treatment of obesity and its associated pathologies are available, with varying degrees of success depending on disease state, personal, lifestyle and genetic factors. However, one of the most effective, and most economically viable therapeutic option for treatment of obesity is physical activity.

Exercise as Frontline Therapy for Obesity and Metabolic Syndrome

Systemic Effects of Exercise

Exercise is a fundamental part of human life and is defined as engaging in structured physical activity in order to improve or maintain fitness [26]. Physical inactivity and a sedentary lifestyle has been associated with an increased risk of developing non-communicable diseases such as obesity, coronary heart disease, cancers, stroke, T2DM, liver disease, and Alzheimer's disease [27]. The effects of exercise have been investigated by a plethora of studies that together suggest significant potential therapeutic effects of exercise in mitigating the development and progression of these chronic diseases. In fact, a systematic review of longitudinal exercise studies reported a negative correlation between physical activity and chronic diseases over time [28]. It is likely that the pleiotropic effects of exercise are mediated in part by exercise-induced systemic adaptations evoked in various organ systems in the body such as the brain, heart, skeletal muscle, and adipose tissue.

Exercise-Induced Protective Effects in Brain

Previous research has illustrated that exercise prevents the decline of cognitive functions associated with age [29, 30], improves neurogenesis [31], and promotes memory enhancement [32]. Low levels of the anti-inflammatory cytokine IL-10 have been linked to aging and neuronal damage [29]. Exercise up-regulates IL-10, and brain-derived neurotrophic factor (BDNF), associated with neurogenesis, in the brains of aged rodents [29, 30]. Exercise has been shown to slow the progression of Alzheimer's disease by eliminating amyloid-beta peptide (the main driver of the disease), suppress cholinergic neuronal cell death, and improve the quality of life of patients with Alzheimer's [33–35]. In Parkinson's disease, resistance exercise decreased the production of nitrogen and reactive oxygen species [36], improved tremor [37] and reduced musculoskeletal deficiencies [38]. Furthermore, exercise mitigated the loss of mitochondrial function and content, and reduced apoptotic signaling in aged rats, albeit the magnitude of the changes was lower in aged compared to young animals [39]. Physical exercise has also been found to rescue metabolomic alterations in the brains of a transgenic mouse model of accelerating aging, the Polg mice [40]. In fact, five months of exercise ameliorated age-associated mitochondrial and physiological dysfunction in Polg mice, induced systemic adaptations and prevented premature mortality [41–43].

Pro-metabolic and Cardiovascular Fitness Associated with Exercise

Endurance exercise is associated with a reduced risk of developing coronary heart disease, hypertension and stroke [44–46]. Exercise-induced induction of antioxidant and nitric oxide levels, and reduction of pro-inflammatory cytokines underlie its protective effect on preventing coronary heart disease [27]. Physically active individuals reportedly have lower triglyceride, cholesterol, and blood pressure levels, improved insulin sensitivity, glucose tolerance and favorable lipoprotein profiles [47, 48]. These adaptations help improve cardiovascular fitness, and also promote the protective effect of exercise on preventing and treating obesity and T2DM. Exercise induces weight loss and a reduction in body mass index (BMI) in obese patients [49–51]. It stimulates the uptake of glucose to the skeletal muscle by increasing the levels of GLUT4 mRNA and protein in an insulin-independent manner [52, 53], thereby rescuing glucose and insulin insensitivity in obese and T2DM patients. Moreover, exercise reduces the levels of hemoglobin A1c (HbA1c), a risk factor associated with T2DM [54]. Another study investigated the effects of the presence or absence of endurance exercise (aerobic and resistance training) in T2DM patients receiving standardized medications. It was observed that patients that underwent endurance exercise compared with sedentary controls did not need medications after a while, cementing the significant potential of exercise to serve as a therapy for T2DM [55, 56]. Additionally, physical activity has been shown to reduce the risk of cancer. In breast cancer patients, physical activity increases the overall survival rate after treatment [57]. Chronic exercise also modulates systemic inflammation [58, 59], and reduces growth factors that promote cancer growth.

The systemic beneficial adaptations that accrue with regular exercise training have been linked with the role of skeletal muscle functioning as an endocrine organ (Fig. 1). Research over the past decade has conclusively shown that contracting muscle releases a myriad of factors including proteins, lipids, mRNA species and DNA metabolites, that play a pivotal role in evoking pro-metabolic adaptations in distal tissues.

Muscle as an Endocrine Organ

Skeletal muscle has been identified as an endocrine organ because it can produce cytokines (referred to as myokines), and is able to communicate with other tissues either in an autocrine, paracrine or endocrine manner [43, 60, 61]. Examples of myokines include IL-6, IL-1b, IL-10, IL-15, TNF-α, toll-like receptor-4 (TLR-4), irisin, meteorin-like protein (METRNL), and BDNF. Several of these myokines possess anti-inflammatory properties, and have been shown to mediate the protective effects of exercise against Alzheimer's, specific cancers, obesity, T2DM and cardiovascular disease [60, 62, 63]. IL-6 is the most common myokine, and possesses anti-inflammatory properties. Exercise induces an increase in circulating

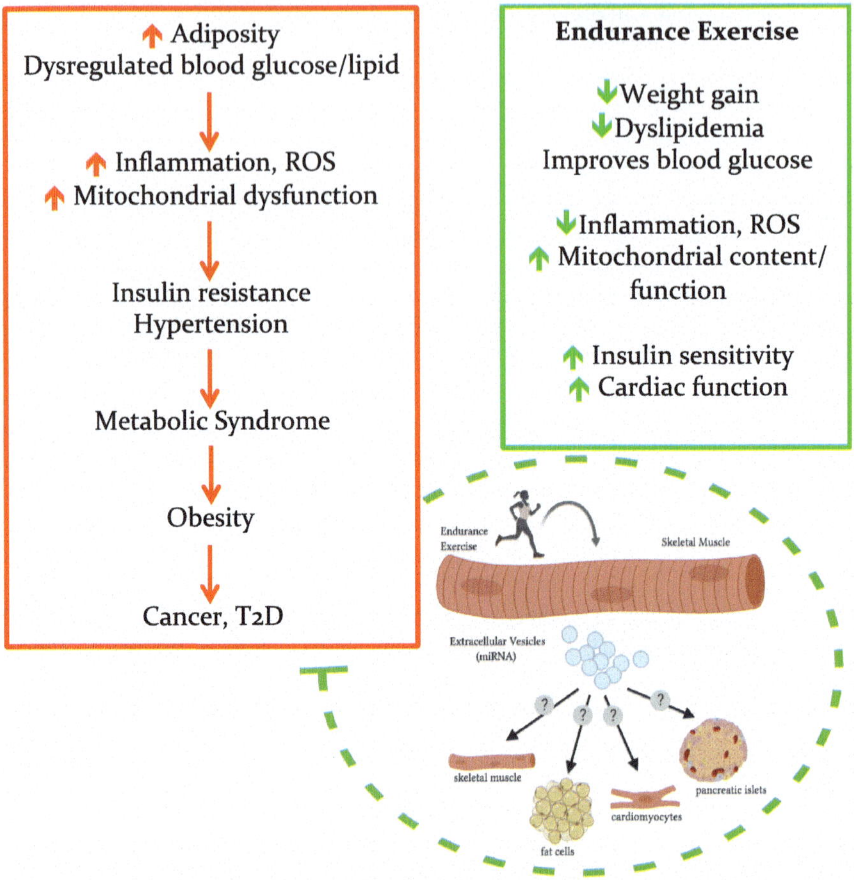

Fig. 1 Endurance exercise modulates the underlying mechanisms that contribute to the development of the metabolic syndrome and obesity. Recent work has illustrated that contracting skeletal muscle confers protective physiological benefits on distal non-exercise tissues such as browning of white adipose cells by releasing myokines. Many of these myokines have been found inside extracellular vesicles (EVs) that mediate cell-to-cell communication. The role of EVs and their enclosed cargo, particularly miRNA species in the treatment of obesity and related pathologies is an exciting new avenue of research

levels of IL-6, which in turn down regulates the production of pro-inflammatory myokines, such as IL-1, TNF-α and TLR-4, and the up regulates the production of anti-inflammatory myokines such as IL-10 [64–66].

Apart from the skeletal muscle, exercise can also have an effect on adipose tissue [67]. Endurance exercise can induce the browning of white adipose tissue (WAT), and this has been associated with reduced risk of T2DM and obesity [60, 68]. The browning of WAT creates a subset of fat cells (called beige or brite cells) that are more metabolically active. An increase in the thermogenic activity of beige cells enhances fuel oxidation, and the total energy expenditure of the body [67]. The myokines that are primarily involved in the induction of WAT browning upon exercise are irisin and METRNL.

Irisin

Irisin was first discovered and characterized by Bostrom et al. [68] in transgenic mice that showed that increased expression of peroxisome proliferator-activated receptor gamma coactivator (PGC-1α) induced the expression of the membrane protein fibronectin type III domain-containing protein 5 (FNDC5) in the skeletal muscle. This membrane protein FNDC5, a precursor of irisin, was cleaved to form irisin, which was then released into the circulation. The transgenic mice were reported to have brown adipocytes within their WAT, and this increased after endurance exercise. Irisin is encoded by the FNDC5 gene, although the presence of irisin in humans has been debated. Some studies have reported an inability to detect the protein, possibly as the antibodies used in the enzyme-linked immunosorbent assay (ELISA) kits lacked specificity [69], or due to the difference in the start codon of FNDC5 gene in human and rodents, being ATA and ATG, respectively [70]. However, a recent study that used mass spectrometry-based analysis reported that irisin is present in humans and is mainly translated from the ATA-codon [71]. Human exercise studies have also shown that endurance exercise increased the levels of plasma irisin [67, 71]. Cumulatively, these reports suggest that endurance exercise increases the circulating concentration of irisin in both mice and humans. During exercise, irisin secreted by the skeletal muscle leads to induction of WAT browning by increasing the mRNA expression of uncoupling protein 1 (UCP1), a marker of fat cell browning [68]. The increase in UCP1 increases energy expenditure, normalizes glucose handling and rescues obesity.

Meteorin-Like Protein (METRNL)

METRNL was first identified by Rao et al. [72], as a circulating factor released from muscle of mice that overexpress PGC-1α4, a spliced isoform of PGC-1α, involved in the regulation of energy expenditure and hypertrophy in the skeletal muscle. METRNL over expression through adenoviral, recombinant protein, or vector transfection-mediated methods, evokes a concomitant increase in the whole body energy expenditure via enhanced thermogenesis. In contrast, blocking METRNL prevents the cold-induced browning of WAT. Both animal and human studies have shown that resistance exercise and cold stimuli significantly increase circulating concentrations of METRNL [67, 73]. During exercise, METRNL is secreted by both the skeletal muscle and adipose tissue. High levels of circulating METRNL promote an up-regulation of genes involved in the browning of WAT, and anti-inflammatory cytokines. Adipocytes contain a heterogeneous population of different immune cell types including eosinophils, macrophages, mast cells etc.

METRNL induced an accumulation of eosinophils in the WAT, which triggered the secretion of IL-13 and IL-4. This, in turn, stimulated the activation of M2 macrophages that are known to be involved in stimulating thermogenesis, and improving glucose homeostasis [72–74].

Other Myokines Involved in WAT Browning

Other myokines that are involved in the browning of WAT include myostatin, fibroblast growth factor-21 (FGF-21) β-aminoisobutyric acid (BAIBA) and lactate [67]. Myostatin has been shown to inhibit the growth and differentiation of skeletal muscle cells [75]. Studies showed that myostatin-knockout mice exhibit a reduction in the accumulation of fat, an increase in muscle mass, and browning of WAT mediated by the up-regulation of UCP1 and PGC-1α genes [76–78]. These studies suggest the importance of myostatin inactivation as a viable anti-obesity therapeutic target. FGF-21 is involved in the metabolism of lipids and browning of WAT by recruiting irisin [79, 80]. METRNL-induced expression of BAIBA in both mice and humans post-exercise is associated with UCP-1 mediated browning of WAT [81]. Lactate is commonly considered a metabolic waste product, as it is the end product of anaerobic glycolysis, and its plasma levels increase correspondingly with increasing exercise intensity. However, lactate has also been identified as a molecule that can transmit signals particularly from skeletal muscle to other tissues such as heart, brain, and liver [67, 82]. Studies in human and mice have shown lactate-dependent induction of WAT browning occurs by increasing the expression of UCP1 and FGF-21 [83].

Altogether, it is clear that exercise-induces release of myokines from skeletal muscle into the circulation. These secreted factors evoke pro-metabolic systemic effects that underpin exercise-mediated rescue of pathophysiological conditions such as obesity and T2DM. Myokines released into the blood can be labile to the environment. This necessitates their release in protective secretory vesicles termed extracellular vesicles.

Extracellular Vesicles and Exercise

Extracellular vesicles (EVs) are lipid membrane-bound structures that are derived from prokaryotic and eukaryotic cells, and play an evolutionary conserved integral role in cellular communication. EVs can be found in most biological fluids including blood, cerebrospinal fluid, saliva, urine, synovial fluid, breast milk, and amniotic fluid [1, 84]. EVs contain different types of biochemical messages, an EV signature that differs based on the cell type of origin, and on physiological conditions. The EV signature is complex consisting of an array of molecular cargo components including proteins, lipids, DNAs, and RNAs, that facilitate intercellular cross talk. The protein cargo of EVs includes CD81, CD63, CD37, major histocompatibility complex class II (MHC II), P-selectin, glycoproteins, integrins and DNA-binding histones among others. The lipid content includes phosphatidylcholine, gangliosides, sphingomyelin and diacylglycerol to name a few. The RNA cargo includes messenger RNA (mRNA), transfer RNA (tRNA), microRNA (miRNA) and ribosomal RNA (rRNA) species [85]. Several comprehensive encyclopedias documenting the growing list of EV molecular cargo exist. These include EVpedia: http://student4.postech.ac.kr/evpedia2_xe/xe/index.php?mid=Home, Vesiclepedia: http://microvesicles.org/, and Exocarta: http://exocarta.org/. EVs are involved in the regulation of number physiological processes including coagulation [86], immune response [87] and cancer progression [84] among many others.

There are three common types of vesicles, classified based on their size, density, and origin. These include exosomes, microvesicles (MVs) and apoptotic bodies (ABs). Exosomes are small cup-shaped vesicles (20–200 nm in diameter and 1.13–1.19 g/mL in density) formed from intraluminal vesicles (ILVs) that are created by the inward pinching of the endosomal membrane. ILVs are packaged within multi-vesicular bodies (MVBs) that fuse with the plasma membrane, and release ILVs as exosomes into the extracellular space. MVs are cup-shaped vesicles (100–1000 nm in diameter), also called ectosomes because they are formed by the direct outward budding of the plasma membrane [87]. ABs are the largest of the EVs (500–5000 nm in diameter, 1.16–1.28 g/ml density), and are formed by plasma membrane blebbing, often observed during the process of apoptosis [84, 88].

Several studies have suggested the possibility of exercise to induce EVs secretion from cells, allowing for the transfer of essential signaling molecules between tissues and cells. Safdar et al. suggested that exercise induces the release of exerkines (myokines released upon exercise), which are enclosed in exosomes [1]. The study also proposed the potential role of these exercise-induced exosomes in treating obesity and T2DM [1]. Whitham et al. reported that exercise-induced the secretion of exosomes and other small vesicles (which contained several myokines) in both human and murine plasma [89]. The study also showed that some of these proteins were transferred via inter-tissue communication to recipient tissues and cells during exercise. Bei et al. [90] found an increase in the plasma EVs in human after exercise. In mice, the authors observed an approximately 1.85-fold increase in serum EVs after three weeks of swimming. This study also showed the protective role of exercise-induced EVs against ischemic injury and suggested exercise-induced EV administration as a viable therapeutic strategy myocardial injury. Guescini et al. observed an increase in EVs and muscle-specific miRNAs in the plasma after acute aerobic exercise in human subjects [91]. Oliveira et al. recently reported EV concentration significantly increased after an acute bout of treadmill exercise in rats [92]. This occurred in tandem with changes in EV RNA cargo: 12 miRNAs and one tRNA were differentially expressed.

EV Cargo: Role of miRNAs

miRNAs are small single-stranded non-coding RNA molecules about 18–24 nucleotides in length that are able to interfere with the translation and degradation of mRNAs by specific complementary binding [93–95]. miRNAs act as regulators of transcriptional and post-transcriptional gene expression, and have been found enclosed within exosomes and EVs. The transfer of EV-miRNAs to recipient cells that take up the EVs, is a highly conserved means of communication between cells and was first described by Valadi et al. [96]. This was supported by other studies that followed after [97–99]. Together these reports showed that miRNAs were transported via exosomes from the donor cells to the recipient cells, and this genetic exchange resulted in reduction and inhibition of several disease conditions. Uptake of EV-miRNA by cells can occur through different mechanisms such as phagocytosis, clathrin-mediated endocytosis or micropinocytosis. After being transported to target cells, miRNAs regulate the expression of target genes in the cell [100]. EV-miRNAs are potentially beneficial, however, they have been linked to some pathological processes including cancer and cardiovascular diseases, and they act as biomarkers for these diseases [93, 94, 101]. Recent studies reported that the miRNA content in EVs of breast and colorectal cancer cell lines were about 5–30% of all RNA cargo [102, 103]. Chen et al. [104] reviewed the role of miRNA-200, -17, -155, -146a, and -210 in lung pathologies. EV-miRNAs also hold considerable potential in stimulating skeletal muscle differentiation and development [105].

Cumulatively, endurance exercise induces an increase in the release of EVs into circulation in both human and rodent models. The EVs released contain myokines, including proteins and miRNAs that are differentially expressed with physical activity. While some studies have investigated the role of different exercise intensities, more work is warranted to carefully delineate the effect of exercise type, intensity, duration, time and recovery period post-exercise to distill the role of contractile activity on EV release and concentration. Furthermore, extensive research into EV molecular cargo such as miRNA species is required to fully decipher their role as circulating biomarkers in physiological and pathophysiological conditions.

Circulating miRNA as Diagnostic Markers

Cell-free miRNAs are highly exposed and can be easily accessible to circulating nucleases (RNAases), which can degrade them. Circulating miRNAs (c-miRNAs) have adopted two mechanisms to protect against degradation by nucleases: formation of specific HDL-binding protein complexes or enclosure within EVs [106]. c-miRNAs have a strong potential of being used as non-invasive diagnostic biomarkers, and have already been linked with a number of diseases such as obesity, cancer and metabolism.

miRNAs Associated with Obesity

An increase in plasma EVs with obesity due to enhanced release from adipose depots, in tandem with several c-miRNAs linked to obesity have been reported [107]. Xie et al. [108] conducted the first study to illustrate the miRNAs that were induced during adipogenesis (miRNA-30, -103, -143, and -422b) showed an inverse pattern of expression in obesity. Ortega et al. [109] reported higher levels of c-miRNA-140-5p, -142-3p and -222, and lower expression of miRNA-532-5p, -125b, -130b, -221, -15a, - 423-5p, and -520c-3p species in obese patients. Furthermore, other studies have also shown differential and deregulated expression profiles of c-miRNAs in obesity [110–112]. Recent studies have associated miRNA-335 and -33 with inflammation and fat accumulation in obesity [113, 114]. Suppressing miRNA-33 mitigates insulin resistance, whereas genetic ablation of miRNA-335 promotes obesity and insulin resistance. A recent review by Ortiz-Dosal et al. [107] comprehensively explored the differential expression profiles of c-miRNAs associated with obesity, and investigated the potential target genes and metabolic pathways involved. The authors identified enhanced levels of 22 c-miRNAs, suppression of 9 c-miRNAs and dysregulation of two c-miRNAs in obesity. Further studies are required to thoroughly decipher the role of c-miRNAs as diagnostic and therapeutic markers in obesity.

miRNAs Linked with Metabolic Syndrome

Metabolic syndrome is linked to pathological conditions such as obesity, diabetes, cancer, cardiovascular disease and nonalcoholic fatty liver disease (NAFLD). It is characterized by hypertension, impaired lipid and glucose profile and oxidative stress. Alterations in circulating miRNA-23a, -27a, -130, -195, -197, -320a, and -509-5p have been associated with metabolic syndrome and are considered to be potential biomarkers for insulin resistance and obesity [115–117]. Others such as miRNA-192, -375, and -15b are increased in type 1 and type 2 diabetic patients [118, 119]. An up-regulation of c-miRNA-18a-5p, -34a-5p, -135a-5p, -195-5p, -320-3p, -674-3p, and -872-5p were observed in chronically stressed rats [120]. Alterations in miRNA-130a, -195, and -92a have been associated with high blood pressure in hypertensive patients [117]. A report linked overexpression of circulating miRNA-1, -21, -133a, and -208 with myocardial infarction [121].

miRNAs Associated with Breast Cancer

Several studies have identified differential expression profiles of c-miRNAs in breast cancer, and their distinct roles in breast cancer initiation and progression.

A miRNA analysis in 76 breast cancer patients and 10 healthy controls revealed that circulating miRNA-21, -125b, -145, and -155 are associated with breast cancer [122]. Another study reported that increasing or decreasing miRNA-21 levels, enhanced or suppressed breast cancer growth, respectively [123]. A genome-wide miRNA analysis in 13 breast cancer patients and 10 healthy controls, reported a significant increase in miRNA-222 was associated with breast cancer [124]. Circulating miRNA-200a and -200c were found to facilitate breast cancer progression [125, 126]. Others reported that c-miRNA-221 and -155 levels were enhanced in triple-negative breast cancer and linked with poor prognosis. Abrogation of c-miRNA-221 and -155 inhibited cell proliferation, angiogenesis and tumor growth, and induced apoptosis [127, 128]. Another recent study suggests that circulating miRNA-21, -190, -200b, and -200c could be potential prognostic markers for metastatic breast cancer [129]. This illustrates the potential role of using c-miRNAs as diagnostic markers of breast cancer, and identifies the transcripts that can be targeted in anti-cancer therapeutic approaches.

miRNAs Linked with Exercise

Endurance exercise training also evokes differential expression profiles of some c-miRNAs. Several miRNAs such as miRNA-1, -133a, -133b, -206, -208, -208b, -378, -486, and -499 are specific to the skeletal muscle, and have been labelled myomiRs [130–134]. Several of these myomiRs are differentially expressed following physical activity, and have been linked to aging, cancer and other pathological states. Expression of myomiRs miRNA-1, -133a, -133b, and -206 is decreased after an acute bout endurance exercise in mice and human. Russell et al. reported miRNA-1 and -29b were enriched, and miRNA-31 suppressed after a short term (10 days) of endurance exercise [135]. A recent study reported alterations in miRNA-21, -16, -93 and -222 levels after 8 weeks of strength and high intensity endurance training in male athletes [136]. It is evident that c-miRNA and myomiR expression levels are regulated with physiological and pathological stimuli. Many of these miRNA transcripts are associated with obesity, cancer, metabolic syndrome and exercise. Given that miRNAs play a crucial rule in regulating translation and gene expression, more work is required to fully understand the regulation and role of c-miRNAs, particularly in the context of EV-miRNA cargo. This could lead to the development of therapeutic interventions to treat obesity and associated metabolic dysfunction, as well as other diseases.

Future Directions and Conclusions

Obesity is a growing problem, associated with several short-term and long-term metabolic and cardiovascular complications. In addition, there is evidence suggesting that excess adiposity during childhood influences growth patterns, and increases risk of obesity and metabolism-related disorders in adulthood [137]. Treatment strategies to treat obesity are multidisciplinary, and include behavioural, physical activity, dietary, pharmacological, and surgical options. Recent evidence has demonstrated that WAT can change its phenotype to a brown-like adipose tissue known as beige/brite adipose tissue. This transition is characterized by an increase in thermogenic capacity mediated by UCP1 [138]. Induction of a thermogenic gene program, and browning of WAT has been implicated in reducing adiposity and insulin resistance following endurance exercise. Given that browning results in thermogenic and metabolically active fat depots, browning of WAT is an attractive therapeutic target for treating obesity [139].

An exercise-based approach to treat obesity is potentially represented by use of myokines that induce browning of WAT. Previous studies have shown that various exercise-responsive myokines, including irisin, FNDC5, and METRNL, positively modulate the browning of WAT. In addition to proteins, c-miRNAs act as molecular instigators of muscle-fat crosstalk that promote WAT browning in response to endurance exercise. Many of the exercise-responsive myokines (both proteins and miRNA species) have been found in EVs that are released from muscle upon physical activity. It is likely that exercise induces browning of WAT using both proteins and miRNA species contained within EVs, in addition to modulating other systemic changes that help mitigate obesity (Fig. 1). Further research investigating the efficacy of isolated exercise-induced EVs as a therapeutic strategy for obesity is warranted. However, it is important to note that specific issues must be addressed when conducting EV-based research. The International Society for Extracellular Vesicles (ISEV) has proposed Minimal Information for Studies of Extracellular Vesicles ("MISEV") guidelines for the EV field in 2018 [140]. In particular, isolation strategies for EVs are variable and can differ based on experimental conditions. To ensure reliability, rigour and reproducibility of results, standard experiments must be performed according to MISEV guidelines before strong conclusions can be made regarding EV-induced therapeutics in obesity and associated pathologies.

References

1. Safdar A, Saleem A, Tarnopolsky MA (2016) The potential of endurance exercise-derived exosomes to treat metabolic diseases. Nat Rev Endocrinol 12(9):504–517
2. Diabetes Canada Clinical Practice Guidelines Expert C, Houlden RL (2018) Introduction. Can J Diab 42(Suppl 1):S1–S5

3. Avery L, Flynn D, van Wersch A, Sniehotta FF, Trenell MI (2012) Changing physical activity behavior in type 2 diabetes: a systematic review and meta-analysis of behavioral interventions. Diabe Care 35(12):2681–2689
4. Flegal KM, Kit BK, Orpana H, Graubard BI (2013) Association of all-cause mortality with overweight and obesity using standard body mass index categories: a systematic review and meta-analysis. JAMA 309(1):71–82
5. Mendrick DL, Diehl AM, Topor LS, Dietert RR, Will Y, La Merrill MA, Bouret S, Varma V, Hastings KL, Schug TT et al (2018) Metabolic syndrome and associated diseases: from the bench to the clinic. Toxicol Sci 162(1):36–42
6. Chung WK (2012) An overview of mongenic and syndromic obesities in humans. Pediatr Blood Cancer 58(1):122–128
7. Han TS, Lean ME (2016) A clinical perspective of obesity, metabolic syndrome and cardiovascular disease. JRSM Cardiovasc Dis 5:2048004016633371
8. de Mello AH, Costa AB, Engel JDG, Rezin GT (2018) Mitochondrial dysfunction in obesity. Life Sci 192:26–32
9. Niemann B, Chen Y, Teschner M, Li L, Silber RE, Rohrbach S (2011) Obesity induces signs of premature cardiac aging in younger patients: the role of mitochondria. J Am Coll Cardiol 57(5):577–585
10. Aballay LR, Eynard AR, Diaz Mdel P, Navarro A, Munoz SE (2013) Overweight and obesity: a review of their relationship to metabolic syndrome, cardiovascular disease, and cancer in South America. Nutr Rev 71(3):168–179
11. Ouchi N, Parker JL, Lugus JJ, Walsh K (2011) Adipokines in inflammation and metabolic disease. Nat Rev Immunol 11(2):85–97
12. Lumeng CN (2013) Innate immune activation in obesity. Mol Aspects Med 34(1):12–29
13. Beberashvili I, Azar A, Abu Hamad R, Sinuani I, Feldman L, Maliar A, Stav K, Efrati S (2019) Abdominal obesity in normal weight versus overweight and obese hemodialysis patients: associations with nutrition, inflammation, muscle strength, and quality of life. Nutrition 59:7–13
14. Stanley TL, Zanni MV, Johnsen S, Rasheed S, Makimura H, Lee H, Khor VK, Ahima RS, Grinspoon SK (2011) TNF-alpha antagonism with etanercept decreases glucose and increases the proportion of high molecular weight adiponectin in obese subjects with features of the metabolic syndrome. J Clin Endocrinol Metab 96(1):E146–E150
15. Committee CCSA (2018) Canadian cancer statistics. Canadian Cancer Society
16. de Oliveira C, Weir S, Rangrej J, Krahn MD, Mittmann N, Hoch JS, Chan KKW, Peacock S (2018) The economic burden of cancer care in Canada: a population-based cost study. CMAJ Open 6(1):E1–E10
17. Stone TW, McPherson M, Gail Darlington L (2018) Obesity and cancer: existing and new hypotheses for a causal connection. EBioMedicine 30:14–28
18. Allott EH, Hursting SD (2015) Obesity and cancer: mechanistic insights from transdisciplinary studies. Endocr Relat Cancer 22(6):R365–R386
19. Gerard C, Brown KA (2018) Obesity and breast cancer—role of estrogens and the molecular underpinnings of aromatase regulation in breast adipose tissue. Mol Cell Endocrinol 466:15–30
20. Cohen DH, LeRoith D (2012) Obesity, type 2 diabetes, and cancer: the insulin and IGF connection. Endocr Relat Cancer 19(5):F27–F45
21. Rinaldi S, Cleveland R, Norat T, Biessy C, Rohrmann S, Linseisen J, Boeing H, Pischon T, Panico S, Agnoli C et al (2010) Serum levels of IGF-I, IGFBP-3 and colorectal cancer risk: results from the EPIC cohort, plus a meta-analysis of prospective studies. Int J Cancer 126 (7):1702–1715
22. Endo H, Hosono K, Uchiyama T, Sakai E, Sugiyama M, Takahashi H, Nakajima N, Wada K, Takeda K, Nakagama H et al (2011) Leptin acts as a growth factor for colorectal tumours at stages subsequent to tumour initiation in murine colon carcinogenesis. Gut 60 (10):1363–1371

23. Cleary MP, Grossmann ME (2009) Minireview: Obesity and breast cancer: the estrogen connection. Endocrinology 150(6):2537–2542
24. Chang SH, Luo S, Thomas TS, O'Brian KK, Colditz GA, Carlsson NP, Carson KR (2017) Obesity and the transformation of monoclonal gammopathy of undetermined significance to multiple myeloma: a population-based cohort study. J Natl Cancer Inst 109(5)
25. Berger NA (2018) Young adult cancer: influence of the obesity pandemic. Obesity (Silver Spring) 26(4):641–650
26. Caspersen CJ, Powell KE, Christenson GM (1985) Physical activity, exercise, and physical fitness: definitions and distinctions for health-related research. Public Health Rep 100 (2):126–131
27. Booth FW, Roberts CK, Laye MJ (2012) Lack of exercise is a major cause of chronic diseases. Compr Physiol 2(2):1143–1211
28. Reiner M, Niermann C, Jekauc D, Woll A (2013) Long-term health benefits of physical activity–a systematic review of longitudinal studies. BMC Public Health 13:813
29. Gomes da Silva S, Simoes PS, Mortara RA, Scorza FA, Cavalheiro EA, da Graca Naffah-Mazzacoratti M, Arida RM (2013) Exercise-induced hippocampal anti-inflammatory response in aged rats. J Neuroinflamm 10:61
30. Littlefield AM, Setti SE, Priester C, Kohman RA (2015) Voluntary exercise attenuates LPS-induced reductions in neurogenesis and increases microglia expression of a proneurogenic phenotype in aged mice. J Neuroinflamm 12:138
31. Wu CW, Chang YT, Yu L, Chen HI, Jen CJ, Wu SY, Lo CP, Kuo YM (2008) Exercise enhances the proliferation of neural stem cells and neurite growth and survival of neuronal progenitor cells in dentate gyrus of middle-aged mice. J Appl Physiol (1985) 105(5):1585–1594
32. Winter B, Breitenstein C, Mooren FC, Voelker K, Fobker M, Lechtermann A, Krueger K, Fromme A, Korsukewitz C, Floel A et al (2007) High impact running improves learning. Neurobiol Learn Mem 87(4):597–609
33. Ang ET, Tai YK, Lo SQ, Seet R, Soong TW (2010) Neurodegenerative diseases: exercising toward neurogenesis and neuroregeneration. Front Aging Neurosci 2
34. Bates KA, Verdile G, Li QX, Ames D, Hudson P, Masters CL, Martins RN (2009) Clearance mechanisms of Alzheimer's amyloid-beta peptide: implications for therapeutic design and diagnostic tests. Mol Psychiatry 14(5):469–486
35. Rolland Y, Pillard F, Klapouszczak A, Reynish E, Thomas D, Andrieu S, Riviere D, Vellas B (2007) Exercise program for nursing home residents with Alzheimer's disease: a 1-year randomized, controlled trial. J Am Geriatr Soc 55(2):158–165
36. Bloomer RJ, Schilling BK, Karlage RE, Ledoux MS, Pfeiffer RF, Callegari J (2008) Effect of resistance training on blood oxidative stress in Parkinson disease. Med Sci Sports Exerc 40(8):1385–1389
37. Schalow G, Paasuke M, Jaigma P (2005) Integrative re-organization mechanism for reducing tremor in Parkinson's disease patients. Electromyogr Clin Neurophysiol 45(7–8):407–415
38. Falvo MJ, Schilling BK, Earhart GM (2008) Parkinson's disease and resistive exercise: rationale, review, and recommendations. Mov Disord 23(1):1–11
39. Ljubicic V, Joseph AM, Adhihetty PJ, Huang JH, Saleem A, Uguccioni G, Hood DA (2009) Molecular basis for an attenuated mitochondrial adaptive plasticity in aged skeletal muscle. Aging (Albany NY) 1(9):818–830
40. Clark-Matott J, Saleem A, Dai Y, Shurubor Y, Ma X, Safdar A, Beal MF, Tarnopolsky M, Simon DK (2015) Metabolomic analysis of exercise effects in the POLG mitochondrial DNA mutator mouse brain. Neurobiol Aging 36(11):2972–2983
41. Safdar A, Annis S, Kraytsberg Y, Laverack C, Saleem A, Popadin K, Woods DC, Tilly JL, Khrapko K (2016) Amelioration of premature aging in mtDNA mutator mouse by exercise: the interplay of oxidative stress, PGC-1alpha, p53, and DNA damage. A hypothesis. Curr Opin Genet Dev 38:127–132
42. Safdar A, Bourgeois JM, Ogborn DI, Little JP, Hettinga BP, Akhtar M, Thompson JE, Melov S, Mocellin NJ, Kujoth GC et al (2011) Endurance exercise rescues progeroid aging

and induces systemic mitochondrial rejuvenation in mtDNA mutator mice. Proc Natl Acad Sci U S A 108(10):4135–4140
43. Safdar A, Khrapko K, Flynn JM, Saleem A, De Lisio M, Johnston AP, Kratysberg Y, Samjoo IA, Kitaoka Y, Ogborn DI et al (2016) Exercise-induced mitochondrial p53 repairs mtDNA mutations in mutator mice. Skelet Muscle 6:7
44. Cornelissen VA, Smart NA (2013) Exercise training for blood pressure: a systematic review and meta-analysis. J Am Heart Assoc 2(1):e004473
45. Reimers CD, Knapp G, Reimers AK (2009) Exercise as stroke prophylaxis. Dtsch Arztebl Int 106(44):715–721
46. Vega RB, Konhilas JP, Kelly DP, Leinwand LA (2017) Molecular mechanisms underlying cardiac adaptation to exercise. Cell Metab 25(5):1012–1026
47. Nystoriak MA, Bhatnagar A (2018) Cardiovascular effects and benefits of exercise. Front Cardiovasc Med 5:135
48. Taylor RS, Brown A, Ebrahim S, Jolliffe J, Noorani H, Rees K, Skidmore B, Stone JA, Thompson DR, Oldridge N (2004) Exercise-based rehabilitation for patients with coronary heart disease: systematic review and meta-analysis of randomized controlled trials. Am J Med 116(10):682–692
49. Goldsmith R, Joanisse DR, Gallagher D, Pavlovich K, Shamoon E, Leibel RL, Rosenbaum M (2010) Effects of experimental weight perturbation on skeletal muscle work efficiency, fuel utilization, and biochemistry in human subjects. Am J Physiol Regul Integr Comp Physiol 298(1):R79–R88
50. Joyner MJ, Pedersen BK (2011) Ten questions about systems biology. J Physiol 589(Pt 5):1017–1030
51. Slentz CA, Duscha BD, Johnson JL, Ketchum K, Aiken LB, Samsa GP, Houmard JA, Bales CW, Kraus WE (2004) Effects of the amount of exercise on body weight, body composition, and measures of central obesity: STRRIDE–a randomized controlled study. Arch Intern Med 164(1):31–39
52. Seki Y, Berggren JR, Houmard JA, Charron MJ (2006) Glucose transporter expression in skeletal muscle of endurance-trained individuals. Med Sci Sports Exerc 38(6):1088–1092
53. Winder WW, Thomson DM (2007) Cellular energy sensing and signaling by AMP-activated protein kinase. Cell Biochem Biophys 47(3):332–347
54. Umpierre D, Ribeiro PA, Kramer CK, Leitao CB, Zucatti AT, Azevedo MJ, Gross JL, Ribeiro JP, Schaan BD (2011) Physical activity advice only or structured exercise training and association with HbA1c levels in type 2 diabetes: a systematic review and meta-analysis. JAMA 305(17):1790–1799
55. Johansen MY, MacDonald CS, Hansen KB, Karstoft K, Christensen R, Pedersen M, Hansen LS, Zacho M, Wedell-Neergaard AS, Nielsen ST et al (2017) Effect of an intensive lifestyle intervention on glycemic control in patients with type 2 diabetes: a randomized clinical trial. JAMA 318(7):637–646
56. Pedersen BK (2019) The physiology of optimizing health with a focus on exercise as medicine. Annu Rev Physiol 81:607–627
57. Chen X, Zheng Y, Zheng W, Gu K, Chen Z, Lu W, Shu XO (2009) The effect of regular exercise on quality of life among breast cancer survivors. Am J Epidemiol 170(7):854–862
58. Bruunsgaard H (2005) Physical activity and modulation of systemic low-level inflammation. J Leukoc Biol 78(4):819–835
59. Pedersen BK, Fischer CP (2007) Beneficial health effects of exercise–the role of IL-6 as a myokine. Trends Pharmacol Sci 28(4):152–156
60. Karstoft K, Pedersen BK (2016) Skeletal muscle as a gene regulatory endocrine organ. Curr Opin Clin Nutr Metab Care 19(4):270–275
61. Pedersen BK (2011) Muscles and their myokines. J Exp Biol 214(Pt 2):337–346
62. Cao L, Choi EY, Liu X, Martin A, Wang C, Xu X, During MJ (2011) White to brown fat phenotypic switch induced by genetic and environmental activation of a hypothalamic-adipocyte axis. Cell Metab 14(3):324–338

63. Kelly AM (2018) Exercise-induced modulation of neuroinflammation in models of Alzheimer's disease. Brain Plast 4(1):81–94
64. Pedersen BK (2009) The diseasome of physical inactivity–and the role of myokines in muscle–fat cross talk. J Physiol 587(Pt 23):5559–5568
65. Pedersen BK, Febbraio MA (2008) Muscle as an endocrine organ: focus on muscle-derived interleukin-6. Physiol Rev 88(4):1379–1406
66. Spielman LJ, Little JP, Klegeris A (2016) Physical activity and exercise attenuate neuroinflammation in neurological diseases. Brain Res Bull 125:19–29
67. Stanford KI, Goodyear LJ (2018) Muscle-adipose tissue cross talk. Cold Spring Harb Perspect Med 8(8)
68. Bostrom P, Wu J, Jedrychowski MP, Korde A, Ye L, Lo JC, Rasbach KA, Bostrom EA, Choi JH, Long JZ et al (2012) A PGC1-alpha-dependent myokine that drives brown-fat-like development of white fat and thermogenesis. Nature 481(7382):463–468
69. Albrecht E, Norheim F, Thiede B, Holen T, Ohashi T, Schering L, Lee S, Brenmoehl J, Thomas S, Drevon CA et al (2015) Irisin—a myth rather than an exercise-inducible myokine. Sci Rep 5:8889
70. Raschke S, Elsen M, Gassenhuber H, Sommerfeld M, Schwahn U, Brockmann B, Jung R, Wisloff U, Tjonna AE, Raastad T et al (2013) Evidence against a beneficial effect of irisin in humans. PLoS ONE 8(9):e73680
71. Jedrychowski MP, Wrann CD, Paulo JA, Gerber KK, Szpyt J, Robinson MM, Nair KS, Gygi SP, Spiegelman BM (2015) Detection and quantitation of circulating human irisin by tandem mass spectrometry. Cell Metab 22(4):734–740
72. Rao RR, Long JZ, White JP, Svensson KJ, Lou J, Lokurkar I, Jedrychowski MP, Ruas JL, Wrann CD, Lo JC et al (2014) Meteorin-like is a hormone that regulates immune-adipose interactions to increase beige fat thermogenesis. Cell 157(6):1279–1291
73. Lee SD, Tontonoz P (2014) Eosinophils in fat: pink is the new brown. Cell 157(6):1249–1250
74. Qiu Y, Nguyen KD, Odegaard JI, Cui X, Tian X, Locksley RM, Palmiter RD, Chawla A (2014) Eosinophils and type 2 cytokine signaling in macrophages orchestrate development of functional beige fat. Cell 157(6):1292–1308
75. McNally EM (2004) Powerful genes–myostatin regulation of human muscle mass. N Engl J Med 350(26):2642–2644
76. McPherron AC, Lee SJ (2002) Suppression of body fat accumulation in myostatin-deficient mice. J Clin Invest 109(5):595–601
77. Shan T, Liang X, Bi P, Kuang S (2013) Myostatin knockout drives browning of white adipose tissue through activating the AMPK-PGC1alpha-Fndc5 pathway in muscle. FASEB J 27(5):1981–1989
78. Whittemore LA, Song K, Li X, Aghajanian J, Davies M, Girgenrath S, Hill JJ, Jalenak M, Kelley P, Knight A et al (2003) Inhibition of myostatin in adult mice increases skeletal muscle mass and strength. Biochem Biophys Res Commun 300(4):965–971
79. Kharitonenkov A, Larsen P (2011) FGF21 reloaded: challenges of a rapidly growing field. Trends Endocrinol Metab 22(3):81–86
80. Lee P, Linderman JD, Smith S, Brychta RJ, Wang J, Idelson C, Perron RM, Werner CD, Phan GQ, Kammula US et al (2014) Irisin and FGF21 are cold-induced endocrine activators of brown fat function in humans. Cell Metab 19(2):302–309
81. Roberts LD, Bostrom P, O'Sullivan JF, Schinzel RT, Lewis GD, Dejam A, Lee YK, Palma MJ, Calhoun S, Georgiadi A et al (2014) beta-Aminoisobutyric acid induces browning of white fat and hepatic beta-oxidation and is inversely correlated with cardiometabolic risk factors. Cell Metab 19(1):96–108
82. Brooks GA (2009) Cell-cell and intracellular lactate shuttles. J Physiol 587(Pt 23):5591–5600
83. Carriere A, Jeanson Y, Berger-Muller S, Andre M, Chenouard V, Arnaud E, Barreau C, Walther R, Galinier A, Wdziekonski B et al (2014) Browning of white adipose cells by intermediate metabolites: an adaptive mechanism to alleviate redox pressure. Diabetes 63 (10):3253–3265

84. El Andaloussi S, Lakhal S, Mager I, Wood MJ (2013) Exosomes for targeted siRNA delivery across biological barriers. Adv Drug Deliv Rev 65(3):391–397
85. Zaborowski MP, Balaj L, Breakefield XO, Lai CP (2015) Extracellular vesicles: composition, biological relevance, and methods of study. Bioscience 65(8):783–797
86. Wei H, Malcor JM, Harper MT (2018) Lipid rafts are essential for release of phosphatidylserine-exposing extracellular vesicles from platelets. Sci Rep 8(1):9987
87. Thery C, Ostrowski M, Segura E (2009) Membrane vesicles as conveyors of immune responses. Nat Rev Immunol 9(8):581–593
88. Crescitelli R, Lasser C, Szabo TG, Kittel A, Eldh M, Dianzani I, Buzas EI, Lotvall J (2013) Distinct RNA profiles in subpopulations of extracellular vesicles: apoptotic bodies, microvesicles and exosomes. J Extracell Vesicles 2
89. Whitham M, Parker BL, Friedrichsen M, Hingst JR, Hjorth M, Hughes WE, Egan CL, Cron L, Watt KI, Kuchel RP et al (2018) Extracellular vesicles provide a means for tissue crosstalk during exercise. Cell Metab 27(1):237–251 e234
90. Bei Y, Xu T, Lv D, Yu P, Xu J, Che L, Das A, Tigges J, Toxavidis V, Ghiran I et al (2017) Exercise-induced circulating extracellular vesicles protect against cardiac ischemia-reperfusion injury. Basic Res Cardiol 112(4):38
91. Guescini M, Canonico B, Lucertini F, Maggio S, Annibalini G, Barbieri E, Luchetti F, Papa S, Stocchi V (2015) Muscle releases alpha-sarcoglycan positive extracellular vesicles carrying miRNAs in the bloodstream. PLoS ONE 10(5):e0125094
92. Oliveira GP Jr, Porto WF, Palu CC, Pereira LM, Petriz B, Almeida JA, Viana J, Filho NNA, Franco OL, Pereira RW (2018) Effects of acute aerobic exercise on rats serum extracellular vesicles diameter, concentration and small RNAs content. Front Physiol 9:532
93. Escudero CA, Herlitz K, Troncoso F, Acurio J, Aguayo C, Roberts JM, Truong G, Duncombe G, Rice G, Salomon C (2016) Role of extracellular vesicles and microRNAs on dysfunctional angiogenesis during preeclamptic pregnancies. Front Physiol 7:98
94. Fernandez-Messina L, Gutierrez-Vazquez C, Rivas-Garcia E, Sanchez-Madrid F, de la Fuente H (2015) Immunomodulatory role of microRNAs transferred by extracellular vesicles. Biol Cell 107(3):61–77
95. Mittelbrunn M, Gutierrez-Vazquez C, Villarroya-Beltri C, Gonzalez S, Sanchez-Cabo F, Gonzalez MA, Bernad A, Sanchez-Madrid F (2011) Unidirectional transfer of microRNA-loaded exosomes from T cells to antigen-presenting cells. Nat Commun 2:282
96. Valadi H, Ekstrom K, Bossios A, Sjostrand M, Lee JJ, Lotvall JO (2007) Exosome-mediated transfer of mRNAs and microRNAs is a novel mechanism of genetic exchange between cells. Nat Cell Biol 9(6):654–659
97. Ding G, Zhou L, Qian Y, Fu M, Chen J, Chen J, Xiang J, Wu Z, Jiang G, Cao L (2015) Pancreatic cancer-derived exosomes transfer miRNAs to dendritic cells and inhibit RFXAP expression via miR-212-3p. Oncotarget 6(30):29877–29888
98. Ong SG, Lee WH, Huang M, Dey D, Kodo K, Sanchez-Freire V, Gold JD, Wu JC (2014) Cross talk of combined gene and cell therapy in ischemic heart disease: role of exosomal microRNA transfer. Circulation 130(11 Suppl 1):S60–S69
99. Vinas JL, Burger D, Zimpelmann J, Haneef R, Knoll W, Campbell P, Gutsol A, Carter A, Allan DS, Burns KD (2016) Transfer of microRNA-486-5p from human endothelial colony forming cell-derived exosomes reduces ischemic kidney injury. Kidney Int 90(6):1238–1250
100. Tkach M, Thery C (2016) Communication by extracellular vesicles: where we are and where we need to go. Cell 164(6):1226–1232
101. Yang Q, Diamond MP, Al-Hendy A (2016) The emerging role of extracellular vesicle-derived miRNAs: implication in cancer progression and stem cell related diseases. J Clin Epigenet 2(1)
102. Cha DJ, Franklin JL, Dou Y, Liu Q, Higginbotham JN, Demory Beckler M, Weaver AM, Vickers K, Prasad N, Levy S et al (2015) KRAS-dependent sorting of miRNA to exosomes. Elife 4:e07197
103. Fiskaa T, Knutsen E, Nikolaisen MA, Jorgensen TE, Johansen SD, Perander M, Seternes OM (2016) Distinct small RNA Signatures in extracellular vesicles derived from breast cancer cell lines. PLoS ONE 11(8):e0161824

104. Chen J, Hu C, Pan P (2017) Extracellular vesicle MicroRNA transfer in lung diseases. Front Physiol 8:1028
105. Wang H, Wang B (2016) Extracellular vesicle microRNAs mediate skeletal muscle myogenesis and disease. Biomed Rep 5(3):296–300
106. Iacomino G, Siani A (2017) Role of microRNAs in obesity and obesity-related diseases. Genes Nutr 12:23
107. Ortiz-Dosal A, Rodil-Garcia P, Salazar-Olivo LA (2019) Circulating MicroRNAs in human obesity: a systematic review. Biomarkers 1–41
108. Xie H, Lim B, Lodish HF (2009) MicroRNAs induced during adipogenesis that accelerate fat cell development are downregulated in obesity. Diabetes 58(5):1050–1057
109. Ortega FJ, Mercader JM, Catalan V, Moreno-Navarrete JM, Pueyo N, Sabater M, Gomez-Ambrosi J, Anglada R, Fernandez-Formoso JA, Ricart W et al (2013) Targeting the circulating microRNA signature of obesity. Clin Chem 59(5):781–792
110. Castano C, Kalko S, Novials A, Parrizas M (2018) Obesity-associated exosomal miRNAs modulate glucose and lipid metabolism in mice. Proc Natl Acad Sci U S A 115(48):12158–12163
111. Heneghan HM, Miller N, McAnena OJ, O'Brien T, Kerin MJ (2011) Differential miRNA expression in omental adipose tissue and in the circulation of obese patients identifies novel metabolic biomarkers. J Clin Endocrinol Metab 96(5):E846–E850
112. Iacomino G, Russo P, Stillitano I, Lauria F, Marena P, Ahrens W, De Luca P, Siani A (2016) Circulating microRNAs are deregulated in overweight/obese children: preliminary results of the I. Family study. Genes Nutr 11:7
113. Otton R, Bolin AP, Ferreira LT, Marinovic MP, Rocha ALS, Mori MA (2018) Polyphenol-rich green tea extract improves adipose tissue metabolism by down-regulating miR-335 expression and mitigating insulin resistance and inflammation. J Nutr Biochem 57:170–179
114. Price NL, Singh AK, Rotllan N, Goedeke L, Wing A, Canfran-Duque A, Diaz-Ruiz A, Araldi E, Baldan A, Camporez JP et al (2018) Genetic ablation of miR-33 increases food intake, enhances adipose tissue expansion, and promotes obesity and insulin resistance. Cell Rep 22(8):2133–2145
115. Deiuliis JA (2016) MicroRNAs as regulators of metabolic disease: pathophysiologic significance and emerging role as biomarkers and therapeutics. Int J Obes (Lond) 40(1):88–101
116. Huang Y, Yan Y, Xv W, Qian G, Li C, Zou H, Li Y (2018) A new insight into the roles of miRNAs in metabolic syndrome. Biomed Res Int 2018:7372636
117. Karolina DS, Tavintharan S, Armugam A, Sepramaniam S, Pek SL, Wong MT, Lim SC, Sum CF, Jeyaseelan K (2012) Circulating miRNA profiles in patients with metabolic syndrome. J Clin Endocrinol Metab 97(12):E2271–E2276
118. Gottmann P, Ouni M, Saussenthaler S, Roos J, Stirm L, Jahnert M, Kamitz A, Hallahan N, Jonas W, Fritsche A et al (2018) A computational biology approach of a genome-wide screen connected miRNAs to obesity and type 2 diabetes. Mol Metab 11:145–159
119. Hernandez-Alonso P, Giardina S, Salas-Salvado J, Arcelin P, Bullo M (2017) Chronic pistachio intake modulates circulating microRNAs related to glucose metabolism and insulin resistance in prediabetic subjects. Eur J Nutr 56(6):2181–2191
120. Zurawek D, Kusmider M, Faron-Gorecka A, Gruca P, Pabian P, Solich J, Kolasa M, Papp M, Dziedzicka-Wasylewska M (2017) Reciprocal MicroRNA expression in mesocortical circuit and its interplay with serotonin transporter define resilient rats in the chronic mild stress. Mol Neurobiol 54(8):5741–5751
121. Zile MR, Mehurg SM, Arroyo JE, Stroud RE, DeSantis SM, Spinale FG (2011) Relationship between the temporal profile of plasma microRNA and left ventricular remodeling in patients after myocardial infarction. Circ Cardiovasc Genet 4(6):614–619
122. Iorio MV, Ferracin M, Liu CG, Veronese A, Spizzo R, Sabbioni S, Magri E, Pedriali M, Fabbri M, Campiglio M et al (2005) MicroRNA gene expression deregulation in human breast cancer. Cancer Res 65(16):7065–7070

123. Frankel LB, Christoffersen NR, Jacobsen A, Lindow M, Krogh A, Lund AH (2008) Programmed cell death 4 (PDCD4) is an important functional target of the microRNA miR-21 in breast cancer cells. J Biol Chem 283(2):1026–1033
124. Wu Q, Wang C, Lu Z, Guo L, Ge Q (2012) Analysis of serum genome-wide microRNAs for breast cancer detection. Clin Chim Acta 413(13–14):1058–1065
125. Tsai HP, Huang SF, Li CF, Chien HT, Chen SC (2018) Differential microRNA expression in breast cancer with different onset age. PLoS ONE 13(1):e0191195
126. Yu SJ, Hu JY, Kuang XY, Luo JM, Hou YF, Di GH, Wu J, Shen ZZ, Song HY, Shao ZM (2013) MicroRNA-200a promotes anoikis resistance and metastasis by targeting YAP1 in human breast cancer. Clin Cancer Res 19(6):1389–1399
127. Kong W, He L, Richards EJ, Challa S, Xu CX, Permuth-Wey J, Lancaster JM, Coppola D, Sellers TA, Djeu JY et al (2014) Upregulation of miRNA-155 promotes tumour angiogenesis by targeting VHL and is associated with poor prognosis and triple-negative breast cancer. Oncogene 33(6):679–689
128. Nassirpour R, Mehta PP, Baxi SM, Yin MJ (2013) miR-221 promotes tumorigenesis in human triple negative breast cancer cells. PLoS ONE 8(4):e62170
129. Papadaki C, Stoupis G, Tsalikis L, Monastirioti A, Papadaki M, Maliotis N, Stratigos M, Mastrostamatis G, Mavroudis D, Agelaki S (2019) Circulating miRNAs as a marker of metastatic disease and prognostic factor in metastatic breast cancer. Oncotarget 10(9):966–981
130. Masi LN, Serdan TD, Levada-Pires AC, Hatanaka E, Silveira LD, Cury-Boaventura MF, Pithon-Curi TC, Curi R, Gorjao R, Hirabara SM (2016) Regulation of gene expression by exercise-related micrornas. Cell Physiol Biochem 39(6):2381–2397
131. Nielsen S, Scheele C, Yfanti C, Akerstrom T, Nielsen AR, Pedersen BK, Laye MJ (2010) Muscle specific microRNAs are regulated by endurance exercise in human skeletal muscle. J Physiol 588(Pt 20):4029–4037
132. Pasiakos SM, McClung JP (2013) miRNA analysis for the assessment of exercise and amino acid effects on human skeletal muscle. Adv Nutr 4(4):412–417
133. Safdar A, Abadi A, Akhtar M, Hettinga BP, Tarnopolsky MA (2009) miRNA in the regulation of skeletal muscle adaptation to acute endurance exercise in C57Bl/6J male mice. PLoS ONE 4(5):e5610
134. Ultimo S, Zauli G, Martelli AM, Vitale M, McCubrey JA, Capitani S, Neri LM (2018) Influence of physical exercise on microRNAs in skeletal muscle regeneration, aging and diseases. Oncotarget 9(24):17220–17237
135. Russell AP, Lamon S, Boon H, Wada S, Guller I, Brown EL, Chibalin AV, Zierath JR, Snow RJ, Stepto N et al (2013) Regulation of miRNAs in human skeletal muscle following acute endurance exercise and short-term endurance training. J Physiol 591(18):4637–4653
136. Horak M, Zlamal F, Iliev R, Kucera J, Cacek J, Svobodova L, Hlavonova Z, Kalina T, Slaby O, Bienertova-Vasku J (2018) Exercise-induced circulating microRNA changes in athletes in various training scenarios. PLoS ONE 13(1):e0191060
137. Shamir R, Phillip M, Turck D (2013) World review of nutrition and dietetics. Nutrition and growth. Introduction. World Rev Nutr Diet 106:1–2
138. Vargas-Castillo A, Fuentes-Romero R, Rodriguez-Lopez LA, Torres N, Tovar AR (2017) Understanding the biology of thermogenic fat: is browning a new approach to the treatment of obesity? Arch Med Res 48(5):401–413
139. Sidossis LS, Porter C, Saraf MK, Borsheim E, Radhakrishnan RS, Chao T, Ali A, Chondronikola M, Mlcak R, Finnerty CC et al (2015) Browning of subcutaneous white adipose tissue in humans after severe adrenergic stress. Cell Metab 22(2):219–227
140. Thery C, Witwer KW, Aikawa E, Alcaraz MJ, Anderson JD, Andriantsitohaina R, Antoniou A, Arab T, Archer F, Atkin-Smith GK et al (2018) Minimal information for studies of extracellular vesicles 2018 (MISEV2018): a position statement of the international society for extracellular vesicles and update of the MISEV2014 guidelines. J Extracell Vesicles 7(1):1535750

Chapter 5
Obesity and Diabetes: Pathophysiology of Obesity-Induced Hyperglycemia and Insulin Resistance

Gaurav Gupta, Ridhima Wadhwa, Parijat Pandey, Sachin Kumar Singh, Monica Gulati, Saurabh Sajita, Meenu Mehta, Avinash Kumar Singh, Harish Dureja, Trudi Collet, Kavita Pabreja, Dinesh Kumar Chellappan and Kamal Dua

Abstract Insulin resistance along with type 2 diabetes (T2D) is some of the complications of obesity. Insulin resistance occurs due to the increased discharge of fatty acids, lipids, and other advancing factors, by the adipose tissue, leading to a myriad of complications. Several other substances namely glycerol, enzymes, inflammatory factors, and hormones are also involved in the overall mechanism. At the point when resistance towards insulin occurs, along with abnormally functioning β-cells, the overall burden of the inability to control elevated glucose levels in the blood ensues. Anomalies in β-cell are thus more critical in the development

G. Gupta (✉)
School of Pharmaceutical Sciences, Jaipur National University, Jagatpura, Jaipur 302017, India
e-mail: gauravpharma25@gmail.com

R. Wadhwa
Faculty of Life Sciences and Biotechnology, South Asian University, Akbar Bhawan, Chanakyapuri 110021, New Delhi, India

P. Pandey
Shri Baba Mastnath Institute of Pharmaceutical Sciences and Research, Baba Mastnath University, Rohtak 124001, India

S. K. Singh · M. Gulati · S. Sajita · M. Mehta
School of Pharmaceutical Sciences, Lovely Professional University, Jalandhar-Delhi G.T. Road (NH-1), Phagwara, Punjab 144411, India

A. K. Singh
Department of Pharmacology, School of Pharmaceutical Education and Research, Jamia Hamdard 110062, New Delhi, India

H. Dureja
Department of Pharmaceutical Sciences, Maharishi Dayanand University, Rohtak 124001, Haryana, India

T. Collet
Inovative Medicines Group, Institute of Health & Biomedical Innovation, Queensland University of Technology, Kelvin Grove, Brisbane, QLD 4059, Australia

P. S. Tappia et al. (eds.), *Pathophysiology of Obesity-Induced Health Complications*, Advances in Biochemistry in Health and Disease 19,
https://doi.org/10.1007/978-3-030-35358-2_5

of T2D and its related symptoms. This information is encouraging investigation of the molecular and hereditary basis of the ailment and new ways to deal with its management and avoidance.

Keywords Obesity · Insulin resistance · Diabetes · β-cells · Genetic

Introduction

Overweight and obesity are characterized by an overabundance gathering of fat tissue to an extent that weakens both physical and psycho-social wellbeing and health. Obesity is viewed as a wellbeing catastrophe in both developing and developed nations [1, 2]. The number of fat people worldwide has achieved 2.1 billion, prompting a blast of obesity-related medical issues related to increased morbidity and mortality [3]. Obese people develop resistance to the cellular activities of insulin, described by a disabled capacity of insulin to repress glucose yield from the liver and to advance glucose take-up in fat and muscle. Insulin obstruction is a key etiological aspect for causing a type 2 diabetes mellitus (T2DM), which has achieved pandemic extents: In the United States, nearly 6% of the current adult populace is diagnosed with this illness [4, 5]. An extra 41 million individuals are prone to be diabetic; with a constellation of resistance to insulin, dyslipidemia and hypertension, which makes the more complicated situation at increased risk for cardiovascular morbidity and mortality [6, 7]. Positive lifestyle changes, while desirable, has demonstrated to be hard to accomplish [8, 9]. Accordingly, a better understanding of the mechanisms at a molecular level underlying insulin resistance will be required to battle the pestilences of T2DM that are strengthened by obesity-linked insulin resistance.

K. Pabreja
Priority Research Centre for Healthy Lungs, The University Of Newcastle, Newcastle, NSW, Australia

D. K. Chellappan
Department of Life Sciences, School of Pharmacy, International Medical University, Bukit Jalil, Kuala Lumpur 57000, Malaysia

K. Dua
Discipline of Pharmacy, Graduate School of Health, University of Technology Sydney, Ultimo, NSW 2007, Australia

Centenary Institute, Royal Prince Alfred Hospital, Camperdown, NSW 2050, Australia

School of Biomedical Sciences and Pharmacy, Priority Research Centre for Healthy Lungs, Hunter Medical Research Institute (HMRI), University of Newcastle, Callaghan NSW 2308, Australia

Incidence

During the last decade, diabetes has become a serious global public health issue and its prevalence has increased dramatically. Diabetes is also one of the major disablers causing additional burden to the existing healthcare scenario. IDF (International Diabetic Federation) predicts that between 2010 and 2030, there will be over 98% jump in diabetics in Africa alone. This will be one of the highest incidence rates in the world. Over 7% of the total diabetics in the world live in Africa. This translates to a whopping 28 million cases. IDF also estimated that by 2030, 366 million deaths worldwide will be due to diabetes [10–12]. The most rapid increase in childhood and adult diabetes is experienced by Mexico, which is now second in Latin America and sixth in the world. Diabetes has been a leading cause of mortality in Mexico since 2000. The overweight and obesity prevalence is approximately 70% in Mexico, which is second after the United States [13–15]. Obesity is the main target for "Global Action Plan for the Prevention and Control of non-communicable disease 2013–2030" [16].

Physiologic Actions of Insulin

Insulin is the main hormone of glucose homeostasis; it invigorates influx of glucose into muscle, synthesis of glycogen in the liver and muscle, and deposition of fat in adipose tissues (Fig. 5.1). Other vital activities of insulin incorporate the improvement of protein synthesis, cell survival and development, aversion of protein catabolism, and anti-inflammatory effects. Insulin-like growth factor (IGF-1) can carry on like insulin, delivering the equivalent valuable effects [17, 18]. Obesity-related T2DM is confirmed by expanded glucose levels in the blood, which result from raised glucose generation in the liver (gluconeogenesis and glycogenolysis) and diminished glucose take-up by muscle [19, 20]. In obesity-promoted T2DM or any pathologic condition related with diminished insulin work, hyperglycemia together with the constricted anabolic and anti-inflammatory impacts prompts muscle protein loss i.e. muscle squandering, increased chances to infection, and higher inflammatory response [21, 22].

Pathophysiology

Pathogenesis of Obesity-Induced Hyperglycemia

One of the most prevalent disorders in the overweight and obese population is diabetes type II. It also accounts for the age-related disorder. The risk of diabetes is related to lifestyle, genetic makeup, and aging. These factors are responsible for

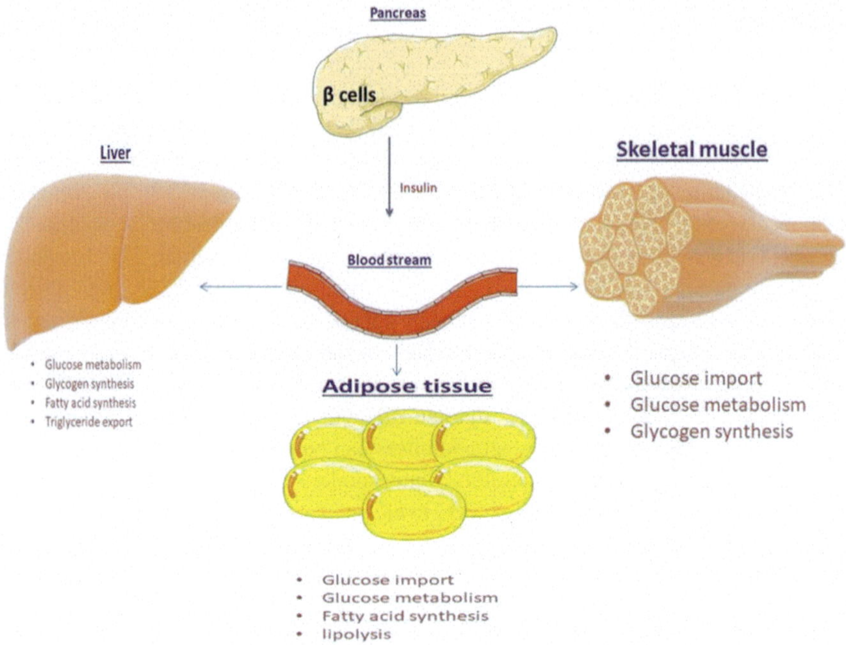

Fig. 5.1 Physiological function of insulin

hyperglycemia, β cell insulin secretion and tissue sensitivity towards insulin. Classically, diabetes type 2 in obesity is explained by insulin resistance and then hyperglycemia leading to β cell death [23, 24]. Insulin resistance occurs from genetic and environmental factors. The hepatic insulin resistance leads to hyperglycemia and diabetes [25]. High plasma insulin causes caloric overload and accumulation of fat in the hepatocytes. This, in turn, results in the failure in glucose production by the liver along with insulin resistance. In addition, it also results in the infiltration of liver fats and lipids which in turn results in the overexpression of β cells, which impedes insulin secretion leading to an eventual rise in the levels of plasma glucose. These elevated plasma sugar levels trigger insulin secretion increasing the rate of hepatic lipogenesis. This, in turn, connects the liver cycle to pancreatic cycle [26]. This has been defined as the insulin resistance in the muscles or otherwise called the twin hypothesis as can be seen in Fig. 5.2.

Pathogenesis of Obesity-Induced Insulin Resistance and Hyperglycaemia

Adipose tissues are insulin sensitive and stores triglycerides via differentiation of preadipocytes to adipocytes, increased glucose and fatty acid uptake from

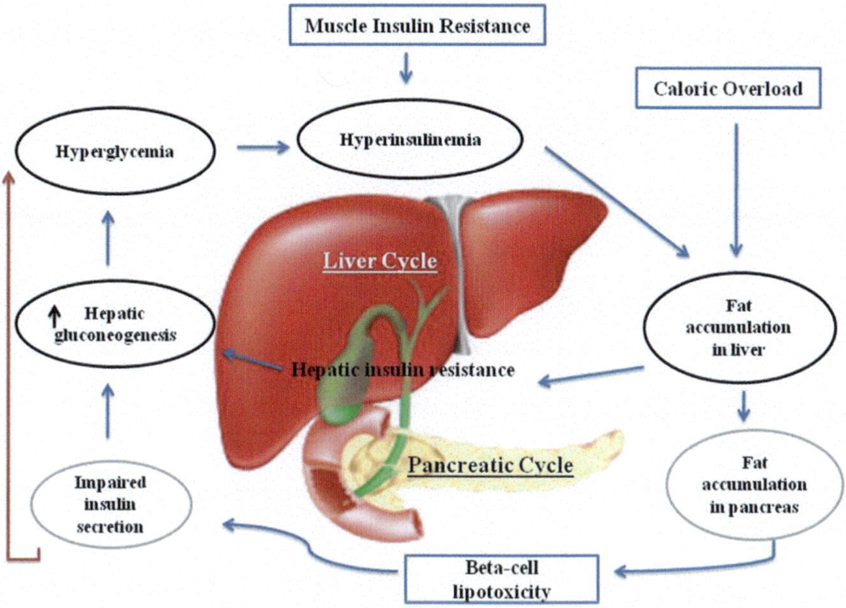

Fig. 5.2 Role of muscle, liver, and pancreas in obesity-induced diabetes type 2

circulation and restrict lipolysis [27]. This occurs by signalling activated by phosphorylation of the insulin receptor leading to binding of PI3K to SH2 domain which can be seen in Fig. 5.3. This generates a secondary messenger for the activation of phosphatidylinositol-(3,4,5)-triphosphate-dependent serine or threonine kinase resulting in glucose transport 4 translocation to the plasma membrane [28]. This is responsible for increased glucose uptake. Furthermore, MAPK also has an indirect metabolic role with anti-lipolytic effect, whereby activates PI3K to stimulate cyclic AMP to reduce fatty acid release from adipocytes [29]. Free fatty acids (FFA) and adipokines from adipose tissues are associated with insulin dysfunction [30]. Thus, FFA and their metabolites activate protein kinases while inhibits nuclear factor κ B kinase β. This has also been reported by Haus and his colleagues, that saturated FFA levels in the plasma are high in obese with diabetes type 2 and causes resistance towards insulin by proinflammatory factors like TNFα [31]. Obesity not only elevates lipolysis but also accumulates macrophages in adipose tissues associated with insulin resistance in obese [32]. Several pro-inflammatory proteins such as IL-18, IL-6, leptin, plasminogen activator inhibitor (PAI)-3 etc. are activated and anti-inflammatory proteins like TNFα, secreted frizzled-related protein 5 (SFRP5) are downregulated which correlates to obesity and metabolic dysfunction [33].

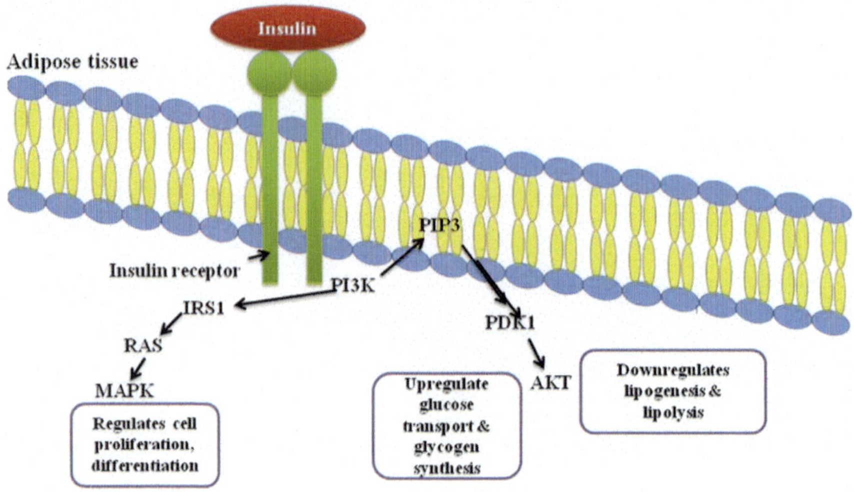

Fig. 5.3 Insulin signaling in adipose tissue

Pathogenesis of Obesity-Induced β-Cell Dysfunction

The progressive decline in β-cell function due to obesity accounts for 86% of youth which is reflected by insulin-producing β-cells to systemic insulin resistance [34]. Obese normoglycemic individuals have elevated β-cell mass and function because of increased glucose and lipid levels [34]. This raised level of nutrients leads to enhanced metabolic activity by β-cells, this successively leads to hyperinsulinemia. Further, prolonged exposure instigates oxidative stress to β-cells [35]. The pancreatic β-cells respond to glucose is dependent on intracellular and extracellular ROS/RNS. This causes an increase in glycolytic flux which stimulates oxidative phosphorylation and ATP production, with increased release of superoxide (O_2^-) anion from electron transport chain. Further, pentose phosphate cycle glucose is converted to a pentose, NADPH is produced from NADPH oxidase (NOX) leading to elevated O_2^- which forms H_2O_2 in the β-cells. Elevated ROS/RNS results in reduced ATP production, DNA damage and advanced glycogen end product (AGEs) [36]. Thus chronic hyperglycaemia causes oxidative stress, and inflammation leading to changes in gene expression responsible for impaired secretion of insulin and apoptosis.

The presence of high saturated fats during circulation competes with glucose for metabolism and tissue uptake. This causes glucolipotoxicity to the beta cells. Genetic variation in the insulin resistant pathways are affected by high body mass indices (BMI) which further attributes to glucose-stimulated insulin secretion (GSIS) due to short exposure of free fatty acids (FFA) [37]. Thus, excessive insulin secretion results in fat storage contributing to obesity. While prolonged exposure compromises glucose metabolism, diminished insulin biosynthesis, dysfunction and loss of β-cells. Also, chronic exposure of FFA hinders Ca^{2+} channels which correspond to increased fat in the islet and intra-pancreatic fat deposits affecting β cell function [38]. Islet inflammation due to metabolism overload also adversely affects the β cell. This recruits immune cells and local synthesis of cytokines leading to β cell dysfunction and insulin resistance [38]. FFA increases palmitic acid in muscles and nuclear factor of κ chain polypeptide gene enhancer for B cells (NFκB) escorts insulin resistance because of β cell death [39].

Current Pharmacotherapy for Obesity-Associated Diabetes

Resistance towards insulin along with altered lipid metabolism are the most common co-morbidities associated with obesity. Paradoxically, some anti-diabetic drugs lead to weight gain after prolonged use. These include sulfonylureas like glibenclamide, glipizide, glimepiride and glyburide, insulin, thiazolidinediones like pioglitazone, meglitinides like nateglinide and repaglinide. Body weight gain increases the risk of diabetes, hypertension and cardiovascular diseases. Hence, the use of some anti-obesity medications in obese diabetic patients is recommended for control of weight, improvement of glycemic control and a decrease in the level of triglycerides [40, 41].

There are six major drugs that are approved by the FDA to reduce obesity in diabetic patients [8]. These include drugs like phentermine, orlistat, extended release (ER) phentermine formulation, lorcaserin, naltrexone as a sustained release (SR) formulation, bupropion as an SR formulation and the injectable version of liraglutide. While orlistat works through a reduction in the absorption of fat, the rest of them act through CNS pathway either by reducing appetite or enhancing satiety. Phentermine is s sympathomimetic amine which is contraindicated in patients suffering from cardiovascular diseases, hypertension, hyperthyroidism or glaucoma. The limited use of phentermine is attributed to these contraindications as most of these conditions happen to be the comorbidities of diabetes [42, 43].

Orlistat is a pancreatic and gastric lipase inhibitor which is contraindicated in patients suffering from malabsorption syndrome or cholestasis. The common adverse effects reported are flatulence, bloating and diarrhea. In a clinical trial involving subjects with abnormal glucose tolerance, the risk of occurrence of T2D was reported to be reduced by 45% when correlated to the group that acted as a placebo [14]. Orlistat is reported to reduce A1C by ~0.3–0.5% when it was used along with oral antidiabetic drugs or any type of insulin [44, 45].

Combination therapy with phentermine/topiramate, a sympathomimetic amine, anorectic and an antileptic drug has been used to achieve weight loss among diabetic patients. It is contraindicated to patients suffering from glaucoma or hyperthyroidism. The findings from the SEQUEL trial revealed that, over the period of 2 years, subjects with T2D showed a 9% loss in their weights, in relation to a 2% loss in their weights among the subjects in the group that acted as the placebo [46, 47]. An added advantage of the therapy is the decreased levels of lipids, decreased blood pressure and an elevated level of cardio-metabolic markers, to an extent that many patients were reported to decrease or completely discontinue hypertension and diabetes therapy. In prediabetic patients, the incidence of T2D is reported to be on a decline by 76% when correlated with the group that acted as the placebo. Another clinical trial reported that patients with T2D showed a reduced 9.4% of baseline weight and 1.6% A1C, compared to a reduction of 2.7% in terms of weight loss along with 1.2% of A1C levels in the group that served as the placebo [45]. The common adverse effect is peripheral neuropathy, which again is a limiting factor since neuropathy is one of the common comorbidities in both types of diabetes.

Lorcaserin is a 5-HT_{2c} receptor agonist. The drug is advised to be administered at a dosage of 10 mg twice a day. It is reported that there was a loss in the baseline bodyweight in subjects treated with this particular drug which was recorded to be up to 5%. These subjects also managed to achieve an improved level of A1C. The values reported were 0.9–1.0% which was correlated to a weight loss percentage of 1.5%. The report also suggested that there was a 0.4% improvement in the levels of A1C from the group that served as a placebo [48, 49]. In another report, subjects with prediabetes who were overweight, when subjected to a management plan involving lorcaserin demonstrated a 38% reduced risk of developing T2D compared to the subjects that were administered the placebo drug [50, 51]. The drug is contraindicated in pregnancy and cannot be recommended in gestational diabetes. Common reported adverse drug events to include headache, dizziness, fatigue, dry mouth, constipation, upper respiratory tract infections, and hypoglycemia.

Naltrexone/bupropion combination therapy includes opioid antagonist and aminoketone antidepressant. Naltrexone is recommended at a dose of 8 mg while bupropion at 90 mg. Pre-clinical studies suggest that the combination increases activity in the pro-opiomelanocortin neurons in the hypothalamus. This is reported to be due to a decrease in the auto-inhibitory effects of β-endorphins resulting from naltrexone. This, in turn, leads to reduced hunger and food consumption [52]. There are a number of randomized and controlled phase 3 clinical trials that have demonstrated a substantial loss in the body weights and recovery of other co-morbid conditions that are connected with overweight and obesity. Results of one of the trials indicate that the patients treated with the combination therapy for one year showed a reduction of 5.9% of initial body weight and 0.6% A1C as compared to a reduction of 2.2% weight and 0.1% A1C in the placebo group [53]. Common side effects include nausea, constipation, headache, vomiting, dizziness, dry mouth and diarrhea [54].

Liraglutide is well known as a drug that acts like glucagon-like peptide-1 (GLP-1). The drug functions as an agonist which is employed in the management of obesity or overweight associated T2D. Among the existing therapies and antidiabetic drugs, liraglutide serves as the single therapy which is used for the symptoms associated with obesity. The usual route of administration of liraglutide is through the subcutaneous route. The drug is administered as an injection through pen equipment. In a 20-week, phase 2, clinical trial, the use of liraglutide was found to result in substantial weight loss [55, 56]. Phase 3 trials of one-year durations showed corroborated findings which demonstrated the continued reduction in the weights of the subjects through a period of at least one year. The mean loss in the subject weights was at 56 weeks [57]. The reports revealed that liraglutide follows a dose-dependent reduction in the subject weights. Another clinical trial in which 846 diabetic patients were involved, who were selected by randomization, were subjected to either liraglutide therapy or a placebo. The findings suggested that the subjects who received liraglutide at 3 mg lost a huge 6.0% from their baseline body weight. In comparison, the weight loss was at 4.7% in the liraglutide group that received 1.8 mg. The weight loss was estimated to be 2.0% in subjects that received the placebo [58, 59]. Some of the common side effects of the drug were observed to be nausea, diarrhea, constipation, vomiting, and headache.

Another approach to management of obesity-related complications is that of the co-administration of probiotics and prebiotics (synbiotics). There are certain synbiotics that are clinically proven to reduce blood glucose levels, hypertension, and body weight. Commonly used synbiotics include Inulin with *Lactobacillus lactis*, *Bifidobacterium bifidium*, *B. longum* (1.5×10^9 CFU/g each) [17] and Inulin with *L. sporogenes* (1×10^7 CFU) [60].

Future Perspectives

Obesity is becoming a major threat to human health and has become an epidemic globally. The problem is associated with many life-threatening conditions including type 2 diabetes mellitus (T2DM), cardiovascular diseases, dyslipidemia, certain types of cancer and hypertension. Obesity-associated T2DM and insulin resistance (IR) are prevalent all around the world. Both hyperglycemia and T2DM develop when the pancreas fails to produce an increased demand for insulin as a result of insulin resistance (IR). IR can develop and may vary at different stages of the development of puberty. In most cases, IR may resolve with the end of puberty which further reduces the chances of post-puberty hyperglycemia [9, 61]. Generally, T2DM can be characterized by both IR and the insufficiency of pancreatic beta-cell.

Obesity is being considered as a key risk factor for developing T2DM as it decreases the sensitivity of organs towards insulin. The major factor associated with decreased functioning of beta cells (responsible for secretion of insulin for facilitating glucose uptake into organs like brain, muscle, adipose tissue, and liver) and IR is saturated fats. As the hyperglycemia persists, these increased saturated fats create a glucolipotoxic state in the body by increasing oxidative stress that is harmful to pancreatic beta cells, this further leads to a reduction in insulin formation and secretion resulting into compromising both function and structure of beta cell [62–64].

The problem has gained special concern due to some recent reports which have reported an approximate drop of 15 years in average life expectancy drops when T2DM develops in adolescents and may lead to chronic complications by the age of

40. The increase in the rates of obesity in recent years has become a costly public health issue. Many epidemiologic studies have reported the link between T2DM and obesity [13, 65]. Once established, obesity is not easy to reverse. Thus, pubertal obesity may be in particular detrimental, for this, early interventions, prior to puberty, may be most helpful in preventing the development of T2DM and its complications [66, 67].

Physical activity can be very effective in preventing or delaying the onset of T2DM and in improving the blood glucose level. Children who have a family history of obesity can increase their insulin sensitivity by including physical activity in their routine, independent of, lifestyle or level of fitness. Though, in a study of T2DM youth named TODAY, it was observed that even an intensive lifestyle intervention with physical activity was not really enough to decrease treatment failure when associated with anti-diabetic drug metformin. The results concluded that once established, T2DM in youth, dietary and exercise changes are very difficult to make [13, 68]. Another study, the Dietary Intervention Study in Children (DISC) determined the effects of high fiber, low-fat diet given during adolescent age. The results observed improved glycemic control as well as blood pressure [69].

Researchers are also working on identifying compounds which can be useful in obesity-induced hyperglycemia and IR. In this connection, a study was conducted to evaluate the effects of freeze-dried jaboticaba peel powder on a number of metabolic parameters in a model of diet-induced obesity. The plant was found to be effective to reduce insulin resistance, as evidenced in the insulin tolerance test, and subsequently confirmed by improved signal transduction through the insulin receptor. Curcumin, the major constituent of turmeric has been reported to have anti-hyperglycemic and insulin sensitizer effects. Thereby, researchers are evaluating the effects of curcumin on hyperglycemic state and insulin resistance in related disorders such as diabetes [70, 71]. Another study conducted especially on postmenopausal women reinforces the hypothesis that a selective modulation of estrogen receptor α activation factor-2 (ER α AF-2), represented a crucial step toward the development of an optimized therapeutic approach to overcome the rising risk of abdominal obesity, metabolic syndrome, and T2DM in this group [72, 73].

Many clinical trials are also undergoing which are evaluating these risk factors and the possible preventive measures for this. Details of which are shown in Table 5.1.

It can be concluded that obesity is associated with hyperglycemia as well as insulin resistance. Therefore, it is important to take preventive steps and control weight gain at early stages of life. For this, we should understand that reducing caloric intake is more effective at achieving initial weight loss than only increasing exercise. The dietary changes alone have been observed to be more effective than physical activity alone. Physical activity remains important for maintaining weight loss but should not be the primary focus of behavioral change for weight loss. Patients on insulin should increase glucose monitoring when starting a new exercise regimen to avoid hypoglycemia during or after exercise. In addition to promoting patients' healthy lifestyle changes, managing medications play an important role for

Table 5.1 List of clinical trial on various drugs for the treatment of obesity-induced diabetes

NCT number	Title	Acronym	Status	Phases	URL
NCT00029848	Obese patients with Type 2 diabetes	RIO-Diabetes	Completed	Phase 3	https://ClinicalTrials.gov/show/NCT00029848
NCT02811484	Efficacy of Exenatide-LAR and Dapagliflozin in overweight/obese, insulin-treated patients with Type 2 diabetes	Dexlar	Withdrawn	Phase 4	https://ClinicalTrials.gov/show/NCT02811484
NCT01667783	Improving diabetes through lifestyle and surgery	IDeaLS	Completed	Phase 2| Phase 3	https://ClinicalTrials.gov/show/NCT01667783
NCT00934570	Activity and Metformin intervention in obese adolescents	REACH	Completed	Phase 4	https://ClinicalTrials.gov/show/NCT00934570
NCT03087032	Liraglutide-bolus versus Glargine-bolus therapy in overweight/obese Type 2 diabetes patients (LiraGooD)	LiraGooD	Recruiting	Phase 4	https://ClinicalTrials.gov/show/NCT03087032
NCT00504673	Comparison of insulin detemir Versus insulin NPH on weight change in overweight and obese with Type 2 diabetes	PREDICTIVE	Completed	Phase 3	https://ClinicalTrials.gov/show/NCT00504673
NCT00478972	Japanese study with rimonabant in obese Type 2 diabetic patients on diet and exercise	SOLO	Terminated	Phase 3	https://ClinicalTrials.gov/show/NCT00478972
NCT00478595	Japanese study with rimonabant in obese Type 2 diabetic patients with oral anti-diabetic drug	SYMPHONY	Terminated	Phase 3	https://ClinicalTrials.gov/show/NCT00478595

(continued)

Table 5.1 (continued)

NCT number	Title	Acronym	Status	Phases	URL
NCT02963922	Effect and safety of liraglutide 3.0 mg in subjects with overweight or obesity and Type 2 diabetes mellitus treated with basal insulin	SCALEâ€ Insulin	Completed	Phase 3	https://ClinicalTrials.gov/show/NCT02963922
NCT00546325	REASSURE: The effect of rimonabant on HbA1c in overweight or obese patients with Type 2 diabetes not adequately controlled on 2 oral antidiabetic agents	REASSURE	Completed	Phase 3	https://ClinicalTrials.gov/show/NCT00546325
NCT01403571	Effectiveness and safety of Salba on weight loss in overweight individuals with Type 2 diabetes	LOSS	Completed	Phase 2	https://ClinicalTrials.gov/show/NCT01403571
NCT02372630	The effect of Linagliptin on inflammation, oxidative stress and insulin resistance in obese Type 2 diabetes subjects	1971	Active, not recruiting	Phase 4	https://ClinicalTrials.gov/show/NCT02372630
NCT03552757	A research study investigating how well Semaglutide works in people with Type 2 diabetes suffering from overweight or obesity	STEP 2	Active, not recruiting	Phase 3	https://ClinicalTrials.gov/show/NCT03552757
NCT00520182	Dietary interventions in Type 2 obese diabetic patients in the community	DIPAC	Unknown status	Phase 3	https://ClinicalTrials.gov/show/NCT00520182
NCT00454597	Effect of the Omentectomy on the hyperglycemia and the resistance	Omentectomy	Completed	Early Phase 1	https://ClinicalTrials.gov/show/NCT00454597

(continued)

Table 5.1 (continued)

NCT number	Title	Acronym	Status	Phases	URL
	to the insulin in patients with morbid obesity				
NCT01601574	Weight loss study for people with Type 2 diabetes	T2D	Completed	Phase 3	https://ClinicalTrials.gov/show/NCT01601574
NCT01713764	A pilot study of the effects of diet and behavioral interventions on health in diabetics	SUCCEED	Completed	Phase 2	https://ClinicalTrials.gov/show/NCT01713764

clinicians in treating patients with type 2 diabetes and obesity. Clinicians should consider altering the diabetes medication regimen and using weight loss medications for these patients.

References

1. Zhou P, Xie W, He S, Sun Y, Meng X, Sun G, Sun X (2019) Ginsenoside Rb1 as an anti-diabetic agent and its underlying mechanism analysis. Cells 8
2. Zhang B, Yang Y, Xiang L, Zhao Z, Ye R (2019) Adipose-derived exosomes: a novel adipokine in obesity-associated diabetes. J Cell Physiol
3. Yu Y, Cai J, She Z, Li H (2019) Insights into the epidemiology, pathogenesis, and therapeutics of nonalcoholic fatty liver diseases. Adv Sci (Weinheim, Baden-Wurttemberg, Germany), 6 1801585
4. Xourgia E, Papazafiropoulou A, Papanas N, Melidonis A (2019) Anti-diabetic treatment leads to changes in gut microbiome. Front Biosci (Landmark edition) 24:688–699
5. Van Herck MA, Weyler J, Kwanten WJ, Dirinck EL, De Winter BY, Francque SM, Vonghia L (2019) The differential roles of t cells in non-alcoholic fatty liver disease and obesity. Front Immunol 10:82
6. Laursen TL, Hagemann CA, Wei C, Kazankov K, Thomsen KL, Knop FK, Gronbaek H (2019) Bariatric surgery in patients with non-alcoholic fatty liver disease—from pathophysiology to clinical effects. World J Hepatol 11:138–149
7. Atawia RT, Bunch KL, Toque HA, Caldwell RB, Caldwell RW (2019) Mechanisms of obesity-induced metabolic and vascular dysfunctions. Front Biosci (Landmark edition) 24:890–934
8. Petrie JR, Guzik TJ, Touyz RM (2018) Diabetes, hypertension, and cardiovascular disease: clinical insights and vascular mechanisms. Can J Cardiol 34:575–584
9. Bigford G, Nash MS (2017) Nutritional health considerations for persons with spinal cord injury. Topics Spinal Cord Inj Rehabil 23:188–206
10. Palomer X, Pizarro-Delgado J, Barroso E, Vazquez-Carrera M (2018) Palmitic and oleic acid: the yin and yang of fatty acids in type 2 diabetes mellitus. Trends Endocrinol Metab: TEM 29:178–190

11. Dombrovska NS, Pleskach OY (2017) Serum content of total adiponectin in the ChNPP accident clean up workers of the ≪iodine period≫ suffering from type 2 diabetes mellitus (literature review and research data). Probl Radiatsiinoi Medytsyny Ta Radiobiolohii 22: 353–371
12. Dalle S, Rossmeislova L, Koppo K (2017) The role of inflammation in age-related sarcopenia. Front Physiol 8:1045
13. Guevara-Aguirre J, Guevara A, Bahamonde M (2018) Insulin resistance depends on GH counter-regulation in two syndromes of short stature. Growth Horm IGF Res Off J Growth Horm Res Soc Int IGF Res Soc 38:44–48
14. Ely BR, Clayton ZS, McCurdy CE, Pfeiffer J, Minson CT (2018) Meta-inflammation and cardiometabolic disease in obesity: can heat therapy help? Temperature (Austin, Tex.) 5:9–21
15. Frankenberg ADV, Reis AF, Gerchman F (2017) Relationships between adiponectin levels, the metabolic syndrome, and type 2 diabetes: a literature review. Arch Endocrinol Metab 61:614–622
16. Whaley-Connell A, Sowers JR (2017) Insulin resistance in kidney disease: is there a distinct role separate from that of diabetes or obesity? Cardiorenal Med 8:41–49
17. Saleem F, Rizvi SW (2017) New therapeutic approaches in obesity and metabolic syndrome associated with polycystic ovary syndrome. Cureus 9:e1844
18. Roder ME (2017) Hyperproinsulinemia in obesity and in type 2 diabetes and its relation to cardiovascular disease. Expert Rev Endocrinol Metab 12:227–239
19. Xiang L, Zhang H, Wei J, Tian XY, Luan H, Li S, Zhao H, Cao G, Chung ACK, Yang C, Huang Y, Cai Z (2018) Metabolomics studies on db/db diabetic mice in skeletal muscle reveal effective clearance of overloaded intermediates by exercise. Anal Chim Acta 1037:130–139
20. Chen XH, Chen PJ, Long Y, Huang QP (2017) [Determination and significance of serum MPO and amylin in adult patients with OSAHS after short-range noninvasive positive pressure ventilation], Lin chuang er bi yan hou tou jing wai ke za zhi = J Clin Otorhinolaryngol Head Neck Surg 31:873–876
21. Sa-Nguanmoo P, Tanajak P, Kerdphoo S, Jaiwongkam T, Wang X, Liang G, Li X, Jiang C, Pratchayasakul W, Chattipakorn N, Chattipakorn SC (2018) FGF21 and DPP-4 inhibitor equally prevents cognitive decline in obese rats, Biomedicine & pharmacotherapy = Biomed Pharmacother 97:1663–1672
22. Lee MJ, Chang BJ, Oh S, Nah SY, Cho IH (2018) Korean red ginseng mitigates spinal demyelination in a model of acute multiple sclerosis by downregulating p38 mitogen-activated protein kinase and nuclear factor-kappaB signaling pathways. J Ginseng Res 42:436–446
23. Taylor R (2013) Type 2 diabetes: etiology and reversibility. Diab Care 36:1047–1055
24. Taylor R (2008) Pathogenesis of type 2 diabetes: tracing the reverse route from cure to cause. Diabetologia 51:1781–1789
25. Pories WJ, Dohm GL (2012) Diabetes: have we got it all wrong?: hyperinsulinism as the culprit: Surgery provides the evidence. Diabetes Care 35:2438–2442
26. Taylor R (2012) Insulin resistance and type 2 diabetes. Diabetes 61:778–779
27. Qatanani M, Lazar MA (2007) Mechanisms of obesity-associated insulin resistance: many choices on the menu. Genes Dev 21:1443–1455
28. Huang X, Liu G, Guo J, Su Z (2018) The PI3 K/AKT pathway in obesity and type 2 diabetes. Int J Biol Sci 14:1483
29. Bagchi D, Preuss HG (2012) Obesity: epidemiology, pathophysiology, and prevention. CRC press
30. Holland WL, Bikman BT, Wang L-P, Yuguang G, Sargent KM, Bulchand S, Knotts TA, Shui G, Clegg DJ, Wenk MR (2011) Lipid-induced insulin resistance mediated by the proinflammatory receptor TLR4 requires saturated fatty acid–induced ceramide biosynthesis in mice. J Clin Invest 121:1858–1870
31. Haus JM, Kashyap SR, Kasumov T, Zhang R, Kelly KR, DeFronzo RA, Kirwan JP (2009) Plasma ceramides are elevated in obese subjects with type 2 diabetes and correlate with the severity of insulin resistance. Diabetes 58:337–343

32. Romeo GR, Lee J, Shoelson SE (2012) Metabolic syndrome, insulin resistance, and roles of inflammation–mechanisms and therapeutic targets. Arterioscler Thromb Vasc Biol 32: 1771–1776
33. Ouchi N, Parker JL, Lugus JJ, Walsh K (2011) Adipokines in inflammation and metabolic disease. Nat Rev Immunol 11:85
34. Bacha F, Gungor N, Lee S, Arslanian SA (2013) Progressive deterioration of β-cell function in obese youth with type 2 diabetes. Pediatr Diab 14:106–111
35. Newsholme P, Haber E, Hirabara S, Rebelato E, Procopio J, Morgan D, Oliveira-Emilio H, Carpinelli A, Curi R (2007) Diabetes associated cell stress and dysfunction: role of mitochondrial and non-mitochondrial ROS production and activity. J Physiol 583:9–24
36. Drews G, Krippeit-Drews P, Düfer M (2010) Oxidative stress and beta-cell dysfunction, Pflügers Archiv-European. J Physiol 460:703–718
37. Genova MP, Atanasova BD, Todorova-Ananieva KN Body mass index and insulin sensitivity/resistance: cross talks in gestational diabetes, normal pregnancy and beyond. In: Body-mass index and health. IntechOpen
38. Ashcroft FM, Rorsman P (2012) Diabetes mellitus and the β cell: the last ten years. Cell 148:1160–1171
39. Eguchi K, Manabe I, Oishi-Tanaka Y, Ohsugi M, Kono N, Ogata F, Yagi N, Ohto U, Kimoto M, Miyake K (2012) Saturated fatty acid and TLR signaling link β cell dysfunction and islet inflammation. Cell Metab 15:518–533
40. Reynoso R, Salgado LM, Calderón V (2003) High levels of palmitic acid lead to insulin resistance due to changes in the level of phosphorylation of the insulin receptor and insulin receptor substrate-1. Springer, Vascular Biochemistry, pp 155–162
41. Tuttle LJ, Bittel DC, Bittel AJ, Sinacore DR (2018) Early-onset physical frailty in adults with diabesity and peripheral neuropathy. Can J Diab 42:478–483
42. Berges-Raso I, Gimenez-Palop O, Gabau E, Capel I, Caixas A, Rigla M (2017) Kallmann syndrome and ichthyosis: a case of contiguous gene deletion syndrome. Endocrinol Diab Metab
43. Lin TY, Lim PS, Hung SC (2018) Impact of misclassification of obesity by body mass index on mortality in patients with CKD. Kidney Int Rep 3:447–455
44. Cheng X, Wang H, Yuan B, Guan P, Wang L (2017) [Prevalence of metabolic syndrome and its family factors for children and adolescents in Chongqing City in 2014], Wei sheng yan jiu = J Hyg Res 46:557–562
45. Browning MG, Khoraki J, DeAntonio JH, Mazzini G, Mangino MJ, Siddiqui MS, Wolfe LG, Campos GM (2005) Protective effect of black relative to white race against non-alcoholic fatty liver disease in patients with severe obesity, independent of type 2 diabetes. Int J Obes 42 (2018):926–929
46. Lee WJ, Almalki O (2017) Recent advancements in bariatric/metabolic surgery. Ann Gastroenterol Surg 1:171–179
47. Yesmin Simu S, Ahn S, Castro-Aceituno V, Yang DC (2017) Ginsenoside Rg5: Rk1 Exerts an Anti-obesity Effect on 3T3-L1 Cell Line by the Downregulation of PPARgamma and CEBPalpha. Iran J Biotechnol 15:252–259
48. Tantipoj C, Sakoolnamarka SS, Supa-amornkul S, Lohsoonthorn V, Deerochanawong C, Khovidhunkit SP, Hiransuthikul N (2017) Screening for type 2 diabetes mellitus and prediabetes using point-of-care testing for HbA1c among Thai dental patients. Southeast Asian J Trop Med Public Health 48:455–465
49. Dowdle SB, Bedard NA, Owens JM, Gao Y, Callaghan JJ (2018) Identifying risk factors for the development of stiffness after revision total knee arthroplasty. J Arthroplasty 33: 1186–1188
50. Mello M, Vasques ACJ, Pareja JC, Oliveira MDS, Novaes FS, Chaim EA, Geloneze B (2017) Effect of biliopancreatic diversion on sleep quality and daytime sleepiness in patients with obesity and type 2 diabetes. Arch Endocrinol Metab 61:623–627

51. Rodriguez A, Guilera N, Mases A, Sierra P, Oliva JC, Colilles C (2018) Management of antiplatelet therapy in patients with coronary stents undergoing noncardiac surgery: association with adverse events. Br J Anaesth 120:67–76
52. Perumpail BJ, Khan MA, Yoo ER, Cholankeril G, Kim D, Ahmed A (2017) Clinical epidemiology and disease burden of nonalcoholic fatty liver disease. World J Gastroenterol 23:8263–8276
53. Hong J, Park S, Menzo EL, Rosenthal R (2018) Midterm outcomes of laparoscopic sleeve gastrectomy as a stand-alone procedure in super-obese patients. Surgery Obes Relat Dis Off J Am Soc Bariatr Surg 14:297–303
54. Capristo E, Panunzi S, De Gaetano A, Raffaelli M, Guidone C, Iaconelli A, L'Abbate L, Birkenfeld AL, Bellantone R, Bornstein SR, Mingrone G (2018) Intensive lifestyle modifications with or without liraglutide 3 mg versus sleeve gastrectomy: a three-arm non-randomised, controlled, pilot study. Diab Metab 44:235–242
55. Valero-Munoz M, Backman W, Sam F (2017) Murine models of heart failure with preserved ejection fraction: a fishing expedition. JACC. Basic Trans Sci 2:770–789
56. Davies MJ, Aronne LJ, Caterson ID, Thomsen AB, Jacobsen PB, Marso SP (2018) Liraglutide and cardiovascular outcomes in adults with overweight or obesity: a post hoc analysis from SCALE randomized controlled trials. Diab Obes Metab 20:734–739
57. Yandrapalli S, Jolly G, Horblitt A, Sanaani A, Aronow WS (2017) Cardiovascular benefits and safety of non-insulin medications used in the treatment of type 2 diabetes mellitus. Postgrad Med 129:811–821
58. Yandrapalli S, Aronow WS (2017) Cardiovascular benefits of the newer medications for treating type 2 diabetes mellitus. J Thoracic Dis 9:2124–2134
59. von Scholten BJ, Davies MJ, Persson F, Hansen TW, Madsbad S, Endahl L, Jepsen CH, Rossing P (2017) Effect of weight reductions on estimated kidney function: post-hoc analysis of two randomized trials. J Diab Complic 31:1164–1168
60. Wangnoo SK, Kumar S, Bhattacharyya A, Tripathi S, Akhtar S, Shetty R, Ghosal S (2016) Liraglutide effect and action in diabetes-In (LEAD-In): A prospective observational study assessing safety and effectiveness of liraglutide in patients with type 2 diabetes mellitus treated under routine clinical practice conditions in India. Indian J Endocrinol Metab 20:838–845
61. Li Z, Liang Y, Xia N, Lai Y, Pan H, Zhou S, Jiang F, He Y (2017) Liraglutide reduces body weight by upregulation of adenylate cyclase 3. Nutr Diab 7:e265
62. Zhu L, Shi J, Luu TN, Neuman JC, Trefts E, Yu S, Palmisano BT, Wasserman DH, Linton MF, Stafford JM (2018) Hepatocyte estrogen receptor alpha mediates estrogen action to promote reverse cholesterol transport during Western-type diet feeding. Mol Metab 8: 106–116
63. Shah VN, Sippl R, Joshee P, Pyle L, Kohrt WM, Schauer IE, Snell-Bergeon JK (2018) Trabecular bone quality is lower in adults with type 1 diabetes and is negatively associated with insulin resistance. Osteoporos Int J Establ Result Coop Between Eur Found Osteoporos Natl Osteoporos Found USA 29:733–739
64. Nouhjah S, Shahbazian H, Shahbazian N, Jahanfar S, Jahanshahi A, Cheraghian B, Mohammadi ZD, Ghodrati N, Houshmandi S (2018) Early postpartum metabolic syndrome in women with or without gestational diabetes: results from life after gestational diabetes Ahvaz cohort study. Diab Metab Syndr 12:317–323
65. Oxenkrug G, van der Hart M, Roeser J, Summergrad P (2017) Peripheral tryptophan—kynurenine metabolism associated with metabolic syndrome is different in Parkinson's and Alzheimer's diseases. Endocrinol Diab Metab J 1
66. Jeznach-Steinhagen A, Ostrowska J, Czerwonogrodzka-Senczyna A, Boniecka I, Gronostajska W (2017) Dietetary recommendation for non-alcoholic fatty liver disease. Polski merkuriusz lekarski: Organ Polskiego Towarzystwa Lekarskiego 43:281–286
67. Aguirre M, Briceno Y, Gomez-Perez R, Zerpa Y, Camacho N, Paoli M (2018) Triglycerides/High density lipoprotein cholesterol ratio as a cardiometabolic risk marker in children and adolescents from Merida city. Venezuela, Endocrinologia, Diab y Nutr 65:74–83

68. Shrivastava U, Fatma M, Mohan S, Singh P, Misra A (2017) Randomized control trial for reduction of body weight body fat patterning, and cardiometabolic risk factors in overweight worksite employees in Delhi, India. J Diab Res 2017:7254174
69. Huang LO, Loos RJF, Kilpelainen TO (2018) Evidence of genetic predisposition for metabolically healthy obesity and metabolically obese normal weight. Physiol Genomics 50:169–178
70. Adak T, Samadi A, Unal AZ, Sabuncuoglu S (2018) A reappraisal on metformin. Regul Toxicol Pharmacol RTP 92:324–332
71. Tobin LM, Mavinkurve M, Carolan E, Kinlen D, O'Brien EC, Little MA, Finlay DK, Cody D, Hogan AE, O'Shea D (2017) NK cells in childhood obesity are activated, metabolically stressed, and functionally deficient. JCI Insight 2
72. Kern PA, Finlin BS, Ross D, Boyechko T, Zhu B, Grayson N, Sims R, Bland JS (2017) Effects of KDT501 on metabolic parameters in insulin-resistant prediabetic humans. J Endocr Soc 1:650–659
73. Duggan C, Baumgartner RN, Baumgartner KB, Bernstein L, George S, Ballard R, Neuhouser ML, McTiernan A (2018) Genetic variation in TNFalpha, PPARgamma, and IRS-1 genes, and their association with breast-cancer survival in the HEAL cohort. Breast Cancer Res Treat 168:567–576

Chapter 6
Obesity and Osteoarthritis: Are Adipokines Bridging Metabolism, Inflammation, and Biomechanics?

Vera Francisco, Clara Ruiz-Fernández, Jesús Pino, Antonio Mera, Miguel Angel Gonzalez-Gay, Francisca Lago, Rodolfo Gómez and Oreste Gualillo

Abstract Osteoarthritis (OA) is a highly prevalent debilitating and painful pathology derived from progressive degeneration of articular joints. Obesity has long been recognized as a significant and potentially preventable risk factor for OA incidence, progression, and disability. Biomechanical loading, together with metabolic and inflammatory imbalances of the joint, strongly contributes to obesity-induced OA pathophysiology. Adipose-tissue derived cytokines—adipokines—have demonstrated roles in modulating pro/anti-inflammatory and anabolic/

V. Francisco · C. Ruiz-Fernández · J. Pino · O. Gualillo (✉)
SERGAS (Servizo Galego de Saude) and IDIS (Instituto de Investigación Sanitaria de Santiago), The NEIRID Group (Neuroendocrine Interactions in Rheumatology and Inflammatory Diseases), Santiago University Clinical Hospital, Xerencia de Xestión Integrada de Santiago de Compostela, Building C, Travesía da Choupana S/N, 15706 Santiago de Compostela, Spain
e-mail: oreste.gualillo@sergas.es

A. Mera
SERGAS (Servizo Galego de Saude), Division of Rheumatology, Santiago University Clinical Hospital, Xerencia de Xestión Integrada de Santiago de Compostela, Travesía da Choupana S/N, 15706 Santiago de Compostela, Spain

M. A. Gonzalez-Gay
Epidemiology, Genetics and Atherosclerosis Research Group on Systemic Inflammatory Diseases, Universidad de Cantabria and IDIVAL, Hospital Universitario Marqués de Valdecilla, Av. Valdecilla, 39008 Santander, Spain

F. Lago
Molecular and Cellular Cardiology Group, SERGAS (Servizo Galego de Saude) and IDIS (Instituto de Investigación Sanitaria de Santiago), Research Laboratory 7, Santiago University Clinical Hospital, Xerencia de Xestión Integrada de Santiago de Compostela, Santiago de Compostela, Spain

R. Gómez
Musculoskeletal Pathology Group, SERGAS (Servizo Galego de Saude) and IDIS (Instituto de Investigación Sanitaria de Santiago), Research Laboratory 9, Santiago University Clinical Hospital, Xerencia de Xestión Integrada de Santiago de Compostela, Santiago de Compostela, Spain

P. S. Tappia et al. (eds.), *Pathophysiology of Obesity-Induced Health Complications*, Advances in Biochemistry in Health and Disease 19,
https://doi.org/10.1007/978-3-030-35358-2_6

catabolic joint balance, with implications in cartilage and bone homeostasis. Mechanical stress may lead to considerable increases in proinflammatory mediators within the joint. Therefore, adipokines emerged as potential candidates to link mechanical, metabolic and inflammatory components of obesity-induced osteoarthritis. Herein we summarize the biology of adipokines in joint tissues, highlighting their implications in the dysregulation of joint homeostasis and, thus OA pathogenesis. Many of the aspects of the adipokine network remain largely unknown and further insights into the intimate mechanisms of adipokines activity will be of great relevance to develop disease-modifying osteoarthritis drugs, especially for obese patients.

Keywords Adipokines · Mechanical load · Inflammation · Obesity · Osteoarthritis

Introduction

Obesity, the greatest public health problem in western countries, has reached epidemic proportions, being still rising in developing countries. Besides representing itself a disabling factor, obesity is strictly associated with other high-incident chronic pathologies, like cardiovascular diseases, type 2 diabetes or arthritis diseases [1]. For long, obesity-focused researches have aimed the identification of risk factors, preventive measures, and treatments. However, public health policies centered on physical activity and diet have demonstrated to be largely ineffective. Additionally, pharmacological approaches have not provided any safe and long-term therapies [1]. Therefore, a deeper understanding of the development of obesity-associated pathologies is needed.

Osteoarthritis (OA) is a painful, debilitating and progressive degenerative disease of the entire joint. It is characterized by molecular (abnormal joint tissue metabolism) anatomic, and/or physiologic derangements, which include cartilage degradation, osteophyte formation, bone remodeling, joint inflammation, and loss of normal joint function [2]. The prevalence of OA significantly increases with age, with a radiographic diagnostic in over 70% of above-65 individuals [3]. Although OA is the most common arthritis form with a high socio-economic impact, currently available treatments demonstrate low efficacy, are frequently associated with adverse effects, and as far to reach disease remission [4]. Moreover, articular cartilage lacks self-healing capacity and OA seems to comprise a group of pathologic processes with a common endpoint, but with a multifactorial etiopathogenesis, such as genetic, molecular and environmental factors, particularly biomechanical stress [3]. Therefore, OA therapy is challenging and deeper knowledge on molecular players contributing to disease onset and development will certainly open new therapeutic routes.

At physiologic conditions, joints are subjected to several loads that, under normal circumstances, have no adverse effects on joint tissues, and are even protective [5]. In obesity, a well known modifiable risk factor of OA incidence, progression and disability [6], increased mechanical loading and "wear-and-tear" stresses have been initially pointed as the main mechanism for the development of obesity-associated OA [7]. However, there is increasing evidence for a multifactorial, systemic link between OA and obesity [8]. White adipose tissue (WAT) is no longer recognized as a simple energy depot, but as an active and inflammatory organ responsible for the production of pleiotropic molecules—adipokines [9]. This new family of proteins has emerged as crucial player in the pathophysiology of immune/inflammatory diseases, including osteoarthritis [10, 11]. Here, we summarize the adipokines biology within the joint, highlighting their role as molecular linkers between metabolism, inflammation, and biomechanics, with implications in the dysregulation of joint homeostasis and, thus OA pathogenesis. Understand the adipokine network and the intimate mechanisms of adipokines activity in OA, may represent innovative therapeutic strategies to develop disease-modifying osteoarthritis drugs, especially for obese patients.

Mechanical Loading and Inflammation

Inflammation has a significant role in altered loading models of OA. After a traumatic injury, the levels of pro-inflammatory cytokines, namely interleukin (IL)-1 and tumor necrosis factor (TNF)-α, were transiently augmented in chondrocytes and articular cartilage [12]. Moreover, mechanical stress induced nitric oxide (NO), prostaglandin E2 (PGE2) and IL-6 secretion [13, 14], while fluid shear stress increased proteoglycan synthesis in chondrocytes [15, 16]. Of note, only chondrocytes embedded in their own ECM, but not that one's embedded in an agarose matrix, show similar increases of pro-inflammatory factors with stress [14, 17, 18], indicating a crucial role for native ECM. In synoviocytes, mechanical stretch also enhances the expression of pro-inflammatory mediators, like cyclooxygenase (COX)-2, PGE2, and IL-1β [19], while in osteoblasts it increases IL-6, COX-2, and IL-8 expression [20]. Altogether, these findings evidenced that altered biomechanical loading is associated with inflammation within the joint that may eventually lead to OA pathogenesis.

Obesity, Biomechanics, and Osteoarthritis

Significant evidence demonstrated that obesity alters gait and joint biomechanics. Accordingly, obese individuals walk slower, with shorter and wider steps, and demonstrated longer stance durations compared with normal-weight subjects [3]. These changes can affect the load-bearing regions of the articular cartilage. Under physiological conditions, chondrocytes maintain a homeostatic equilibrium between the catabolic and anabolic processes, leading to the slow turnover of the cartilage extracellular matrix (ECM). An imbalance in chondrocyte metabolism towards catabolic processes leads to the progressive cartilage degeneration. This shift is influenced by soluble mediators, such as growth factors and cytokines, local matrix composition, and biophysical factors, including mechanical stresses [3]. An altered joint loading, either single (acute impact event) or repetitive (cumulative contact stress), can lead to OA pathophysiology by altering the composition, structure, metabolism, and mechanical properties of joint tissues, in particular cartilage and subchondral bone [3]. Impact load augments tissue hydration and cellular activity and leads to subchondral bone remodeling and ECM splitting [21], all features of early OA. Moreover, joint instability in OA animal models increase hydration, collagen disruption, and matrix turnover followed by a decrease in the tissue stiffness in tension, compression, and shear [3]. Furthermore, an increase of both articular cartilage and synovial fluid biomarkers correlated with joint histological damage was verified [3].

Although mechanical loading is an important factor relating to obesity and OA, there are cumulative data pointing to a systemic link between these pathologies [8]. A small reduction of body weight (5 kg), and in particular body fat (2.4%) is associated with an over 50% decrease in the OA risk or progression [22]. Furthermore, epidemiological studies verified that the risk of developing hand OA, a non-weight bearing joint, is about twofold in obese patients, compared with normal-weight subjects [23]. Furthermore, extreme obese animals due to impaired leptin signaling (leptin-deficient (ob/ob), or leptin receptor-deficient (db/db) mice), demonstrated similar incidence rates of knee OA compared to controls [24]. These findings suggest that adipose-tissue derived factors, could have an important role linking obesity and OA.

Leptin

Leptin, from the Greek root *leptos*, meaning "thin", is a 16 kDa non-glycosylated cytokine-like hormone of 146 amino acids encoded by the obese gene and predominantly produced in WAT [25]. Leptin was initially described as a neuroendocrine regulator of weight homeostasis by inducing anorexigenic neuropeptides and suppressing orexigenic factors on hypothalamic nuclei but also stimulates thermogenesis and energy expenditure by modulating insulin activity and lowering blood glucose levels [26, 27]. Thus, the central leptin resistance, due to the impairment of leptin transportation, signaling and targeting specific neural circuits, is proposed as the primary risk factor for obesity [26]. Leptin exerts its biological activity through activation of leptin receptors (Ob-R or LEPR), which are found with at least six alternative splicing forms, including a long isoform (Ob-Rb), four short isoforms (Ob-Ra, Ob-Rc, Ob-Rd and Ob-Rf) and a soluble isoform (Ob-Re) [26, 28]. The long form receptor Ob-Rb has the full intracellular domain that allows the transduction of leptin signal through JAK and STAT signaling pathways [29]. Additionally, Ob-Rb could also activate ERK1/2, p38 MAPK, JNK, PKC, and PI3 K pathways [26]. Suppressor of cytokine signaling-3 (SOCS-3) and protein tyrosine kinase 1B (PTP1B) acted as negative regulators of leptin signaling [26, 28]. Leptin receptors are widely expressed in peripheral tissues and thus, leptin is considered a pleiotropic hormone implicated in several physiologic processes, namely insulin secretion, bone metabolism, and immune responses (both innate and adaptive) [30]. Consequently, this hormone has been considered as a potential linker between obesity and OA. Accordingly, obese individuals have enhanced levels of leptin—hyperleptinemia, and higher OA prevalence [31].

Although WAT is the major source of leptin, IPFP and synovial membrane (or synovium) also synthesized leptin, being its synovial fluid (SF) concentration equivalent, or even higher than in serum, which indicates a local production of leptin during OA or factors affecting its clearance [32, 33]. In fact, OA cartilage and osteophytes demonstrated increased synthesis and secretion of leptin when compared to normal cartilage [34]. Accordingly, the SF and serum leptin levels in OA patients were higher compared with controls and a gender-specific correlation has been described [35, 36]. Of note, the expression of suppressor of cytokine signaling-3 (SOCS-3), a negative regulator of leptin signaling, is decreased in the cartilage of obese OA patients [37], and the expression of leptin and Ob-Rb is significantly increased in the advanced defective area of OA cartilage compared to adjacent intact regions [36]. Serum leptin has also been associated with radiographic severity in OA patients, particularly in females [38, 39]; however, there are conflicting results [40]. Nevertheless, the ratio of SF to plasma leptin has been pointed as a more suitable biomarker than plasma leptin alone to evaluate the knee OA severity [41].

Most of the previous data revealed a pro-inflammatory and catabolic action for leptin in cartilage metabolism by acting alone or in synergy with other pro-inflammatory mediators, therefore contributing to OA pathophysiology [30, 42]. Leptin could augment IL-8 and IL-6 production in fibroblast-like synoviocytes, being involved Ob-Rb, JAK2, insulin receptor substrate-1 (IRS-1), PI3K, Akt, NF-kB, and AP-1 signaling pathways [43, 44]. Leptin also regulates the production of pro-inflammatory factors by immune cells [30]; namely, it induces IL-6, IL-8, and CCL3 expression in CD4+ T cells from OA patients, but nor from healthy individuals [45]. Furthermore, leptin induces Th1 phenotype and could increase leukocyte and monocyte infiltration into the joint by up-regulation of VCAM-1, thus promoting and perpetuating inflammation in OA joints [30, 46]. In human and murine chondrocytes, as well as in cartilage explants, leptin, in synergy with IL-1 or IFN-γ, induced the expression of pro-inflammatory factors, such as NO, NOS2, PGE2, COX-2, IL-6 and IL-8 [47–50]. It was demonstrated that JAK2, PI3K, MEK1, ERK1/2, p38 MAPK, and JNK were involved in leptin-regulated NOS2 expression [47–50].

Leptin can also promote OA-associated cartilage destruction by directly inducing the expression of MMPs. In particular, it augments MMP-1, -3, and -13 expression in human OA cartilage, likely through activation of NF-κB, protein kinase C (PKC), and JNK signaling pathways [51]. Whereas in bovine cartilage, leptin, alone or in combination with IL-1, promotes cartilage collagen release as well as MMP-1, and 13 expression, via STAT1, STAT3, STAT5, MAPK, Akt, and NF-κB pathways [43]. In vivo, injection of recombinant murine leptin in knee joints of rats, augments the expression of ADAMTS-4, and -5, MMP-2, and 9, cathepsin D and collagen II, while basic fibroblast growth factor (bFGF) is decreased [52]. Nevertheless, there are some studies describing a protective action of leptin in OA by exerting anabolic functions in articular cartilage through induction of growth factors expression, such as transforming growth factor (TGF)-β and insulin-like growth factor (IGF) [35]. Altogether, these data indicated that leptin may have a dual effect on OA development. Likely, augmented catabolic and pro-inflammatory activity together with impaired anabolic and anti-inflammatory functions may lead to an overall detrimental effect of leptin. Therefore, further investigation addressing this metabolic balance of leptin in OA would be important [43].

MicroRNAs (miRNAs), small single-stranded non-coding segments of RNA, are increasingly recognized as regulatory molecules involved in pathologic processes, including OA, inflammation, and obesity. The levels of miR-27, which directly targets the 3'-untranslated region of leptin, were found to be diminished in OA chondrocytes and the injection of miR-27 lentiviral overexpression vector in OA rats resulted in decreased levels of IL-6 and -8, as well as MMP-9 and -13, therefore suggesting a protective action of miR-27 in OA, likely through leptin targeting [53].

Besides synoviocytes and chondrocytes, osteoblasts are also significant leptin targeted cells. OA subchondral osteoblasts produce twofold leptin and Ob-Rb compared to normal ones, which could contribute to osteoblast proliferation and differentiation. Implicated mechanisms included the modulation of alkaline phosphatase (ALP), osteocalcin (OC), collagen type I and TGF-β1 production (osteoblasts metabolic markers), as well as ERK1/2 and p38 MAPK pathways [54]. Likewise, leptin also induced the proliferation of nucleus pulposus cells, leading to a cell cluster formation and proliferation of fibro-cartilaginous tissue [55]. Considering that intervertebral disc degeneration (IVDD) appears like OA relatively to increased cell proliferation, it was hypothesized that leptin could induce OA synoviocytes and osteoblasts proliferation, therefore contributing to osteophyte formation [43]. Moreover, chondrogenic progenitor cells (CPCs), seed cells crucial to cartilage homeostasis and replacement of damage tissue, change their differentiation pattern in the presence of leptin. In particular, this adipokine reduces CPC migratory ability and chondrogenic potential, while it induces CPC senescence and osteogenic transformation [56].

Based on these results, upregulated levels of leptin may be a risk factor for OA pathophysiology by influencing a pro-inflammatory environment, cartilage catabolic activity, as well as cartilage and bone remodeling. Hence, leptin has been indicated as a sensitive biomarker for predicting OA severity, pain and cartilage damage [57].

Adiponectin

Adiponectin (also called Acrp30, GBP28, apM1, or AdipoQ) is a 244-residue protein, structurally homologous to collagen type VIII and X, and complement factor C1q, that is encoded by ADIPOQ gene located on chromosome 3q27, a locus correlated with diabetes and cardiovascular disease susceptibility [58]. Several adiponectin molecular forms could be found, including globular (gAPN), full-length (fAPN), low MW (LMW), medium MW (MMW), high MW (HMW), and serum albumin bonded LMW (Alb-LMW) adiponectin [58, 59]. All of them have different biological activities that are mainly mediated by two specific adiponectin receptors: AdipoR1, mainly found in skeletal muscle, and AdipoR2, predominantly present in liver. The signal transduction of adiponectin receptors involves the activation of AMPK, PPAR-α and PPAR-γ [58]. Additionally, T-cadherin has also been identified as a receptor for high molecular adiponectin multimers [60]. In morbidly obese patients, circulating adiponectin levels tend to be low and augmented with weight loss and treatment with thiazolidinediones (PPAR agonists), which enhance insulin sensitivity [60, 61]. Accordingly, adiponectin has been described as an endogenous insulin sensitizer by stimulation of fatty acid oxidation and glucose uptake in the muscle, and reduction of glucose synthesis in the liver [60]. Furthermore, there is increasing evidence revealing the importance of adiponection in inflammation-related pathologies, such as cardiovascular disease, type 2 diabetes, metabolic syndrome, rheumatoid arthritis and OA [58], likely due to modulation of immune system. In particular, adiponectin inhibited proinflammatory responses by polarizing macrophages M1 to M2, Th1/Th17 to Th2/Tregs, and inhibiting TLR4-mediated NF-κB activation [62].

Based on both clinical and experimental data, adiponectin have implicated in the pathophysiology of OA. However, the exact effect of adiponectin in OA, whether protective or not, is still contradictory. It was verified that serum and plasma adiponectin levels were augmented in OA patients compared to healthy controls [63], but some studies showed that the difference is not statistically significant after age, gender, and BMI adjustments [43]. Furthermore, adiponectin levels were associated with OA radiological severity and erosive OA, being until considered an OA biomarker [57], while other studies reported no association [64]. Remarkably, SF adiponectin levels were almost 100 times lower than those in plasma, and have been positively correlated with aggrecan degradation markers, whereas negatively correlated with radiographic disease severity and knee OA pain [43]. Moreover, the percentages of hexamer and HMW adiponectin forms per total adiponectin are lower in SF than plasma, whereas the trimer form is higher [43]. Additionally, it was demonstrated that total adiponectin, rather than HMW, was positively associated with knee OA, while negatively correlated with radiographic progression in hand OA [43]. These data indicated the importance of considering the different adiponectin isoforms when evaluating it as a marker for OA risk, activity or progression.

Within the joint, adiponectin could be expressed by synoviocytes, IPFP, osteophytes, cartilage and bone tissues [32]. Moreover, AdipoR1 and AdipoR2 are differentially expressed in OA chondrocytes compared with normal counterparts [65, 66]. Regarding adiponectin activity in the joint, it could downregulate IL-1β-induced MMP13 expression and to upregulate TIMP-2 in human chondrocytes, clearly revealing a chondroprotective action in OA progression by exerting anabolic and anti-inflammatory roles. However, most of the studies established adiponectin as a catabolic and pro-inflammatory mediator in OA. Accordingly, adiponectin serum levels or released by OA cartilage are positively associated with degradation biomarkers, such as cartilage oligomeric matrix protein (COMP), MMP-3, and -13, and mPGES [43]. In OA cartilage and human primary chondrocytes, adiponectin augmented NO, NOS2, IL-6, IL-8, MCP-1, MMP-1, -3, -9, and 13 [67]. Moreover, in human and murine chondrocytes, adiponectin increased VCAM-1 expression, thus enhancing and perpetuating cartilage-degradation in inflamed OA joints [46]. In OA synoviocytes, adiponectin dose-dependently increased PGE2 production [68] and augments the expression of NOS2, IL-6, MMP-1, -3, and MMP-13 via AdipoR1/5'-AMPK, MAPK, and NF-κB signaling pathways, exerting inflammatory and matrix degradation functions during OA [43]. Additionally, adiponectin also induces ICAM-1 expression in human OA synoviocytes, through liver kinase B1 (LKB1)/CaMKII, AMPK, c-Jun, and AP-1 pathways, with subsequent monocyte adhesion to OA synoviocytes [43].

Adiponectin also induced osteoblast proliferation and differentiation, through a dose- and time-dependent increase of ALP activity, osteocalcin, and type I collagen production, as well as matrix mineralization [69]. AdipoR/JNK pathway seems to mediate the adiponectin action on osteoblast proliferation, while AdipoR/p38 signaling is involved in adiponectin-induced osteoblast differentiation [69]. Nevertheless, there are contradictory results demonstrating the inhibition of osteoblasts proliferation and promotion of apoptosis by adiponectin [70]. It was evidenced that mechanical loading augments adiponectin and its receptors in skeletal muscle, while exercise increase serum adiponectin in high-fat-sedentary rats compared to controls, with potential effect in preventing bone loss [71]. Therefore, these results indicated that adiponectin altered bone metabolism and biomechanical properties. But, further studies will be of relevance to clarify the exact action of adiponectin in joint cartilage and bone metabolism, and its contribution to OA pathogenesis. Of note, the different adiponectin isoforms have demonstrated diverse biological effects [43]. As total adiponectin has demonstrated contradictory activities in OA, further evaluation of the relative contribution of each isoform to inflammatory response and matrix degradation in OA joint will be highly valuable [72].

Lipocalin-2

Lipocalin-2 (LCN2; also known as neutrophil gelatinase-associated lipocalin, 24p3, p25, migration-stimulating factor inhibitor, human neutrophil lipocalin, α-1-microglobulin—related protein, siderocalin or uterocalin) is a glycoprotein encoded by a gene located at the chromosome locus 9q34.11 [73]. It circulates as a 25 kDa monomer, a 46 kDa homodimer or in a covalent complex with MMP-9, thus blocking MMP-9 auto-degradation [74]. Given the presence of hydrophobic ligand binding pocket, the members of the lipocalin family also bind and transport steroids, fatty acids, LPS, iron, and in the case of LCN2, siderophores. LCN2 binds at least to two surface receptors: LCN2 receptor (also called SLC22A17, 24p3R, or NGALR), a brain-type organic cation transporter that binds mouse LCN2; and megalin (also known as LRP2, or glycoprotein GP330), a multiligand scavenger receptor that binds human LCN2 [11]. Originally identified in human neutrophil granules and mouse kidney cells, LCN2 is mainly produced by WAT, but is also expressed in immune cells, liver, spleen, and chondrocytes [73]. LCN2 actions include induction of apoptosis in hematopoietic cells, modulation of inflammation and metabolic homeostasis, being evidenced as important player in obesity-related diseases [73]. Moreover, LCN2 have been revealed as a sensor of mechanical loading, inflammatory status and catabolic stimuli of the joint, with possible involvement in OA pathophysiology [43].

LCN2 was widely described as a regulator of immune system. In particular, it binds to enterobactin, a Gram-negative bacteria siderophore responsible for iron transport into the cell. Consequently, bacterial iron stores depletion occurs and a bacteriostatic effect is observed [73]. These LCN2 properties have been implicated in gastrointestinal tract protection against pathogens [75]. Moreover, key inflammatory transcription factors, including NF-κB, STAT1, STAT3, and C/EBP, have been shown to transactivate LCN2 expression, indicating this adipokine as inflammatory regulator [76, 77]. Recently, E74-like factor 3 (ELF3), a transcription factor induced by inflammatory cytokines, was identified as LCN2 expression regulator in human chondrocytes [77]. Accordingly, LCN2 plays anti-inflammatory actions in regulating M1/M2 macrophage polarization via NF-κB/STAT3 loop activation [78], while, in adaptive immunity, LCN2 was described to induce human leukocyte antigen G (a tolerogenic mediator) on $CD4^+$ T cells, and to up-regulate Treg cells expansion in healthy individuals [73]. Supporting the role of LCN2 in immune system response, plasma LCN2 concentrations have been correlated with several inflammatory and metabolic parameters [67].

In obese patients, LCN2 expression is augmented and could be reversed by thiazolidinediones treatment [79]. LCN2 levels were also elevated in OA synovial fluid and cartilage [73, 80], thus suggesting a possible involvement in OA development. In fact, LCN2 affected cartilage degradation by blocking MMP-9 auto-degradation and by reducing chondrocyte proliferation [67]. At the same time, IL-1β, LPS, dexamethasone, adipokines (leptin and adiponectin), and osteoblast conditioned medium (paracrine stimulation), increased LCN2 expression in chondrocytes [73, 81]. Of note, NO promote LPS-mediated LCN2 expression in chondrocytes, suggesting a feedback loop regulating its expression [82]. Furthermore, LCN2 expression could be induced by glucocorticoids, alone or in combination with IL-1, via corticoids receptors, PI3 K, ERK1/2 and JAK2 pathways [83] in chondrocytes. In osteoblasts, the absence of mechanical loading induced LCN2 expression, likely contributing to bone metabolism through stimulation of pro-osteoclastogenic factors, receptor activator of NF-κB ligand and IL-6, and inhibition of the anti-osteoclastogenic factor, osteoprotegerin [73]. In these cells, the LCN2 expression is also increased by inflammatory cytokines (TNF-α and IL-17) [73].

Altogether, these results indicated LCN2 responsiveness to inflammation and mechanical load within the joint, leading to alterations in cartilage, subchondral bone and bone-cartilage crosstalk underlying OA pathogenesis. Nevertheless, recently, it was verified that LCN2 overexpression in chondrocytes did not affect the expression of cartilage matrix molecules or matrix-degrading enzymes [84]. Moreover, LCN2 overexpression in mouse cartilage did not cause OA pathogenesis, and LCN2 KO mice showed no alteration in cartilage destruction, in the destabilization of the medial meniscus-induced OA model [84]. Although LCN2 can contribute to OA pathophysiology, it appears to be not sufficient by itself to induce OA cartilage destruction in mice. Further studies are demanded to fully clarify the role of LCN2 in OA development in humans.

Progranulin

Progranulin [PGRN; also known as acrogranin, granulin-epithelin precursor (GEP), proepithelin, GP88, or PC-cell-derived growth factor] is a cysteine-rich secreted glycoprotein encoded by the GRN gene, located on chromosome 17q21.32 [85]. It is a 68–88 kDa protein that can undergo enzymic proteolysis originating small homologous subunits—granulins or epithelins [86]. Evidence that PGRN is internalized through endocytosis suggests the existence of a PGRN cell surface receptor. However, a unique PGRN receptor has not yet been identified. By now, some types of cell surface receptors for PGRN, but not for individual granulins, have been proposed: sortilin, which mediates extracellular PGRN uptake and extracellular levels in CNS [87, 88]; death receptor 3, which is involved in several inflammatory disorders [89]; EphA2, a member of the large family of Ephrin receptor tyrosine kinases [90]; and TNFR1 and TNFR2 (TNF receptors), which have been correlated with PGRN anti-inflammatory activity [91]. TNF-α demonstrated a higher affinity to TNFR1, ubiquitously expressed and trigger of proinflammatory signaling pathways, than for TNFR2, mainly expressed in hematopoietic cells and related to anabolic responses [91]; while PGRN exhibited comparable affinity to both receptors, being higher than TNF-α [91]. Taking account these observations, a minimal engineered PGRN-protein atsttrin was developed as TNF/TNFR signalling antagonist [91]. A wide range of cells was reported to express PGRN, including epithelial cells, macrophages, chondrocytes and adipocytes. Recently identified as an adipokine, PGRN has been implicated in inflammation, obesity and rheumatic diseases, being a potential therapeutic target and biomarker in inflammatory diseases [85].

As an anti-inflammatory protein, PGRN also suppresses the production of CXCL9 and CXCL10,through the TNFR1 pathway [89, 92]. Moreover, PGRN induced Treg populations and IL-10 production [89, 92]. On the other hand, PGRN can be degraded by several proteinases, like MMP-9, -12 and -14, ADAMT S7, elastase, and proteinase-3, generating granulins, which have pro-inflammatory activity [89, 92].

Recent studies reveal that PGRN is implicated in OA pathogenesis. OA patients present higher expressions of PGRN in cartilage, synovial and IPFP samples [93]. PGRN expression is also elevated during chondrocyte differentiation in vitro [93].

Moreover, aged PGRN-KO mice develop OA-like phenotype and both recombinant PGRN and PGRN-derived atsttrin could effectively prevent the OA onset and progression [89]. In OA, PGRN also exhibits anti-inflammatory properties by promoting anabolic metabolism via TNFR2, and by inhibiting IL-1β-mediated catabolic metabolism (suppression of NOS2, COX-2, MMP-13 and VCAM-1) through TNFR1 binding. By blocking TNF-α mediated NF-κB activation, PGRN inhibited MMPs and ADAMTS expression, and thus, cartilage degradation [89, 93]. Additionally, PGRN is involved in chondrocytes differentiation and proliferation as well as in endochondral ossification of growth plate during development [94]. Confirming the protective role of PGRN in OA, intra-articular injection of mesenchymal stem cells expressing recombinant atsttrin, prevented the progression of degenerative changes in a surgically induced preclinical OA mouse model [95]. Furthermore, intra-articular injection of etanercept (a fusion-soluble TNFR2 protein that inhibits TNF-α and has therapeutic activity in rheumatoid arthritis patients) blocks PGRN binding to TNFR2 and causes a more severe joint destruction in a preclinical OA mouse model [86]. Hence, PGRN identification and atsttrin development as TNF-α/TNFR pathway antagonist, may lead to new therapeutic interventions for OA.

Conclusions

There are several factors that contribute to obesity-induced degeneration of joint tissues and subsequent development of OA (Fig. 6.1). Obesity, one of the primary risk factors for osteoarthritis, is associated with a state of low-grade chronic inflammation. Both local and systemic pro-inflammatory mediators and cytokines have been evidenced as key players in the progressive degeneration of joint tissues and OA development. Furthermore, mechanical stress of the joint (abnormal, altered or injurious loading) locally augmented the expression of pro-inflammatory factors, which could, at least in part, lead to an imbalance towards catabolic processes. However, the precise relationship between biomechanical factors and inflammation are not yet fully understood and further studies are necessary.

The data summarized here indicated that dysfunctional white adipose tissue together with interactions between joint mechanical overloading and local and/or systemic inflammation may prompt the pathogenesis and the development of osteoarthritis. In this context, adipokines, pleiotropic molecules synthesized and up-regulated by adipocytes as well as by chondrocytes and other cell types from the joints (including immune infiltrating cells), are evidenced as crucial promoters and sustainers of inflammatory and degradative processes associated with joint homeostasis unbalance and OA development. Nevertheless, many aspects of the adipokine network, especially the interplay between inflammatory paths and mechanical and metabolic processes in the cartilage and bone, remain largely unknown.

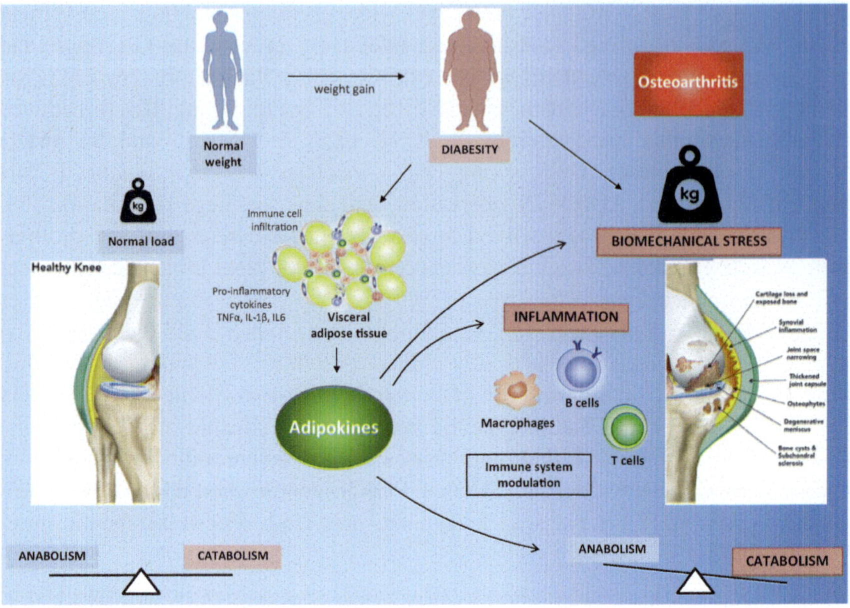

Fig. 6.1 Contributory factors responsible for obesity-induced degeneration of joint tissues and development of OA

Doubtless, further insights into the intimate mechanisms regulating peripheral and central adipokines activity might be of great relevance for future osteoarthritis therapeutic strategies.

Acknowledgements OG and FL are Staff Personnel of Xunta de Galicia (Servizo Galego de Saude, SERGAS) through a research-staff stabilization contract (ISCIII/SERGAS). VF is a "Sara Borrell" Researcher funded by ISCIII and FEDER (CD16/00111). RG is a "Miguel Servet" Researcher funded by Instituto de Salud Carlos III (ISCIII) and FEDER. CRF is a pre-doctoral research scholar funded by ISCIII and FEDER (Exp. 18/00188). OG, MAGG and RG are members of RETICS Programme, RD16/0012/0014 (RIER: Red de Investigación en Inflamación y Enfermedades Reumáticas) via Instituto de Salud Carlos III (ISCIII) and FEDER. FL is a member of CIBERCV (Centro de Investigación Biomédica en Red de Enfermedades Cardiovasculares).

The work of OG and JP (PI17/00409), RG (PI16/01870 and CP15/00007) and FL (PI15/00681 PI18/00821 and CB16/11/00226) was funded by Instituto de Salud Carlos III and FEDER. OG is a beneficiary of a project funded by Research Executive Agency of the European Union in the framework of MSCA-RISE Action of the H2020 Programme (Project number 734899). RG and OG are beneficiaries of a project funded by Mutua Madrileña 2018.). The funders had no role in study design, data collection, and analysis, decision to publish, or preparation of the manuscript. OG is also beneficiary of a grant from Xunta de Galicia, Consellería de Economia, Emprego e Industria, Axencia Galega de Innovación, Grant n° IN607B2019/10.

References

1. Zhang Y, Liu J, Yao J et al (2014) Obesity: pathophysiology and intervention. Nutrients 6 (11):5153–5183
2. Loeser RF, Goldring SR, Scanzello CR, Goldring MB (2012) Osteoarthritis: a disease of the joint as an organ. Arthritis Rheum 64(6):1697–1707
3. Guilak F (2011) Biomechanical factors in osteoarthritis. Best Pract Res Clin Rheumatol 25 (6):815–823
4. Glyn-Jones S, Palmer AJR, Agricola R et al (2015) Osteoarthritis. Lancet 386(9991):376–387
5. Guilak F, Fermor B, Keefe FJ et al (2004) The role of biomechanics and inflammation in cartilage injury and repair. Clin Orthop Relat Res 423:17–26
6. Blagojevic M, Jinks C, Jeffery A, Jordan KP (2010) Risk factors for onset of osteoarthritis of the knee in older adults: a systematic review and meta-analysis. Osteoarthr Cartil 18(1):24–33
7. Gabay O, Hall DJ, Berenbaum F, Henrotin Y, Sanchez C (2008) Osteoarthritis and obesity: experimental models. Jt Bone Spine 75(6):675–679
8. Aspden RM (2011) Obesity punches above its weight in osteoarthritis. Nat Rev Rheumatol 7 (1):65–68
9. Vieira-Potter VJ (2014) Inflammation and macrophage modulation in adipose tissues. Cell Microbiol 16(10):1484–1492
10. Tilg H, Moschen AR (2006) Adipocytokines: mediators linking adipose tissue, inflammation and immunity. Nat Rev Immunol 6(10):772–783
11. Francisco V, Pino J, Gonzalez-Gay MA et al (2018) Adipokines and inflammation: is it a question of weight? Br J Pharmacol 175(10):1569–1579
12. Pickvance EA, Oegema TR, Thompson RC (1993) Immunolocalization of selected cytokines and proteases in canine articular cartilage after transarticular loading. J Orthop Res 11(3):313–323
13. Das P, Schurman DJ, Smith RL (1997) Nitric oxide and G proteins mediate the response of bovine articular chondrocytes to fluid-induced shear. J Orthop Res 15(1):87–93
14. Lee DA, Frean SP, Lees P, Bader DL (1998) Dynamic mechanical compression influences nitric oxide production by articular chondrocytes seeded in agarose. Biochem Biophys Res Commun 251(2):580–585
15. Mohtai M, Gupta MK, Donlon B et al (1996) Expression of interleukin-6 in osteoarthritic chondrocytes and effects of fluid-induced shear on this expression in normal human chondrocytesin vitro. J Orthop Res 14(1):67–73
16. Smith RL, Donlon BS, Gupta MK et al (1995) Effects of fluid-induced shear on articular chondrocyte morphology and metabolismin vitro. J Orthop Res 13(6):824–831
17. Fermor B, Weinberg JB, Pisetsky DS, Misukonis MA, Banes AJ, Guilak F (2001) The effects of static and intermittent compression on nitric oxide production in articular cartilage explants. J Orthop Res 19(4):729–737
18. Fermor B, Weinberg JB, Pisetsky DS, Misukonis MA, Fink C, Guilak F (2002) Induction of cyclooxygenase-2 by mechanical stress through a nitric oxide-regulated pathway. Osteoarthr Cartil 10:792–798
19. Takao M, Okinaga T, Ariyoshi W et al (2011) Role of heme oxygenase-1 in inflammatory response induced by mechanical stretch in synovial cells. Inflamm Res 60(9):861–867
20. Sanchez C, Pesesse L, Gabay O et al (2012) Regulation of subchondral bone osteoblast metabolism by cyclic compression. Arthritis Rheum 64(4):1193–1203
21. Radin EL, Martin RB, Burr DB, Caterson B, Boyd RD, Goodwin C (1984) Effects of mechanical loading on the tissues of the rabbit knee. J Orthop Res 2:221–234
22. Felson DT, Zhang Y, Anthony JM, Naimark A, Anderson JJ (1992) Weight loss reduces the risk for symptomatic knee osteoarthritis in women. Ann Intern Med 116(7):535
23. Yusuf E, Nelissen RG, Ioan-Facsinay A et al (2010) Association between weight or body mass index and hand osteoarthritis: a systematic review. Ann Rheum Dis 69(4):761–765

24. Griffin TM, Huebner JL, Kraus VB, Guilak F (2009) Extreme obesity due to impaired leptin signaling in mice does not cause knee osteoarthritis. Arthritis Rheum 60(10):2935–2944
25. Green ED, Maffei M, Braden VV et al (1995) The human obese (OB) gene: RNA expression pattern and mapping on the physical, cytogenetic, and genetic maps of chromosome 7. Genome Res 5(1):5–12
26. Zhou Y, Rui L (2014) Leptin signaling and leptin resistance. Front Med 7(2):207–222
27. Cohen B, Novick D, Rubinstein M (1996) Modulation of insulin activities by leptin. Science 274(5290):1185–1188
28. Münzberg H, Morrison CD (2015) Structure, production and signaling of leptin. Metabolism 64(1):13–23
29. Frühbeck G (2006) Intracellular signalling pathways activated by leptin. Biochem J 393(1):7–20
30. Francisco V, Pino J, Campos-Cabaleiro V et al (2018) Obesity, fat mass and immune system: role for leptin. Front Physiol 9:640
31. Richter M, Trzeciak T, Rybka JD et al (2017) Correlations between serum adipocytokine concentrations, disease stage, radiological status and total body fat content in the patients with primary knee osteoarthritis. Int Orthop 41(5):983–989
32. Presle N, Pottie P, Dumond H et al (2006) Differential distribution of adipokines between serum and synovial fluid in patients with osteoarthritis. Contribution of joint tissues to their articular production. Osteoarthr Cartil 14(7):690–695
33. Conde J, Scotece M, López V et al (2013) Differential expression of adipokines in infrapatellar fat pad (IPFP) and synovium of osteoarthritis patients and healthy individuals. Ann Rheum Dis 73(3):631–633
34. Vuolteenaho K, Koskinen A, Moilanen E (2014) Leptin—a link between obesity and osteoarthritis. Applications for prevention and treatment. Basic Clin Pharmacol Toxicol 114 (1):103–108
35. Dumond H, Presle N, Terlain B et al (2003) Evidence for a key role of leptin in osteoarthritis. Arthritis Rheum 48(11):3118–3129
36. Simopoulou T, Malizos KN, Iliopoulos D et al (2007) Differential expression of leptin and leptin's receptor isoform (Ob-Rb) mRNA between advanced and minimally affected osteoarthritic cartilage; effect on cartilage metabolism. Osteoarthr Cartil 15(8):872–883
37. Vuolteenaho K, Koskinen A, Moilanen T, Moilanen E (2012) Leptin levels are increased and its negative regulators, SOCS-3 and sOb-R are decreased in obese patients with osteoarthritis: a link between obesity and osteoarthritis. Ann Rheum Dis 71(11):1912–1913
38. Karvonen-Gutierrez CA, Harlow SD, Mancuso P, Jacobson J, De Leon CFM, Nan B (2013) Association of leptin levels with radiographic knee osteoarthritis among a cohort of midlife women. Arthritis Care Res 65(6):936–944
39. Zhang P, Zhong ZH, Yu HT, Liu B (2015) Significance of increased leptin expression in osteoarthritis patients. PLoS ONE 10(4):e0123224
40. Massengale M, Lu B, Pan JJ, Katz JN, Solomon DH (2012) Adipokine hormones and hand osteoarthritis: radiographic severity and pain. PLoS ONE 7(10):e47860
41. Staikos C, Ververidis A, Drosos G, Manolopoulos VG, Verettas D-A, Tavridou A (2013) The association of adipokine levels in plasma and synovial fluid with the severity of knee osteoarthritis. Rheumatology 52(6):1077–1083
42. Scotece M, Mobasheri A (2015) Leptin in osteoarthritis: focus on articular cartilage and chondrocytes. Life Sci 140:75–78
43. Tu C, He J, Wu B, Wang W, Li Z (2019) An extensive review regarding the adipokines in the pathogenesis and progression of osteoarthritis. Cytokine 113:1–12
44. Pearson MJ, Herndler-Brandstetter D, Tariq MA et al (2017) IL-6 secretion in osteoarthritis patients is mediated by chondrocyte-synovial fibroblast cross-talk and is enhanced by obesity. Sci Rep 7(1):3451
45. Scotece M, Pérez T, Conde J et al (2017) Adipokines induce pro-inflammatory factors in activated CD4+ T cells from osteoarthritis patient. J Orthop Res 35(6):1299–1303

46. Conde J, Scotece M, López V et al (2012) Adiponectin and leptin induce VCAM-1 expression in human and murine chondrocytes. PLoS ONE 7(12):e52533
47. Otero M, Gomez Reino JJ, Gualillo O (2003) Synergistic induction of nitric oxide synthase type II: in vitro effect of leptin and interferon-gamma in human chondrocytes and ATDC5 chondrogenic cells. Arthritis Rheum 48(2):404–409
48. Otero M, Lago R, Lago F, Reino JJG, Gualillo O (2005) Signalling pathway involved in nitric oxide synthase type II activation in chondrocytes: synergistic effect of leptin with interleukin-1. Arthritis Res Ther 7(3):R581–R591
49. Otero M, Lago R, Gómez R, Lago F, Gomez-Reino JJ, Gualillo O (2007) Phosphatidylinositol 3-kinase, MEK-1 and p38 mediate leptin/interferon-gamma synergistic NOS type II induction in chondrocytes. Life Sci 81(19–20):1452–1460
50. Vuolteenaho K, Koskinen A, Kukkonen M et al (2009) Leptin enhances synthesis of proinflammatory mediators in human osteoarthritic cartilage–mediator role of NO in leptin-induced PGE2, IL-6, and IL-8 production. Mediators Inflamm 2009:345838
51. Koskinen A, Vuolteenaho K, Nieminen R, Moilanen T, Moilanen E (2011) Leptin enhances MMP-1, MMP-3 and MMP-13 production in human osteoarthritic cartilage and correlates with MMP-1 and MMP-3 in synovial fluid from oa patients. Clin Exp Rheumatol 29(1):57–64
52. Bao JP, Chen WP, Feng J, Hu PF, Shi ZL, Wu LD (2010) Leptin plays a catabolic role on articular cartilage. Mol Biol Rep 37(7):3265–3272
53. Zhou B, Li H, Shi J (2017) miR-27 inhibits the NF-κB signaling pathway by targeting leptin in osteoarthritic chondrocytes. Int J Mol Med 40(2):523–530
54. Mutabaruka M-S, Aoulad Aissa M, Delalandre A, Lavigne M, Lajeunesse D (2010) Local leptin production in osteoarthritis subchondral osteoblasts may be responsible for their abnormal phenotypic expression. Arthritis Res Ther 12(1):R20
55. Zhao CQ, Liu D, Li H, Jiang LS, Dai LY (2008) Expression of leptin and its functional receptor on disc cells: contribution to cell proliferation. Spine (Phila. Pa. 1976) 33(23):858–864
56. Zhao X, Dong Y, Zhang J et al (2016) Leptin changes differentiation fate and induces senescence in chondrogenic progenitor cells. Cell Death Dis 7(4):e2188
57. Poonpet T (2014) Adipokines: biomarkers for osteoarthritis? World J Orthop 5(3):319
58. Liu M, Liu F (2014) Regulation of adiponectin multimerization, signaling and function. Best Pract Res Clin Endocrinol Metab 28(1):25–31
59. Sun Y, Xun K, Wang C et al (2009) Adiponectin, an unlocking adipocytokine: review. Cardiovasc Ther 27(1):59–75
60. Kadowaki T, Yamauchi T (2005) Adiponectin and adiponectin receptors. Endocr Rev 26 (3):439–451
61. Maeda N, Takahashi M, Funahashi T et al (2001) PPARgamma ligands increase expression and plasma concentrations of adiponectin, an adipose-derived protein. Diabetes 50(9):2094–2099
62. Luo Y, Liu M (2016) Adiponectin: a versatile player of innate immunity. J. Mol. Cell Biol. 8 (2):120–128
63. Laurberg TB, Frystyk J, Ellingsen T et al (2009) Plasma adiponectin in patients with active, early, and chronic rheumatoid arthritis who are steroid- and disease-modifying antirheumatic drug-naive compared with patients with osteoarthritis and controls. J Rheumatol 36(9):1885–1891
64. Fioravanti A, Cheleschi S, De Palma A et al (2017) Can adipokines serum levels be used as biomarkers of hand osteoarthritis? Biomarkers 1–6 [Epub ahead of print]
65. Kang EH, Lee YJ, Kim TK et al (2010) Adiponectin is a potential catabolic mediator in osteoarthritis cartilage. Arthritis Res Ther 12(6):R231
66. Wang Q, Cai J, Wang J et al (2014) Down-regulation of adiponectin receptors in osteoarthritic chondrocytes. Cell Biochem Biophys 70(1):491–497
67. Francisco V, Pérez T, Pino J et al (2017) Biomechanics, obesity, and osteoarthritis. The role of adipokines: when the levee breaks. J Orthop Res 36(2):594–604

68. Zuo W, Wu Z-H, Wu N et al (2011) Adiponectin receptor 1 mediates the difference in adiponectin- induced prostaglandin E2 production in rheumatoid arthritis and osteoarthritis synovial fibroblasts. Chin Med J (Engl) 124(23):3919–3924
69. Luo X-H, Guo L-J, Yuan L-Q et al (2005) Adiponectin stimulates human osteoblasts proliferation and differentiation via the MAPK signaling pathway. Exp Cell Res 309(1):99–109
70. Kajimura D, Lee HW, Riley KJ et al (2013) Adiponectin regulates bone mass via opposite central and peripheral mechanisms through FoxO1. Cell Metab 17(6):901–915
71. Tang L, Gao X, Yang X et al (2016) Ladder-Climbing training prevents bone loss and microarchitecture deterioration in diet-induced obese rats. Calcif Tissue Int 98(1):85–93
72. Toussirot É, Binda D, Gueugnon C, Dumoulin G (2012) Adiponectin in autoimmune diseases. Curr Med Chem 19(32):5474–5480
73. Abella V, Scotece M, Conde J et al (2015) The potential of lipocalin-2/NGAL as biomarker for inflammatory and metabolic diseases. Biomarkers 20(8):565–571
74. Villalvilla A, García-Martín A, Largo R, Gualillo O, Herrero-Beaumont G, Gómez R (2016) The adipokine lipocalin-2 in the context of the osteoarthritic osteochondral junction. Sci Rep 6:29243
75. Moschen AR, Adolph TE, Gerner RR, Wieser V, Tilg H (2017) Lipocalin-2: a master mediator of intestinal and metabolic inflammation. Trends Endocrinol Metab 28(5):388–397
76. Fujino R-S, Tanaka K, Morimatsu M, Tamura K, Kogo H, Hara T (2006) Spermatogonial cell-mediated activation of an IkappaBzeta-independent nuclear factor-kappaB pathway in Sertoli cells induces transcription of the lipocalin-2 gene. Mol Endocrinol 20(4):904–915
77. Conde J, Otero M, Scotece M et al (2016) E74-like factor 3 and nuclear factor-κB regulate lipocalin-2 expression in chondrocytes. J Physiol 21(21):6133–6146
78. Guo H, Jin D, Chen X (2014) Lipocalin 2 is a regulator of macrophage polarization and NF-κB/STAT3 pathway activation. Mol Endocrinol 28(10):1616–1628
79. Wang B-W, Hung H-F, Chang H et al (2007) Mechanical stretch enhances the expression of resistin gene in cultured cardiomyocytes via tumor necrosis factor. Am J Physiol Hear Circ Physiol 293:2305–2312
80. Gupta K, Shukla M, Cowland JB, Malemud CJ, Haqqi TM (2007) Neutrophil gelatinase-associated lipocalin is expressed in osteoarthritis and forms a complex with matrix metalloproteinase 9. Arthritis Rheum 56(10):3326–3335
81. Conde J, Scotece M, Gomez R, Lopez V, Gomez-Reino JJ, Gualillo O (2011) Adipokines and osteoarthritis: novel molecules involved in the pathogenesis and progression of disease. Arthritis 2011:203901 (2011)
82. Gómez R, Scotece M, Conde J et al (2013) Nitric oxide boosts TLR-4 mediated lipocalin 2 expression in chondrocytes. J Orthop Res 31(7):1046–1052
83. Conde J, Lazzaro V, Scotece M et al (2017) Corticoids synergize with IL-1 in the induction of LCN2. Osteoarthr Cartil 25(7):1172–1178
84. Choi WS, Chun JS (2017) Upregulation of lipocalin-2 (LCN2) in osteoarthritic cartilage is not necessary for cartilage destruction in mice. Osteoarthr Cartil 25(3):401–405
85. Abella V, Pino J, Scotece M et al (2017) Progranulin as a biomarker and potential therapeutic agent. Drug Discov Today 22(10):1557–1564
86. Wei J-L, Buza Iii J, Liu C-J (2016) Does progranulin account for the opposite effects of etanercept and infliximab/adalimumab in osteoarthritis? J Orthop Res 34(1):12–14
87. Hu F, Padukkavidana T, Vægter CB et al (2010) Sortilin-mediated endocytosis determines levels of the frontotemporal dementia protein, progranulin. Neuron 68(4):654–667
88. Carrasquillo MM, Nicholson AM, Finch N et al (2010) Genome-wide screen identifies rs646776 near sortilin as a regulator of progranulin levels in human plasma. Am J Hum Genet 87(6):890–897
89. Jian J, Li G, Hettinghouse A, Liu C (2018) Progranulin: a key player in autoimmune diseases. Cytokine 101:48–55
90. Neill T, Buraschi S, Goyal A et al (2016) EphA2 is a functional receptor for the growth factor progranulin. J Cell Biol 215(5):687–703

91. Tang W, Lu Y, Tian Q-Y et al (2011) The growth factor progranulin binds to TNF receptors and is therapeutic against inflammatory arthritis in mice. Science 332(6028):478–484
92. Wei J, Hettinghouse A, Liu C (2016) The role of progranulin in arthritis. Ann N Y Acad Sci 1383(1):5–20
93. Abella V, Scotece M, Conde J et al (2016) The novel adipokine progranulin counteracts IL-1 and TLR4-driven inflammatory response in human and murine chondrocytes via TNFR1. Sci. Rep. 6(1):20356
94. Feng JQ, Guo F-J, Jiang B-C et al (2010) Granulin epithelin precursor: a bone morphogenic protein 2-inducible growth factor that activates Erk1/2 signaling and JunB transcription factor in chondrogenesis. FASEB J. 24(6):1879–1892
95. Xia P, Wang X, Lin Q, Li X (2015) Efficacy of mesenchymal stem cells injection for the management of knee osteoarthritis: a systematic review and meta-analysis. Int Orthop 39 (12):2363–2372

Chapter 7
Understanding the Initiation and Progression of Diet-Induced Obesity and Associated Pathophysiology: Lessons Learned from a Rat Model

David A. Hart, Walter Herzog, Jaqueline L. Rios, Raylene A. Reimer and Kelsey H. Collins

Abstract In human populations, the development and progression of obesity and its consequences on host systems is complex. Risk for obesity can reside in an individual's genome, in epigenetic modifications, and in dietary and activity level variables. In addition to host system factors, the diet can also influence the gut microbiota leading to alterations to the host and such alterations can contribute to pathophysiological sequelae in the host. In order to better understand the impact of

D. A. Hart (✉) · W. Herzog · J. L. Rios · R. A. Reimer
Human Performance Laboratory, Faculty of Kinesiology,
c/o McCaig Institute for Bone & Joint Health, University of Calgary,
3330 Hospital Drive NW, T2N 4N1 Calgary, AB, Canada
e-mail: hartd@ucalgary.ca

W. Herzog
e-mail: wherzog@ucalgary.ca

J. L. Rios
e-mail: jaquelinelourdes.rio@ucalgary.ca

R. A. Reimer
e-mail: reimer@ucalgary.ca

D. A. Hart · W. Herzog
Department of Surgery, McCaig Institute for Bone & Joint Health,
University of Calgary, Calgary, AB, Canada

D. A. Hart
Bone & Joint Health Strategic Clinical Network, Alberta Health Services,
Edmonton, AB, Canada

R. A. Reimer
Department of Biochemistry& Molecular Biology, University of Calgary,
Calgary, AB, Canada

K. H. Collins
Department of Orthopaedic Surgery, Washington University at St. Louis,
St. Louis, MO, USA
e-mail: kelseycollins@wustl.edu

P. S. Tappia et al. (eds.), *Pathophysiology of Obesity-Induced Health Complications*, Advances in Biochemistry in Health and Disease 19,
https://doi.org/10.1007/978-3-030-35358-2_7

diet on obesity development, our group has studied the consequences of a high-fat/high-sucrose (HFS) diet on physiological systems of Sprague-Dawley male rats over short term and long term exposure to the diet. Medium (12 weeks) to long term (28 weeks) exposure to the diet leads to development of knee, shoulder, and to a lesser degree, hip damage, alterations to some skeletal muscles, development of insulin resistance and type 2 diabetes, development of features of systemic metabolic syndrome, fatty liver disease, and alterations to the vitreous humor of the eye. Very short time exposure (days to 4 weeks) leads to early, but fluctuating changes to serum cytokine profiles, changes to some skeletal muscles, and the onset of knee joint damage. Alterations to the gut microbiota are evident following medium to longer term exposure to the diet, but not during the shorter time frame. Exposure of rats on the HFS diet to either a modest exercise protocol or an oligofructose prebiotic initiated at the same time as the HFS diet is initiated, completely prevented development of joint damage at 12 weeks. Thus, the development of obesity, and at least some of its pathophysiological sequelae in this model, are modifiable by low cost, minimally invasive interventions. Such findings provide an opportunity to determine whether some of the consequences of exposure to the HFS diet develop in parallel or serially, and to identify potential points in the process that are reversible. Current studies are focused on addressing such questions.

Keywords Obesity · Metabolic syndrome · Skeletal muscle · Joint integrity · Gut microbiome · Insulin resistance · Diet-induced obesity · Genetic models · Rat models · Exercise · Prebiotics

Background

The issue of obesity and its consequences has taken on renewed importance due to the prevalence of the condition, particularly in Western and developed nations [1]. In many such environments, the current incidence of obesity is greater than 35% [1]. For many countries, the emergence of obesity as a critical negative factor in the health and well-being of their populations has largely occurred over the past 40 years or so, with a continuing rise in incidence over that period of time. Also of concern is the rising rate of obesity in childhood and adolescence [2].

For many individuals with obesity, direct and indirect consequences of obesity pose severe risk for loss of health and quality of life. The former includes diabetes, cardiovascular disease, accelerated loss of cognition, respiratory diseases, hypertension, liver disease, joint diseases, sarcopenia, and loss of ocular integrity. The relationships between these risks for disease, cause and effect relationships, and why some individuals have higher risk for some of the consequences than others, have been difficult to ascertain due to the heterogeneity of *Homo sapiens,* as well as time-related factors, such as the length of time an individual has lived with obesity, the extent of the metabolic derangement, and likely, their chronological age. Thus, some forms of obesity may be related to, or the result of, genetics [3], epigenetics

[4], age-related factors (menopause in females, andropause in males) [5], other sex-related factors, activity levels [6, 7], and diet considerations [8, 9]. By some accounts, there are more than 200 genetic contributors to obesity [10], a number which may be staggering to unravel to better understand the mechanisms underpinning the condition. However, such genetic, and likely transgenerational inheritance of epigenetic contributors, are risk factors and many of them are not manifested unless other variables come into play, such as environmental factors. This conclusion is based on the fact that inheritance of genetic risk is likely a slow process, plausibly taking millennia, and potentially serving survival advantage during the evolution of *Homo sapiens* when food insecurity was more common. While epigenetic alterations leading to enhanced risk for obesity may have a shortened time frame compared to genetic risks, they likely do not account for the rapid and continuing rise in the prevalence of obesity over the past 40 years, referred to by some [1, 11, 12] as an epidemic. Thus, environmental factors may be influencing these increases in obesity incidence on a background of genetic and epigenetic heterogeneity.

As dissection of such relationships in humans is limited from a number of perspectives (i.e. heterogeneity, ethics, etc.), many researchers have turned to preclinical models to gain insights regarding intrinsic and extrinsic factors leading to obesity induction and progression, and the associated pathophysiological responses.

Rodent Models

To better understand the interplay between genetics and epigenetics variables and environmental factors, rodent models offer several advantages, including genetic manipulation, assessment of epigenetic modifications induced by environmental stressors, variation in nutrition via defined diets, invasive procedures can be used, the sequences of pathophysiological responses can be assessed, the effectiveness of defined interventions can be assessed, and models with defined genetic backgrounds can be used for the studies. However, one cannot lose sight of the fact that such advantages of some preclinical models (i.e. those with highly restricted genetic backgrounds) may also limit the effective translation of the findings to a heterogeneous human population.

Established mouse models have taken advantage of spontaneous genetic mutations, such as a lack of the leptin gene [13], or the leptin receptor gene [14], as well as other mutations or genetic modifications in specific metabolic pathways [15]. In addition, inbred mice have also been used with obesity-inducing diets containing high fat or high fat plus high sucrose to better mimic some of the attributes of Western diets consumed by many humans [16]. In addition, inbred mice exposed to an obesity-inducing diet have had obesity and its associated pathophysiology positively, albeit limitedly, impacted by interventions, such as a defined exercise program [17]. However, many of the studies in the literature using mouse models have focused on well-established obesity, and not the inductive phases of the disease to better understand the sequence of events and the temporal aspects of the biological systems impacted. Furthermore, many reports focus on a specific host target system (i.e. bone adaptations; cartilage damage; muscle repair complications) rather than taking an integrated systems biology approach [8].

Available rat models can also exhibit defined genetic [18, 19] or epigenetic alterations [20], and can also offer the advantage of not being completely inbred to better understand variation and commonalities in response to changing environmental factors such as diet. Furthermore, different obesity-inducing diets have been used with rat models (i.e. high fat or high fat-high sucrose) to assess the host systems affected [8]. Again, most studies have assessed the effects of the diet using longer exposure to the diet resulting in established (8–12 weeks) or chronic (24–26 weeks) obesity. Furthermore, nearly all studies, either mouse or rat have used sexually mature or adult animals for the studies. Few studies were focused on addressing the induction of childhood obesity in spite of the fact that in humans this is a relevant population that is growing rapidly, and one that has significant implications for long term health risks [21].

Thus, rodent models offer the opportunity to assess a number of aspects of obesity induction, and the effectiveness of both surgical [22] and non-surgical interventions [23–26] on modifying specific components of the pathophysiological consequences of obesity.

Characterization of Pathophysiological Responses to Diet-Induced Obesity in Sprague-Dawley Rats

Rodent models have provided a number of insights into obesity onset and progression. Some of the mouse models are genetic in nature, and therefore represent a specific subset of alterations that result in obesity [27]. Alternatively, other models use modified diets to induce obesity, either as high fat diets [28] or diets containing high fat and high sugar [8]. Furthermore, some of the rodent models use inbred strains of mice, such as the *ob/ob* mouse, which offer some advantages, but also some disadvantages as extrapolation of the findings to more genetically heterogeneous populations is problematic.

Constitutive genetic models (i.e. not inducible using a drug, like doxycycline or tamoxifen) also can exert their impact from the time of birth, and therefore, the young animals are influenced by the mutations from birth through lactation, weaning, puberty, and maturation. Thus, the metabolic alterations associated with the genetic mutations, such as leptin or leptin receptor expression/function, from the time of birth and their consequences could influence growth, establishment of normal hormonal relationships when going through puberty, and the functional relationship with the gut microbiota. Of note, inducible systems, like the Cre/Lox inducible genetic recombination system, can allow for temporal control of the induction of a genetic defect of interest, allowing for normal growth and development of the animal [29]. This is another key tool from which genetic features can be probed in the case of obesity.

In contrast, dietary alterations can be initiated in either male or female rodents when the animals are young (weanlings) or when more mature after puberty onset and following skeletal maturity (mice), or maturation (rats). Thus, such models offer the advantage of being able to study host and microbiota responses at an early age, at maturity, or during aging ± menopause equivalency. Bearing in mind that a variety of genetic and epigenetic contributors may influence which animals are predisposed to respond or become resistant to obesity, using genetically heterogeneous populations, such as rat strains (i.e. Sprague-Dawley rats), the influence of genetic variation on responsiveness to the dietary insult can be examined, as well as sex differences in response patterns.

A recent development in the study of nutrient effects on host responsiveness is that of transgenerational effects, mediated via epigenetic changes [30]. For instance, human populations subjected to starvation can impart a risk for obesity on their offspring via epigenetic alterations [31, 32]. Similarly, epigenetic alterations regarding obesity can occur in the offspring of rat mothers exposed to diet-induced obesity prior to pregnancy [33].

Thus, there are multiple ways to develop obesity in rodent models. Understanding the nuances of the pathophysiology in such models, and the ability to successfully intervene to address the obesity and the physiological consequences under a variety of circumstances will likely provide insights into human obesity development and progression, hopefully leading to effective treatment that can thwart the current epidemic and its impact on both health and the health care system.

Pathophysiology in the Adult Sprague-Dawley High Fat-High Sucrose (HFS) Model of Diet-Induced Obesity

As outlined above, there are a number of rodent models available to study obesity, its consequences, and its contributors. However, it is beyond the scope of this chapter to address many of them, and therefore, the focus will be on the model we have been characterizing over the past several years, namely the male Sprague-Dawley (SD) rat model using a defined high fat-high sucrose (HFS) obesity-inducing dietary challenge.

Many reports have used diet-induced obesity in rats, usually feeding the diet from adulthood for 8–12 weeks or 26–28 weeks [8, 34–36]. Male rats, 10–12 weeks of age were started on the HFS diet for 12 weeks, and then the animals were divided into the top third, which gained the most weight and had the highest fat mass (HFS Responders), and the bottom third who gained the least (HFS Resistant) ($p < 0.05$). In these animal subpopulations, a number of host systems

Table 7.1 Changes in host systems and relationship to knee osteoarthritis damage (Mankin score) in animals challenged with 12-weeks of the HFS diet

Analyte	Tissue	Direction in DIO versus chow	DIO-P versus DIO-R	P-value	Relationship to Mankin score
–	Body fat	↑	↑	0.0001	Moderate, positive
–	Body mass	↑	↑	0.0001	Moderate, positive
IL-6	Synovial fluid	↑	↑ (p = 0.026)	0.287	Very weak; NS
Leptin	Synovial fluid	↑	≈	0.001	Moderate, positive
Leptin	Serum	↑	≈	0.001	Weak, positive
IL-1β	Synovial fluid	↑	≈	0.003	Weak, positive
TNF-α	Serum	↑	≈	0.023	Weak, positive
MCP-1	Serum	↑	≈	0.004	Weak, positive

Relations are defined as: r value of 0.00–0.19 as "very weak"; 0.2–0.39 "weak", 0.40–0.59 "moderate"; 0.60–0.79 "strong"; 0.80–1.0 "very strong" per Evans 1996 [38]; *NS* Not Significant

were altered together, and many were not different between obesity prone or obesity resistant animals. (examples summarized in Table 7.1, original data can be found in [8]. Obesity, and corresponding chronic-low grade inflammation, is associated with the onset and progression of knee OA. The origin of this inflammation is poorly understood. Here, the effect of high fat, high sucrose (HFS) diet induced obesity (DIO) on local (synovial fluid), and systemic (serum) inflammation is evaluated after a 12-week obesity induction and a further 16-week adaptation period. For 12-weeks of obesity induction, n = 40 DIO male Sprague-Dawley rats consumed a HFS diet while the control group (n = 14) remained on chow. DIO rats were allocated to prone (DIO-P, top 33% based on weight change) or resistant (DIO-R, bottom 33%) groups at 12-weeks. Animals were euthanized at 12- and after an additional 16-weeks on diet (28-weeks). At sacrifice, body composition and knee joints were collected and assessed. Synovial fluid and sera were profiled using cytokine array analysis. At 12-weeks, DIO-P animals demonstrated increased Modified Mankin scores compared to DIO-R and chow (p = 0.026), and DIO-R had higher Mankin scores compared to chow (p = 0.049). While numerous systemic and limited synovial fluid inflammatory markers were increased at 12-weeks in DIO animals compared to chow, by 28-weeks there were limited systemic differences but marked increases in local synovial fluid inflammatory marker concentrations. Metabolic OA may manifest from an initial systemic inflammatory disturbance. Twelve weeks of obesity induction leads to a unique inflammatory profile and induction of metabolic OA which is altered after a further 16-weeks of obesity and HFS diet intake, suggesting that obesity is a dynamic, progressive process [37].

In addition, in the rats on the HFS diet, the gut microbiota were also extensively altered compared to those on the control chow diet [36]. Interestingly, regarding joint damage, not all articulating joints were affected the same, with the knees and shoulders being affected more than the hips [39]. Furthermore, aspects of eye integrity were found to be altered, with the vitreous humor (VH) and VH cells (VHC) transformed into a more inflammatory state [40]. As the eye is a local immune privileged site (discussed in [40, 41]), such changes may lead to, or contribute to, development of pathology in the eye.

In contrast, when male SD rats were started on the HFS diet at 10–12 weeks of age and then kept on the diet for 28 weeks, the $HFS_{resistant}$ caught up to the $HFS_{responsive}$ animals for all aspects of the obesity and its consequences, including body mass, fat mass, insulin resistance, skeletal muscle alterations, knee joint damage, and serum + synovial fluid biomarker alterations [39, 42–44], with the changes summarized in Table 7.2. In addition, gut microbiota alterations became stabilized, with increased serum bacterial lipopolysaccharide levels detected [44].

Thus, initially there is detectable heterogeneity in the rate of responsiveness of male SD rats to the HFS diet, potentially indicating that there may be genetic variables that contribute to the responsive phenotype, but over time, when the obesity becomes more chronic, potential genetic factors contributing to resistance are overcome, either directly by the diet, or indirectly due to the early consequences of the diet. Thus, based on this experimental design, exposure to the HFS diet for an extended period of time (12 weeks) in male adult SD rats "sets the stage" for development of a pattern of pathophysiological consequences leading to fully developed systemic (T2D, metabolic syndrome) and local (knee joint, bone marrow, muscle) impact, and also alterations to the gut microbiota [37]. We have concluded that skeletal muscles are likely central to the development of the pathophysiological consequences of the HFS diet (reviewed in [8]). Inflammation can arise in response to a variety of stimuli, including infectious agents, tissue injury, autoimmune diseases, and obesity. Some of these responses are acute and resolve, while others become chronic and exert a sustained impact on the host, systemically or locally. Obesity is now recognized as a chronic low-grade, systemic inflammatory state that predisposes to other chronic conditions including metabolic syndrome (MetS). Although obesity has received considerable attention regarding its pathophysiological link to chronic cardiovascular conditions and type 2 diabetes, the musculoskeletal (MSK) complications (i.e., muscle, bone, tendon, and joints) that result from obesity-associated metabolic disturbances are less frequently interrogated. As musculoskeletal diseases can lead to the worsening of MetS, this underscores the imminent need to understand the cause and effect relations between the two, and the convergence between inflammatory pathways that contribute to MSK damage. Muscle mass is a key predictor of longevity in older adults, and obesity-induced sarcopenia is a significant risk factor for adverse health outcomes. Muscle is highly plastic, undergoes regular remodeling, and is responsible for the majority of total body glucose utilization, which when impaired leads to insulin resistance. Furthermore, impaired muscle integrity, defined as persistent muscle loss, intramuscular lipid accumulation, or connective tissue deposition, is a

Table 7.2 Changes in host systems and relationship to knee osteoarthritis damage (Mankin score) in animals challenged with 28-week HFS diet, adapted from [36]

Analyte	Tissue	Direction in DIO versus chow	DIO-P versus DIO-R	P-value	Relationship to Mankin score
–	Body fat	↑	≈	0.001	Moderate, positive
–	Body mass	↑	≈	0.001	NS
Lipopolysaccharide	Serum	↑	≈	0.008	NS
Lactobacillus spp.	Gut microbiota	↓	≈	0.001	Weak, negative
Methanobrevibacter spp.	Gut microbiota	≈	≈	NS	Weak, positive
IL-6	Synovial fluid	↑	≈	0.05	NS
Leptin	Synovial fluid	↑	≈	0.022	Moderate, positive
Leptin	Serum	↑	↑	0.017	Moderate, positive
IL-1α	Synovial fluid	↑	↑	0.020	Weak, positive
IL-1β	Synovial fluid	≈	≈	NS	Weak, NS
TNF-α	Synovial fluid	↑	≈	0.003	NS
TNF-α	Serum	≈	≈	NS	NS
MCP-1	Synovial fluid	↑	≈	0.002	NS
MCP-1	Serum	≈	≈	NS	NS

Relations are defined as: r value of 0.00–0.19 as "very weak"; 0.2–0.39 "weak", 0.40–0.59 "moderate"; 0.60–0.79 "strong"; 0.80–1.0 "very strong" per Evans 1996 [38]; *NS* Not Significant

hallmark of metabolic dysfunction. In fact, many common inflammatory pathways have been implicated in the pathogenesis of the interrelated tissues of the musculoskeletal system (e.g., tendinopathy, osteoporosis, and osteoarthritis). Despite these similarities, these diseases are rarely evaluated in a comprehensive manner. The aim of this review is to summarize the common pathways that lead to musculoskeletal damage and disease that result from and contribute to MetS. We propose the overarching hypothesis that there is a central role for muscle damage with chronic exposure to an obesity-inducing diet. The inflammatory consequence of diet and muscle dysregulation can result in altered tissue repair and an imbalance toward negative adaptation, resulting in regulatory failure and other musculoskeletal tissue damage. The commonalities support the conclusion that musculoskeletal pathology with MetS should be evaluated in a comprehensive and integrated manner to understand risk for other MSK-related conditions [8], specifically in the case of subsequent development of musculoskeletal damage. Thus, timing of assessment for the sequence of pathophysiological events is critical, particularly if one is trying to identify "transition points" that may be amenable to reversing the obesity and specific aspects of its consequences through development of targeted interventions.

Pathophysiology of Short-Term Exposure of Adult SD Rats to an HFS Diet

Based on the results of the above discussed studies, 12 weeks of exposure to an HFS diet leads to a number of alterations in both host systems, as well as the gut microbiota. However, if timing is critical to understanding of the sequence of events, then it is important to perform short-term studies to determine whether some of the pathophysiological sequelae occur sequentially (and possibly interact to yield the ultimate phenotype), or occur in parallel from the initiation of the exposure to the HFS diet. Therefore, male 10–12 week old SD rats were exposed to the HFS diet and sacrificed 1, 3, 7, 14, or 28 days after HFS diet initiation [8, 45].

Based on the results obtained, some host responses to the HFS diet could be detected by 3 days on the diet, while other systems affected by longer term exposure were not. For some responses, such as the serum biomarker profile responses detected using a 27-plex protein array and the Luminex platform, there was detectable but variable responses from day 3 to day 14 which then stabilized somewhat by day 28. Thus, it would appear that the initial responses of the host are to respond in a cyclic manner with cycles of being impacted by the dietary insult, initiation of resistance to the insult, again being impacted by the dietary insult until some of the resistance mechanisms are overridden and the system finally adapts to the dietary insult leading to development of obesity and some of its more pronounced sequelae (insulin resistance likely at the level of skeletal muscles [45], serum biomarker changes indicative of metabolic syndrome, joint damage and abnormal adipose deposition to name a few) (summarized in Table 7.3).

Another feature that resulted from these short-term studies was that not all skeletal muscles were affected equally [45]. Specifically, muscles more dependent on glycolysis (e.g. VL, a predominantly fast twitch-fibred muscle) were affected early, while the soleus muscle (a predominantly slow twitch-fibred muscle), that is more dependent on fatty acid metabolism [47], was not affected. Thus, development of insulin resistance while on the HSF diet, which is reported to involve skeletal muscles [48], does not appear to affect all muscles equally, and may be dependent on the type of muscle, and the functioning of the muscles (reviewed in [8]). Inflammation can arise in response to a variety of stimuli, including infectious agents, tissue injury, autoimmune diseases, and obesity. Some of these responses are acute and resolve, while others become chronic and exert a sustained impact on the host, systemically or locally. Obesity is now recognized as a chronic low-grade,

Table 7.3 Changes in host systems and relationship to body fat in animals challenged with 4-weeks of a HFS diet, adapted from [46]

Analyte	Tissue	Direction in DIO versus chow	P-value	Relationship to body fat
–	Body mass	↑	<0.05	Moderate, positive
Oil Red O (Intramuscular lipid)	Vastus lateralis	↑	<0.05	Weak, positive
IL-6	Serum	↑	0.005	
Leptin	Serum	↑	<0.10	
Bacteroides/ Prevotella spp.	Gut microbiota	↓	<0.05	
Lactobacillus spp.	Gut microbiota	↓	<0.05	
Clostridium cluster I	Gut microbiota	↑	<0.05	
Clostridium cluster XI	Gut microbiota	↑	<0.05	
Clostridium cluster IV	Gut microbiota	↓	<0.05	
Clostridium cluster XIV	Gut microbiota	↓	<0.05	

Relations are defined as in Tables 7.1 and 7.2

systemic inflammatory state that predisposes to other chronic conditions including metabolic syndrome (MetS). Although obesity has received considerable attention regarding its pathophysiological link to chronic cardiovascular conditions and type 2 diabetes, the musculoskeletal (MSK) complications (i.e., muscle, bone, tendon, and joints) that result from obesity-associated metabolic disturbances are less frequently interrogated. As musculoskeletal diseases can lead to the worsening of MetS, this underscores the imminent need to understand the cause and effect relations between the two, and the convergence between inflammatory pathways that contribute to MSK damage. Muscle mass is a key predictor of longevity in older adults, and obesity-induced sarcopenia is a significant risk factor for adverse health outcomes. Muscle is highly plastic, undergoes regular remodeling, and is responsible for the majority of total body glucose utilization, which when impaired leads to insulin resistance. Furthermore, impaired muscle integrity, defined as persistent muscle loss, intramuscular lipid accumulation, or connective tissue deposition, is a hallmark of metabolic dysfunction. In fact, many common inflammatory pathways have been implicated in the pathogenesis of the interrelated tissues of the musculoskeletal system (e.g., tendinopathy, osteoporosis, and osteoarthritis). Despite these similarities, these diseases are rarely evaluated in a comprehensive manner. The aim of this review is to summarize the common pathways that lead to musculoskeletal damage and disease that result from and contribute to MetS. We propose the overarching hypothesis that there is a central role for muscle damage with chronic exposure to an obesity-inducing diet. The inflammatory consequence of diet and muscle dysregulation can result in altered tissue repair and an imbalance toward negative adaptation, resulting in regulatory failure and other musculoskeletal tissue damage. The commonalities support the conclusion that musculoskeletal pathology with MetS should be evaluated in a comprehensive and integrated manner to understand risk for other MSK-related conditions [8].

Furthermore, it was also clear from these studies that not all animals in these short-term exposure groups responded to the HFS dietary insult the same, and thus, there was considerable animal-to-animal variation even though they were all males of a similar age, had been raised on a similar chow diet prior to the HFS dietary insult, and were housed under similar conditions. Thus, again, this variation may be a reflection of some of the genetic heterogeneity in the SD population. Of note, many of these initial host changes appear to occur in concert with detectable alterations to the gut microbiota (unpublished observations). For example, tighter clustering by principal component analysis was observed between gut microbiota relative abundance with prolonged exposure to HFS [45]. Therefore, there are complex interactions between the host and gut microbiota, and the roles they are playing in the development of progress of pathophysiological events in response to the HFS diet.

While these short-term HFS exposure studies are not comprehensive, it is clear that to better understand the variables contributing to the obesity phenotypes observed after longer term exposure (e.g. 12–28 weeks), additional shorter term studies, such as those of 1, 2, 4, 6, 8 weeks duration may provide further understanding of the sequence of events, dysregulation, and tissue damage occurring after exposure to an obesity-inducing diet. Such understanding may again provide insights into potential "transition points" that could be sites for targeted interventions to prevent obesity progression, and ultimately enhance reversal.

Targeting Prevention of the Pathophysiological Consequences of the HFS Obesity-Inducing Diet

Improved understanding of the induction, establishment, and entrenchment of obesity and its consequences following exposure to a HFS diet will potentially lead to opportunities to identify points to target with specific interventions. To initiate such intervention studies, we started with a focus on characterization of interventions that would prevent the development of obesity and its consequences in the SD model. The first two interventions chosen for study were moderate exercise (impacting host systems including skeletal muscles J.L. Rios, PhD. Thesis, University of Calgary, 2019) and a prebiotic (directly impacting the gut microbiota) previously shown to be effective in both the rat model [23, 24, 49] and human populations [50, 51]. However, such prebiotics can also potentially impact some host systems via synthesis and host uptake of short chain fatty acids (e.g. butyrate, propionate) that can subsequently impact host systems such as centres in the brain [52].

Somewhat surprisingly, initiation of moderate exercise, prebiotic treatment, and exercise + prebiotic treatment, started at the time of HFS initiation with male SD rats at 10–12 weeks of age for a term of 12 weeks, all lead to a complete prevention of knee joint damage development in this model [53]. In addition, all treatment regimens were effective in preventing development of insulin resistance, and other manifestations of the HFS diet (J. L. Rios, Ph.D. Thesis, University of Calgary, 2019 [53]). These findings with the exercise protocol are consistent with a recent report indicating that exercise can prevent some diet-induced changes in a mouse model [54]. However, that study did not investigate the potential of the prebiotic or a combination of the interventions.

Thus, at the 12-week time point of feeding the HFS diet, both the dietary manipulation of gut microbiota and the exercise protocol appear to be impacting the host equally to prevent development of pathophysiological consequences to the diet. Based on the short-term studies discussed above, one might have expected that the exercise protocol, affecting host systems such as skeletal muscle integrity might have been more effective than the prebiotic, but this was not the case. Therefore, by 12 weeks on the diet, any early differences in the impact of the two interventions may have become obscured, a conclusion which reinforces the importance of doing shorter term studies to elucidate important details in the mechanistic progression of metabolic alterations induced by the HFS diet. Of note, by 12 weeks on the diet the combination of the two interventions was as effective as each intervention alone, and thus, they may either share some mechanistic commonalities, their mechanisms intersect at some common central point, or their mechanisms differ, but there are multiple pathways to prevent the impact of the HFS diet on host systems and development of pathophysiology. Again, insights into these options may be elucidated by shorter term studies. Such insights may be critical to enhance effective translation of the findings to subpopulations of humans with obesity.

Future Directions

The above discussed studies are encouraging regarding the elaboration of improved understanding of the pathophysiological impact of an HFS diet on male rats, and the results also provide opportunities to build on the information provided to address additional questions. These include: (1) Are the effects of the HFS diet similar or different when female SD are investigated? It is known that regulation of fat and sugar metabolism in males and females is different [55], as well as predisposition to the intergenerational transmission of obesity in males compared to females [24], so this is an important question to address. (2) Are the responses to the diet dependent on the age of the animals? Thus, are older animals and weanlings affected similarly as young post-puberty adults? Recent preliminary studies indicate that male SD rats exposed to the HFS diet as weanlings exhibit a different phenotype than the young adults regarding the development of knee joint damage [56], so age at the time of exposure may play a significant role in pathophysiology development, and exposure prior to onset of puberty may lead to a unique metabolic set point which could make reversal more challenging and different from adult onset. (3) Which pathophysiological changes are due to the HF versus the HS components of the diet, and does the combination (HFS) lead to a more complex phenotypic set of pathophysiological consequences? It is well-known that using both a high fat and high sugar diet can induce more extreme pro-inflammatory dysregulation when compared to high-fat diets alone [16]. The HFS diet was chosen to reflect elements of the "fast food" western diet, and thus is a "worst case" scenario for diet-induced obesity. Therefore, it will be important to determine whether one can dissect specific elements of the pathophysiologic sequelae of the diet by separating it into components? (4) Can timing of the interventions (i.e. exercise versus prebiotic) after initiation of the HFS diet exposure lead to an altered phenotypic response pattern in the rats? Can this response pattern reverse specific elements of metabolic dysregulation and provide insights into changes occurring in parallel or serially, and which are more associated with the gut microbiota alterations versus host systems? Since the two interventions studied thus far are simple, non-invasive and cost-effective, do they continue to exhibit effectiveness when initiated at various times after the HFS diet exposure is started, and if they lose their effectiveness differentially and in a phenotype-specific manner, can that provide additional clues to mechanistic implications?

Finally, it will be important to evaluate the translatability of effective rat interventions to human populations/subpopulations with obesity. Exercise and prebiotics are two potential candidates as they may not be species-restricted, at least to the same extent as possible pharmacologic interventions, where species-specific differences in target molecules preclude or interfere with effective translation.

A number of the above outlined directions to build on what has been learned are currently in progress in the authors' laboratories.

Acknowledgements The authors thank Ruth Seerattan for excellent technical support for the studies. The authors acknowledge the financial support from the AHS Strategic Clinical Network Program (DAH), CIHR (WH & RAR), the CAPES fund of Brazil (JLR), and NIH T32 NIDDK, Alberta Innovates Health Solutions, Canadian Institutes of Health Research, Killam Trusts (KHC).

Conflict of Interest The authors declare they have no conflicts of interest to disclose.

References

1. Hruby A, Hu FB (2015) The epidemiology of obesity: a big picture. Pharmacoeconomics [Internet]. 2015 Jul [cited 2019 Jan 4] 33(7):673–689. http://www.ncbi.nlm.nih.gov/pubmed/25471927
2. Kumar S, Kelly AS (2017) Review of childhood obesity. Mayo Clin Proc [Internet]. 2017 Feb [cited 2019 Jan 13] 92(2):251–265. http://www.ncbi.nlm.nih.gov/pubmed/28065514

3. Xia Q, Grant SFA (2013) The genetics of human obesity. Ann N Y Acad Sci [Internet]. 2013 Apr [cited 2019 Jan 13] 1281(1):178–190. http://www.ncbi.nlm.nih.gov/pubmed/23360386
4. Rosen ED, Kaestner KH, Natarajan R, Patti M-E, Sallari R, Sander M et al (2018) Epigenetics and epigenomics: implications for diabetes and obesity. Diabetes [Internet]. 2018 Oct 1 [cited 2019 Jan 13] 67(10):1923–1931. http://www.ncbi.nlm.nih.gov/pubmed/30237160
5. Lamberts SWJ, van den Beld AW, van der Lely A-J (1997) The endocrinology of aging. Science (80–). 278:419–423
6. Westerterp KR (1999) Assessment of physical activity level in relation to obesity: current evidence and research issues. Med Sci Sports Exerc [Internet]. 1999 Nov [cited 2019 Jan 13] 31(11 Suppl):S522–525. http://www.ncbi.nlm.nih.gov/pubmed/10593522
7. Hoos MB, Gerver WJM, Kester AD, Westerterp KR (2003) Physical activity levels in children and adolescents. Int J Obes [Internet]. 2003 May 17 [cited 2019 Jan 13] 27(5):605–609. http://www.nature.com/articles/0802246
8. Collins KH, Herzog W, MacDonald GZ, Reimer RA, Rios JL, Smith IC et al (2018) Obesity, metabolic syndrome, and musculoskeletal disease: common inflammatory pathways suggest a central role for loss of muscle integrity. Front Physiol 9
9. Johnson AR, Makowski L (2015) Nutrition and metabolic correlates of obesity and inflammation: clinical considerations. J Nutr [Internet]. 2015 May 1 [cited 2019 Jan 13] 145 (5):1131S–1136S. https://academic.oup.com/jn/article/145/5/1131S/4589947
10. McPherson R (2007) Genetic contributors to obesity. Can J Cardiol [Internet]. 2007 Aug [cited 2019 Jan 13] 23(Suppl A):23A–27A. http://www.ncbi.nlm.nih.gov/pubmed/17668084
11. Swinburn BA, Sacks G, Hall KD, McPherson K, Finegood DT, Moodie ML et al (2011) The global obesity pandemic: shaped by global drivers and local environments. Lancet [Internet]. 2011 Aug 27 [cited 2014 Jan 9] 378(9793):804–814. http://www.sciencedirect.com/science/article/pii/S0140673611608131
12. Roberto CA, Swinburn B, Hawkes C, Huang TT-K, Costa SA, Ashe M et al (2015) Patchy progress on obesity prevention: emerging examples, entrenched barriers, and new thinking. Lancet 385(9985):2400–2409
13. Castracane VD, Henson MC (2006) The obese (ob/ob) mouse and the discovery of leptin. In: Leptin [Internet]. Springer US, Boston, MA [cited 2019 Jan 16], pp 1–9. http://link.springer.com/10.1007/978-0-387-31416-7_1
14. Björnholm M, Münzberg H, Leshan RL, Villanueva EC, Bates SH, Louis GW et al (2007) Mice lacking inhibitory leptin receptor signals are lean with normal endocrine function. J Clin Invest [Internet]. 2007 May [cited 2019 Jan 16] 117(5):1354–1360. http://www.ncbi.nlm.nih.gov/pubmed/17415414
15. Efeyan A, Comb WC, Sabatini DM (2015) Nutrient-sensing mechanisms and pathways. Nature [Internet]. 2015 Jan 15 [cited 2019 Jan 16] 517(7534):302–310. http://www.nature.com/articles/nature14190
16. Schemmel R, Mickelsen O, Tolgay Z (1969) Dietary obesity in rats: influence of diet, weight, age, and sex on body composition. Am J Physiol [Internet]. 1969 Feb [cited 2013 Dec 14] 216 (2):373–379. http://www.ncbi.nlm.nih.gov/pubmed/5766993
17. Jung AP, Luthin DR (2010) Wheel access does not attenuate weight gain in mice fed high-fat or high-CHO diets. Med Sci Sports Exerc [Internet]. 2010 Feb [cited 2019 Jan 16] 42(2):355–360. http://www.ncbi.nlm.nih.gov/pubmed/19927024
18. Lutz TA, Woods SC (2012) Overview of animal models of obesity. Curr Protoc Pharmacol [Internet]. 2012 Sep [cited 2019 Jan 16]; Chapter 5: Unit 5.61. http://www.ncbi.nlm.nih.gov/pubmed/22948848
19. Vogel H, Kraemer M, Rabasa C, Askevik K, Adan RAH, Dickson SL (2017) Genetic predisposition to obesity affects behavioural traits including food reward and anxiety-like behaviour in rats. Behav Brain Res [Internet]. 2017 Jun 15 [cited 2019 Jan 16] 328:95–104. https://www.sciencedirect.com/science/article/pii/S0166432816311779

20. Seki Y, Williams L, Vuguin PM, Charron MJ (2012) Minireview: epigenetic programming of diabetes and obesity: animal models. Endocrinology [Internet]. 2012 Mar [cited 2019 Jan 16] 153(3):1031–1038. http://www.ncbi.nlm.nih.gov/pubmed/22253432
21. World Health Organization (2015) Interim report of the commission on ending childhood obesity [Internet]. 2015. http://www.who.int/end-childhood-obesity/commission-ending-childhood-obesity-interim-report.pdf?ua=1
22. Oberbach A, Schlichting N, Heinrich M, Lehmann S, Till H, Mohr FW et al (2014) Weight loss surgery improves the metabolic status in an obese rat model but does not affect bladder fibrosis associated with high fat diet feeding. Int J Obes [Internet]. 2014 Aug 29 [cited 2019 Jan 16] 38(8):1061–1067. http://www.ncbi.nlm.nih.gov/pubmed/24166068
23. Bomhof MR, Saha DC, Reid DT, Paul HA, Reimer RA (2014) Combined effects of oligofructose and Bifidobacterium animalis on gut microbiota and glycemia in obese rats. Obesity (Silver Spring) [Internet]. 2014 Mar [cited 2014 Oct 20] 22(3):763–771. http://doi.wiley.com/10.1002/oby.20632
24. Paul HA, Collins KH, Nicolucci AC, Urbanski SJ, Hart DA, Vogel HJ et al (2019) Maternal prebiotic supplementation reduces fatty liver development in offspring through altered microbial and metabolomic profiles in rats. FASEB J [Internet]. 2019 Jan 10 [cited 2019 Jan 14] fj.201801551R. https://www.fasebj.org/doi/10.1096/fj.201801551R
25. Lee JR, Tapia MA, Nelson JR, Moore JM, Gereau GB, Childs TE et al (2019) Sex dependent effects of physical activity on diet preference in rats selectively bred for high or low levels of voluntary wheel running. Behav Brain Res [Internet]. 2019 Feb 1 [cited 2019 Jan 16] 359:95–103. http://www.ncbi.nlm.nih.gov/pubmed/30392852
26. Chen C-N (Joyce), Liao Y-H, Lin S-Y, Yu J-X, Li Z-J, Lin Y-C et al (2017) Diet-induced obesity accelerates blood lactate accumulation of rats in response to incremental exercise to maximum. Am J Physiol Integr Comp Physiol [Internet]. 2017 Nov 1 [cited 2019 Jan 16] 313 (5):R601–607. http://www.ncbi.nlm.nih.gov/pubmed/28855180
27. Robinson SW, Dinulescu DM, Cone RD (2000) Genetic models of obesity and energy balance in the mouse. Annu Rev Genet [Internet]. 2000 Dec [cited 2019 Jan 16] 34(1):687–745. http://www.ncbi.nlm.nih.gov/pubmed/11092843
28. Hariri N, Thibault L (2010) High fat diet-induced obesity in animal models. Nutr Res Rev [Internet]. 2010 Dec [cited 2013 Feb 20] 23(2):270–299. http://www.ncbi.nlm.nih.gov/pubmed/20977819
29. McLellan MA, Rosenthal NA, Pinto AR (2017) Cre-*lox* P-mediated recombination: general principles and experimental considerations. In: Current protocols in mouse biology [Internet]. Wiley, Hoboken, NJ, USA [cited 2019 Jan 19], pp 1–12. http://www.ncbi.nlm.nih.gov/pubmed/28252198
30. Lillycrop KA, Burdge GC (2011) Epigenetic changes in early life and future risk of obesity. Int J Obes [Internet]. 2011 Jan 15 [cited 2019 Jan 16] 35(1):72–83. http://www.nature.com/articles/ijo2010122
31. Radford EJ, Ito M, Shi H, Corish JA, Yamazawa K, Isganaitis E et al (2014) In utero effects. In utero undernourishment perturbs the adult sperm methylome and intergenerational metabolism. Science [Internet]. 2014 Aug 15 [cited 2019 Jan 16] 345(6198):1255903. http://www.ncbi.nlm.nih.gov/pubmed/25011554
32. Youngson NA, Morris MJ (2013) What obesity research tells us about epigenetic mechanisms. Philos Trans R Soc Lond B Biol Sci [Internet]. 2013 Jan 5 [cited 2019 Jan 16] 368(1609):20110337. http://www.ncbi.nlm.nih.gov/pubmed/23166398
33. Mulligan CM, Friedman JE (2017) Maternal modifiers of the infant gut microbiota: metabolic consequences. J Endocrinol 235(1):R1–12
34. Collins KH, Reimer RA, Seerattan RA, Leonard TR, Herzog W (2015) Using diet-induced obesity to understand a metabolic subtype of osteoarthritis in rats. Osteoarthr Cartil 23(6)
35. Collins KH, Hart DA, Reimer RA, Seerattan RA, Herzog W (2016) Response to diet-induced obesity produces time-dependent induction and progression of metabolic osteoarthritis in rat knees. J Orthop Res 34(6)

36. Collins KH, Paul HA, Reimer RA, Seerattan R-A, Hart DA, Herzog W (2015) Relationship between inflammation, the gut microbiota, and metabolic osteoarthritis development: studies in a rat model. Osteoarthr Cartil 23(11):1989–1998
37. Collins KH, Hart DA, Reimer RA, Seerattan RA, Herzog W (2016) Response to diet-induced obesity produces time-dependent induction and progression of metabolic osteoarthritis in rat knees. J Orthop Res 34(6):1010–1018
38. Evans JD (1996) Straightforward statistics for the behavioral sciences [Internet]. Brooks/Cole Pub. Co., Pacific Grove; 1996 [cited 2019 Feb 7], 600 pp. https://www.worldcat.org/title/straightforward-statistics-for-the-behavioral-sciences/oclc/32465263
39. Collins KH, Hart DA, Seerattan RA, Reimer RA, Herzog W (2017) High-fat/high-sucrose diet-induced obesity results in joint-specific development of osteoarthritis-like degeneration in a rat model. Submitt to Bone Jt Res; Manuscript ID: BJR-2017-0201
40. Collins KH, Herzog W, Reimer RA, Reno CR, Heard BJ, Hart DA (2017) Diet-induced obesity leads to pro-inflammatory alterations to the vitreous humour of the eye in a rat model. Inflamm Res
41. Niederkorn JY (2012) Ocular immune privilege and ocular melanoma: parallel universes or immunological plagiarism? Front Immunol [Internet]. 2012 Jun 13 [cited 2017 Jun 15] 3:148. http://journal.frontiersin.org/article/10.3389/fimmu.2012.00148/abstract
42. Collins KH, Hart D, Smith I, Issler A, Reimer R, Seerattan R et al (2017) Acute and chronic changes in rat soleus muscle after high-fat high-sucrose diet. Physiol Rep 5(10):e13270
43. Collins KH, Hart DA, Reimer RA, Seerattan RA, Banker CW, Sibole SC et al (2016) High-fat high-sucrose diet leads to dynamic structural and inflammatory alterations in the rat vastus lateralis muscle. J Orthop Res [Internet]. 2016 Mar 17 [cited 2016 Mar 21] 34(12):2069–2078. http://www.ncbi.nlm.nih.gov/pubmed/26990324
44. Collins KH, Paul HA, Reimer RA, Seerattan RA, Hart DA, Herzog W (2015) Relationship between inflammation, the gut microbiota, and metabolic osteoarthritis development: Studies in a rat model. Osteoarthr Cartil 23(11)
45. Collins KH, Paul HA, Hart DA, Reimer RA, Smith IC, Rios JL et al (2016) A high-fat high-sucrose diet rapidly alters muscle integrity, inflammation and gut microbiota in male rats. Sci Rep 6
46. Collins KH, Paul HA, Hart DA, Reimer RA, Smith IC, Rios JL et al (2016) A high-fat high-sucrose diet rapidly alters muscle integrity, inflammation and gut microbiota in male rats. Sci Rep [Internet]. 2016 Nov 17 [cited 2016 Nov 17] 6:37278. http://dx.doi.org/10.1038/srep37278
47. Ciapaite J, van den Berg SA, Houten SM, Nicolay K, van Dijk KW, Jeneson JA (2015) Fiber-type-specific sensitivities and phenotypic adaptations to dietary fat overload differentially impact fast-versus slow-twitch muscle contractile function in C57BL/6J mice. J Nutr Biochem [Internet]. 2015 Feb [cited 2015 May 11] 26(2):155–164. http://www.sciencedirect.com/science/article/pii/S0955286314002174
48. Kraegen EW, James DE, Storlien LH, Burleigh KM, Chisholm DJ (1986) In vivo insulin resistance in individual peripheral tissues of the high fat fed rat: assessment by euglycaemic clamp plus deoxyglucose administration. Diabetologia [Internet]. 1986 Mar [cited 2015 Aug 3] 29(3):192–198. http://www.ncbi.nlm.nih.gov/pubmed/3516775
49. Paul HA, Collins KH, Bomhof MR, Vogel HJ, Reimer RA (2018) Potential impact of metabolic and gut microbial response to pregnancy and lactation in lean and diet-induced obese rats on offspring obesity risk. Mol Nutr Food Res
50. Nicolucci AC, Hume MP, Martínez I, Mayengbam S, Walter J, Reimer RA (2017) Prebiotics reduce body fat and alter intestinal microbiota in children who are overweight or with obesity. Gastroenterology [Internet]. 2017 Sep [cited 2017 Oct 22] 153(3):711–722. http://www.ncbi.nlm.nih.gov/pubmed/28596023
51. Parnell JA, Reimer RA (2012) Prebiotic fibres dose-dependently increase satiety hormones and alter Bacteroidetes and Firmicutes in lean and obese JCR:LA-cp rats. Br J Nutr [Internet]. 2012 Feb [cited 2014 Oct 6] 107(4):601–613. http://www.pubmedcentral.nih.gov/articlerender.fcgi?artid=3827017&tool=pmcentrez&rendertype=abstract

52. Bourassa MW, Alim I, Bultman SJ, Ratan RR (2016) Butyrate, neuroepigenetics and the gut microbiome: can a high fiber diet improve brain health? Neurosci Lett [Internet]. 2016 Jun 20 [cited 2019 Jan 16] 625:56–63. https://www.sciencedirect.com/science/article/pii/S0304394016300775
53. Rios JL, Bomhof MR, Reimer RA, Hart DA, Collins KH, Herzog W (2019) Protective effect of prebiotic and exercise intervention on knee health in a rat model of diet-induced obesity. Sci Rep 9:3893. https://doi.org/1038/s41598-019-40601-x
54. Evans CC, LePard KJ, Kwak JW, Stancukas MC, Laskowski S, Dougherty J et al (2014) Exercise prevents weight gain and alters the gut microbiota in a mouse model of high fat diet-induced obesity. Federici M, editor. PLoS One [Internet]. 2014 Mar 26 [cited 2019 Jan 16] 9(3):e92193. http://www.ncbi.nlm.nih.gov/pubmed/24670791
55. Varlamov O, Bethea CL, Roberts CT, Jr (2014) Sex-specific differences in lipid and glucose metabolism. Front Endocrinol (Lausanne) [Internet]. 2014 [cited 2019 Jan 16] 5:241. http://www.ncbi.nlm.nih.gov/pubmed/25646091
56. Collins KH, MacDonald GZ, Hart DA, Seerattan RA, Rios JL, Reimer RA et al (2019) Adult versus weanling exposure to a high fat high sucrose diet leads to differences in development of joint damage: potential impact of timing on establishment of metabolic and functional set points. J Sport Heal Sci (in press)

Chapter 8
Immune Modulation and Macrophage Polarization in the Pathogenesis of Pancreatic Dysfunction and Obesity

Nuray Yazihan and Sevginur Akdas

Abstract The prevalence of obesity and metabolic syndrome continue to increase by years and now obesity is a growing public health problem. World Health Organization (WHO) reports that more than 1.9 billion adults were overweight in 2016, 39% of them obese and 8.5–12.2% of adults aged 18 years and older had diabetes. Nutritional imbalance (over- or under-nutrition), lifestyles or environmental factors can affect the hemostasis of the body. Hormonal, inflammatory, oxidative and nutritional status determines the activation of cellular and systemic pathways in the body. Chronic disturbances of the nutritional status of the body especially obesity increases the prevalence of chronic non-communicable diseases (NCD) (WHO Diabetes [1]). Body weight and metabolism are determined by a complex orchestration of the function of several cells, organs and tissues. Underlying mechanisms of obesity and insulin resistance are still unknown. Immune cells within the metabolism related organs also likely contribute to systemic control of glucose, lipid. Increased production of local and systemic adipokines and cytokines, polarization of macrophages, T helper subtype changes could contribute to pathologies linking obesity to diabetes, both by decreasing insulin sensitivity, by compromising β-cell function and disturbing adipose tissue metabolism and distribution. Recent studies show that oxidative stress, systemic chronic inflammation or dysregulation of immune system contribute to development of NCD. Metabolic syndrome and diabetes have similar pathophysiological mechanisms with other NCD. For this reason, relevant interventions to modulate oxidative and inflammatory status of the body will make meaningful improvements in the mortality and morbidity associated with NCD. Chronic nutrition imbalance and obesity promote low-grade inflammation. "Metainflammation" is a term used for chronic inflammation of organs consists of the gastrointestinal system (including

N. Yazihan · S. Akdas
Interdisciplinary Food, Metabolism and Clinical Nutrition Department, Institute of Health Sciences, Ankara University, Ankara, Turkey

N. Yazihan (✉)
Internal Medicine, Pathophysiology Department, Faculty of Medicine, Ankara University, Morfoloji Building, Sihhiye, Ankara, Turkey
e-mail: nurayyazihan@yahoo.com

P. S. Tappia et al. (eds.), *Pathophysiology of Obesity-Induced Health Complications*, Advances in Biochemistry in Health and Disease 19,
https://doi.org/10.1007/978-3-030-35358-2_8

liver), muscle and adipose tissue. Macrophages are guardians of the tissues. They regulate immune responses, homeostasis in the different physiological and pathological conditions. Clarification of the underlying mechanisms of chronic inflammation will provide an explanation for mechanisms of obesity and the associated complications and will supply information for new therapeutic approaches. This noteworthy perception is allowing us to more assuredly define the role that macrophages involve in health and in obesity and how inflammatory mediators behave as signaling molecules in this pathway. Additionally, on a molecular level, we are beginning to figure out how such factors as nutritional, metabolic status, hormonal changes, lifestyle, genetic and epigenetic factors interrelate and terminate in different phenotypes and characteristics, and which interventions may modulate immune functions. Therefore, this chapter will review the metabolic regulation of plasticity of macrophages and role of inflammation and macrophage polarization in obesity, diabetes and pathogenesis.

Keywords Macrophage polarization · Immunomodulation · Metainflammation · Obesity · Diabetes mellitus

Macrophage Phenotypes

Macrophages are important components of an organisms microenvironment and it's defense against pathogens. They have a wide range of functions in the body's homeostasis, defense and regeneration.

Macrophages, whose main functions are phagocytosis, also play a central role in the natural and acquired immune responses. Mononuclear phagocytes originate from the bone marrow, circulate in the bloodstream and become active in some tissues and mature. In addition to monocyte derived macrophages there are specialized resident macrophages in the tissues. Red pulp and marginal zone macrophages in the spleen, microglia in the brain, osteoclasts and peritoneal macrophages are some of resident macrophages of tissues. Tissue-specific macrophages are replaced from bone marrow-derived monocytes when needed in different organs. Macrophages are the main effector cells of the late period of the natural immune response. They control and limit the inflammation, sustain tissue homeostasis and promote subsequent repair.

Macrophages exhibit functional plasticity during the many other functions of tissue homeostasis, such as continuity of the tissue, protection from pathogens, destruction of harmful substances and waste products and repair of tissue. Plasticity of macrophage differentiation is necessary to accomplish these divergent activities. Heterogeneity of macrophages is required for natural immunity, as well as in both general or specific pathophysiological responses.

Beside its tissue-specific types, it is shown that macrophages have phenotypic heterogeneity. Adaptation and metabolic programming of macrophage tissue/body conditions determines and controls plasticity and differentiation of macrophages into these subtypes.

Like T helper cells, macrophages also polarize to distinct phenotypes expressing unique cell surface molecules and secreting discrete sets of cytokines and chemokines. Macrophages classified as M1 (classical) and M2 (alternative) subtypes which are mainly induced by T helper-1 (Th-1) and Th-2 cells. The classical M1 phenotype supports proinflammatory Th1 responses driven by cytokines such as TNF, IL-1, IL-6, IL-12 and IL-23, while the alternate M2 phenotype is generally supportive of anti-inflammatory processes driven by IL-10. Innate immunity or TLR activation via LPS and IFN-γ, involve in M1 polarization. M2 polarization is usually induced by IL-4 or IL-13. Higher levels of IRF5 expression was shown in M1 phenotype and important in many autoimmune diseases [2, 3].

The classification of M2 subtype further classified into subsets; M2a, M2b, M2c and M2d based on the type of stimulation and the subsequent expression of surface molecules and cytokines. The macrophages in atherosclerotic plaques can be termed as Mox, Mheme, MHb and M4. If any injury or hemorrhage occurs around the plaque, macrophage subtypes related to hemoglobin and heme become activated. These macrophages start the repairing mechanism, have significant anti-inflammatory capacity and decrease foam cell formation. Macrophages found in the hemorrhagic zones of human atherosclerotic lesions differentiate to Mhb by haemoglobin–haptoglobin complexes and to Mheme by haemoxygenase 1 (HO1) and the scavenger receptor CD163. Mheme have increased levels of cholesterol efflux transporters to reduce cholesterol storage in macrophages which is important for anti-atherogenic activity. M1 and M4 macrophages promote, while M2 and Mheme macrophages counteract foam cell formation, thus having opposite effect on atherosclerosis progression [2, 4–6].

Cholesterol, LPS and proinflammatory cytokines TNF-α, and IL-1 induce macrophages to M1 phenotype via TLR4 and NFkb-related inflammatory pathways. When macrophages are exposed to oxidized phospholipids, they differentiate to Mox phenotype. Mox macrophages are regulated by oxidized phospholipids. Nrf2, HO-1, Txnrd1, and Srxn1 genes regulate their antioxidant properties. Chemokine CXCL4 or platelet factor 4 induce macrophages to M4 phenotype. M4 macrophages differentiate in response to the chemo-attractant CXCL4, thus showing pro-inflammatory and pro-atherogenic effects. But the M4 phenotype is different from others and has the ability to change its phenotype to M1, M2. M4 phenotype is very common in atherosclerotic areas as seen in Fig. 8.1 [2, 4, 5, 7].

Nowadays it is accepted that macrophages have important roles both in induction, progression and control of inflammation and tissue regeneration. Both M1 and M2 phenotypes take part in these processes. Biomarkers of phenotypes, and specific biomolecules can be used for diagnosis, as well as for follow up of many diseases. Time-dependent control of macrophage phenotypes can be the target of the treatment and new therapeutic interventions can be developed especially in atherosclerosis, oxidative, inflammatory and hemorrhagic damage [8–10]. Sodhi et al. showed that

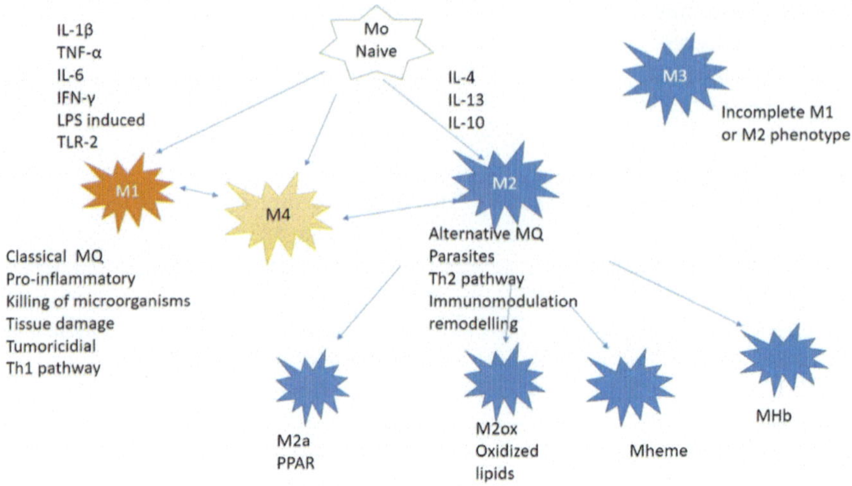

Fig. 8.1 Subtypes and functions of macrophages

induction of HO-1 is important for attenuation of oxidative damage and lipid accumulation in hepatocytes through Sirtuin1 (SIRT1) pathway in the fructose-induced obesity model [11]. Induction of HO-1 by hemin protects high-fat diet mice from obesity-induced inflammation. Hemin treatment induces M2 phenotype macrophage marker genes while it also reduces M1 marker genes. Increased IL-4, Mrc1 and Clec 10a gene expression whereas decreased JNK, NF-kb, CD274 and TNF-α expressions were also observed. It is reported that induction of HO-1 with hemin may be a useful treatment for adipocyte inflammation. Induction of the PPARγ and STAT6 pathways involved in macrophage modulation by hemin [12].

Phenotypic changes and polarization are under the control of many factors. Monocyte-driven and resident macrophages show different responses to external and internal stimuli. Biomolecules secreted from AT have the capacity to change phenotypes of the macrophages in AT. Osteopontin is an adipokine with proinflammatory action. Schuch et al. investigated the effects of osteopontin on bone marrow-derived and AT macrophages in obesity-induced osteopontin deficient (Spp1(−/−)) and wild-type C57BL/6 (WT) mice [13]. They found that Spp1 (−/−) with high-fat diet mice express more M1 and less M2 markers compared to wild type mice. The cytokine secretion levels of Spp1(−/−) with high-fat diet mice were found similar to wild type. LPS-induced cytokine responses and phagocytic activities of human monocyte-derived macrophages and altered M1/M2-related marker expression were reported in the presence of osteopontin in the medium.

Also, exposure to Th2 cytokines IL-4, IL-13, GM-CSF increases proliferation of AT macrophages whereas Th1 cytokines inhibit. AT macrophages of obese mice have increased sensitivity to IL-4 by increased phosphorylation of STAT6. The researchers found that IL-6 is necessary for IL-4 and IL-13 responses [14].

Control of Macrophage Metabolism

Metabolic regulation and intracellular energy sources are important for the determination of phenotypic stage of macrophages and functions of immune cells. Previously it has been shown that M1 macrophages have a more prominent glycolytic rate whereas M2 macrophages have fatty acid metabolism and oxidative phosphorylation. Adipose tissue O2 level is important for the regulation of cellular metabolism. HIF-1 is expressed as a response to hypoxia and metabolic stress. Boutens et al. showed that glycolytic pathways are activated during metabolic stress and hypoxia by induction of HIF-1, deletion of HIF-1 did not affect the inflammatory status, cytokine production and insulin resistance after 8th week of high feed diet, it is important for controlling of other metabolic properties of macrophages during the early stages of obesity [15]. They found adipose tissue macrophages of obese mice and coculture of bone marrow derived macrophages with adipose tissues of obese mice have more prominent glycolysis and OXPHOS pathways compared to lean derived adipose tissue macrophages and exposed to lean adipose tissue.

Similar findings were reported by Fujisaka et al. They found that M1 type macrophages of adipose tissue function as HIF-1 dependent and it is independent in obese mice. IL-1β, IL-6, Nos2 expressions are under control of hypoxia and HIF-1α in cultured macrophage cells, but IL-4 and depending on the condition TNF-α and some other cytokines are not affected by HIF-1 α deficiency [16].

HIF-1α overexpressed Lsl-HIF1 dPA mice are shown to have higher expression of IL-6, IL-1β, TNF-α, and Cd11cexpression while lower expression of Arg1, CD206 and Chi313 in peritoneal macrophages. Macrophages of HIF-1α overexpressed mice show M1 polarization, which can be seen in Fig. 8.2 [17].

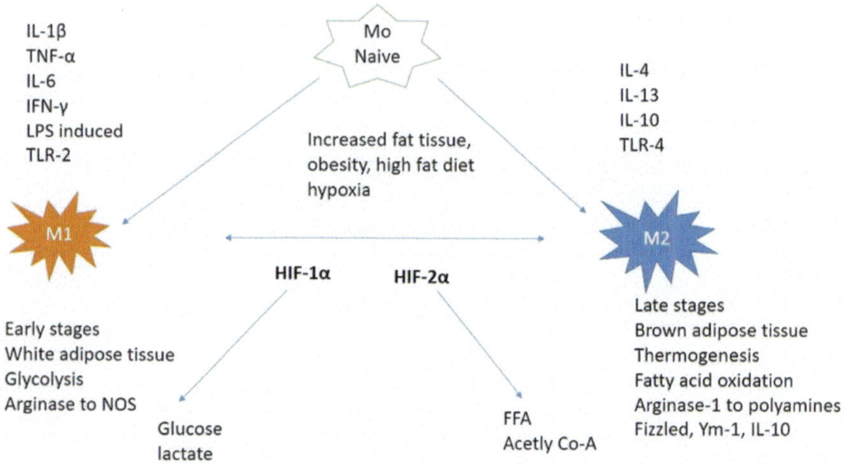

Fig. 8.2 Regulation of macrophage polarization. M: macrophage, IL: interleukin, TNF: tumor necrotizing factor, IFN: interferon, LPS: lipopolysaccharide, TLR: toll-like receptor, NOS: nitric acid oxidase synthase, HIF: hypoxia-inducible factor, FFA: free fatty acid

Infection or inflammation on the microenvironment decrease tissue oxygen and nutrients. HIFs are important for the adaptation of cells to deprivation. HIFs activate a stream of genes that maintain the survival of the cells. Oxygen sensing proteins and specific inhibitors regulate the HIFs' activities. Prolyl hydroxylase domain enzymes (PHD1-3) function as oxygen sensors are activators of HIF-1α and asparaginyl hydroxylase enzyme factor is an inhibitor of HIF-1α. These two inducers and inhibitor require Fe^{2+}, ascorbic acid and the tricarboxylic acid (TCA) cycle intermediary α-ketoglutarate. During hypoxia, HIF-1α activates glycolysis that produces lactate and sufficient supply of energy and is what is needed to the function of innate immune cells [18]. Takeda et al. showed that HIF-1α and HIF-2 are important for M1 and M2 phenotyping [19]. Both HIF-1α and HIF-2 is important for T cell functions and metabolism. IL-4 is a strong activator of HIF-2 in immune cells [20]. Choe et al. reported that HIF-2 is also important for direct regulation of inflammatory responses in adipose tissue in obese patients [21]. HIF-2 suppresses proinflammatory cytokines and NOS activity that promotes M2 polarization through induction of arginase 1, which can be seen in Fig. 8.2.

Glutamine and α-ketoglutarate are other effective inducers of M1/M2 phenotypes. Glutaminolysis, increased fatty acid oxidation produce α-ketoglutarate that induces alternative activation of macrophages via Jmjd3-dependent epigenetic reprogramming of M2 genes. α-ketoglutarate/succinate ratio controls the M2/M1 ratio. Succinate synthesized from glutamine-dependent anerplerosic reactions of TCA and γ-aminobutyric acid shunt. Glutamine promotes polarization of M1 macrophage. In many pathological conditions such as diabetic and cancer, patients have impaired glutamine metabolism and M2 polarization [22, 23].

Pancreas and Immune System

Innate lymphoid cells, resident macrophages and immune cells from bone marrow with dendritic cells are especially found in all tissues. Resident immune cells are important for self-defence and homeostasis of the tissues. Similarly, host or resident immune cells originated from the embryonic liver and then from bone marrow. The functional and steady state of pancreatic macrophages in different pathological conditions define the fates of the diseases. Calderon et al. evaluated the embryonic source of the resident macrophages in pancreatic islets and in the inter-acinarstroma [24]. They found IL-1β and TNF-α expressed macrophages originated from hematopoiesis and myeloid source. Steady-state macrophages of pancreatic islets are shown as M1 type. They found that over 98% of CD45 islet resident macrophages express CD11c, MHC-II, F4/80, CD11b, CD64, lysozyme, CX3CR1, and CD68 without mannose receptor 1 (Mrc1; CD206) and macrophage galactose-type C-type lectin (Mgl1/2; CD301).

The functional difference of pancreas as exocrine (acinar glands) and endocrine (islet) glands is due to a distinct contact with bloodstream and digestive system and are the main determinants of the self-defense system of the pancreatic tissue.

Different results related with the immune system of the pancreas are reported from different research groups. Ferris et al. (2017) showed that resident immune cells of the pancreas are on the proinflammatory state with M1 macrophage phenotype predominance. These cells function as a barrier for the pancreas. The interaction between resident macrophages and antigen presenting cells is one of the main starting points of autoimmunity in the gland. CD103(+) dendritic cells together with CD8(+) cells are regulated by Batf3 transcription factor [25].

Now it is known that resident macrophages have important roles in the pathogenesis of nutritional pancreatic dysfunction. A short term, high fat diet induces macrophage infiltration and islet cell proliferation without inducing insulin resistance. Islet cells are the main target of autoimmune reactions in the diabetes. Immune cells around the islet cells can regulate insulin secretion [26, 27]. Torres-Castro et al. showed that augmented inflammation via M1 pathway in human monocyte cell culture model that higher levels of CD11c and lower CD206 expression in hyperglycemic conditions which is shown previously in obese patients [28].

Inhibition of proinflammatory cytokine production from pancreatic islets prevents diabetic nephropathy and other complications. It is shown that cannabinoid-1 receptor (CB1R) activation takes part in islet inflammation and deletion of CB1R prevents islet cell damage, decrease IL-β TNF-α, CCL2, and interferon regulatory factor (IRF5) induced proinflammatory signaling in macrophages invaded pancreatic islets. Knockdown of IRF5 prevents genetically diabetic rats from pancreatic islet loss [29].

Moganti et al. showed that hyperglycemia induces a mixture of M1 and M2 related cytokine responses. Following an acute TNF-α secretion, IL-1β and IL-1Ra activation continue. M2 macrophages express a lower level of CCL18 but still have significant markers of M2 [30].

There are inconsistencies in the status of macrophages in diabetic pathogenesis. Cucak et al. reported that most of the immune cells of the pancreatic islet are islet-derived macrophage subsets expressed moderate MHC-II, high galectin-3, and low CD80/CD86 levels rather than antigen presenting cells [31]. After the early stages of diabetes development M1 polarization then changes to M2 phenotype and profibrotic characters added to proinflammation.

Targeting of inflammation started to get attention in both type I and II diabetes. The alterations in the metabolism of arachidonic acid and lipoxygenase (LO) pathways are shown previously in the pathogenesis of diabetic complications. 12-LO is most common form of lipoxygenases in pancreas and transform polyunsaturated fatty acids into different types of eicosanoids, such as leukotrienes. Deletion of 12-LO shown to intercept occurrence of hyperglycemia in high fat diet and streptozotocin induced diabetes models. Anti-inflammatory applications against specific proinflammatory molecules can be successful in the management of type 2 diabetes [32, 33].

One of the most common anti-diabetic agents, metformin, has been shown to be effective for control of chronic inflammatory status of obesity via modulation of macrophage polarization to M2 phenotype by AMPK pathway [34].

Obesity and Macrophages

Obesity is the first stage of metabolic syndrome and type II diabetes formation. With increasing infiltration of proinflammatory immune cells into adipose tissue, low-grade chronic inflammation starts. Crosstalk between innate and alternative immunity determines the fate of the insulin response and resistance balance. The control and status of the inflammation differ in the lean and obese state of adipose tissue (AT). The balance of lipogenesis and lipolysis is under the control of the hormonal status of the body, which is also linked to adipokines of the AT. Eosinophils, mast cells, innate and adaptive type lymphoid cells, and mainly adipose tissue macrophages (ATMs) are the principal and most common immune cell population in the AT, especially in obesity. With the activation of M2 phenotype; Th2 related cytokines IL-4, IL-5, IL-13, IL-17 and IL-33 become more prominent [35, 36].

Adipose tissue (AT) can be classified as white, beige and brown according to their physical appearance and function. AT is an active organ of the body responses to metabolic changes of the body and environmental stress conditions. Excess energy deposits in AT and thermogenesis starts for adaptation to cold. It behaves as an endocrine organ with a capacity to regulate metabolic homeostasis and secretion of bioactive molecules. Sensitivity to hormones and body location, the phenotype of AT is a determining factor of regulation of adipokine secretion, adipocyte formation and meta-inflammation formation [37]. White adipose tissue is found subcutaneous or viscerally with a large uniocular shape, it functions as lipid deposition. WAT responds to hormonal stimuli as well as to insulin. Brown adipose tissue [38] is multi-ocular and functions as a heat producer. The balance between white and brown AT is important for metabolic consumption of fat tissue.

It is reported that cold stimulation starts the crosstalk between brain and fat tissue. Brown adipose tissue [38] is the main actor for the cold response and control of heat production, adaptation to thermogenesis and energy production in the mitochondria. Glycolytic pathways and lipolysis start with uncoupling protein-1 (UCP-1) and catecholamines secreted because of meta-inflammation [39, 40]. Fischer et al. reported that there was no catecholamine secretion, from adipose tissue [41]. Rajasekaran et al. showed that monocyte chemoattractant protein-1 (MCP-1) is important for M2 polarization and switching of WAT cells to BAT and promoting energy expenditure [42].

Neurotransmitters are important for regulation of lipogenesis and lipolysis balance. Within the calcitonin gene family, bioactive peptide adrenomedullin 2 expression level is found to be inversely correlated with weight gain. Adrenomedullin 2 promotes the genesis of white adipose tissue and also increases thermogenesis in brown adipose tissue via activation of M2 macrophages. Adrenergic stimulation induces prolactin expression on THP-1 monocytes which has been shown to induce lipogenesis and M1 polarization [43, 44].

Controversial studies are presented on the subject of macrophage polarization and obesity. Hu et al. reported that inhibition of arginase one of the M2 phenotype marker controls macrophage activation and infiltration to adipose tissue in high-fat diet (HFD) induced obesity model [45]. N(ω)-hydroxy-nor-l-arginine inhibitor nor-NOHA decreased proinflammatory cytokines chemoattractant protein-1 (MCP-1), TNF-α, and IL-6 and increased the anti-inflammatory cytokine IL-10 in co-cultured 3T3-L1 adipocytes and RAW 264.7 [46].

Increasing body mass and obesity results in cardiovascular complications in patients. Sex, age, body mass index, smoking and serum lipid levels are risk factors of cardiovascular diseases. Inflammation of AT was found correlated only with subcutaneous inflammation that was determined by the differences of markers of phagocytic pro-inflammatory ($CD14^+$ $CD16^+$ $CD36^{high}$), anti-inflammatory ($CD14^+$ $CD16^-$ $CD163^+$) and transitional subsets of macrophages [47].

Nitric oxide (NO) is one of the main pathways in polarization of macrophages, decreasing insulin resistance and related inflammation. Arginase takes part both in M1 and M2 phenotypes. Inducible NO synthase (iNOS) is more prominent in M1, endothelial NO synthase (eNOS) in M2 type. eNOS protects hepatic structures against high fat-induced liver damage. Nitrate increases fatty acid β-oxidation in adipocytes by increasing oxygen consumption. Series of transcriptional and proteomic analyses showed that dietary nitrate can modulate mitochondrial functions, increases the expression of thermogenic genes in brown adipose tissue. Nitrate activates conversion of WAT to BAT by inducing expression of BAT-specific gene expression in WAT. In a type 2 diabetic rat model it is shown that long term (3 months) sodium nitrite (50 mg/L) in drinking water administration increases adipose tissue NO and GMP levels. Sodium nitrite is protective against type 2 diabetes mellitus by converting WAT cells to beige/brite cells which have antiobesity and antidiabetic capacity [48–51].

Diabetes and Inflammation

Diabetes mellitus is defined as "a chronic, metabolic disease characterized by elevated levels of blood glucose (or blood sugar), which leads over time to serious damage to the heart, blood vessels, eyes, kidneys, and nerves" [1].

Long term dysregulation of metabolism disturbs immune cell functions and immune cell distribution. $CD8^+$ effector T cells are the main actors of adipose tissue inflammation. Subsets of $CD4^+$ helper cells, T regulatory (Treg) cells are usually reduced in obese adipose tissue that can secrete inhibitory cytokines. The proliferation of adipose cells increases the stress and oxygen need of the adipose tissue. Immune system dysfunction occurs in pancreatic islets, spleen, liver, adipose tissue and all over the body. Lymphocytes, neutrophils, monocyte/macrophages, resident immune cells and mast cells are actors of the chronic low-level inflammation in all effected tissues.

The innate immune response is the first line of self-defense against internal and external stimuli. Increased diabetic stress is due to metabolic disturbances such as hyperglycemia, hyperlipidemia, advanced end-stage glycosylation products (AGEs) and oxidative stress activates macrophages. M1-type innate immune response starts in the early stages. Increased production of cytokines and chemokines, complements, oxLDL, AGEs and end products of cellular damage promote further inflammation. M1/M2 immunomodulation and mast cell involvement regulate the progress of the problem. Especially in renal damage, immune system modulation is the chief of the complex orchestra. In diabetic kidney, AGEs, ROS and oxLDL in the circulation activate cytokines, chemokines and complement systems. Invading of neutrophils and mast cells in the glomerulus, tubular and mesangial areas increase damage leading to fibrosis and loss of renal function.

Mast cells are accepted as one of the MASTer of the inflammation in many cases. Mast cells are fascinating cells, they take the names from their rich content "Mastzellen", meaning "well-fed cells" in 1877. We know that mast cells are heterogeneous cells with divergent functions. They originate from bone marrow $CD34^{+}$, $CD117^{+}$ stem cells and complete maturation in tissues. Due to having a variety of components, they have diversified effects on various physiological and pathological conditions. Similar to resident macrophages, they act as sentinels of the microvilli and modulates the inflammation, damage and repair processes [52, 53]. The mast cell stabilizers used in allergic reactions are found to be effective in high-fat diet induced obesity and diabetes in mice [54]. Mast cell regulation can be used as a supplementary treatment for diabetes and obesity. Inhibition of mast cell-related cytokines is effective for prevention of body weight gain, correction of glucose intolerance, insulin resistance [55]. Atopic lesions, allergic reactions are related with mast cell and Th2 cell activation. These cells activate alternative responses or M2 polarization of macrophages [56].

Complement system mediators especially C3a and C5a, behave as a proinflammatory and chemotactic factor for immune cells. Recent studies focused on the role of complement activation in the pathogenesis of diabetes and related complications. It is well known that C3a activation is related with disturbed glucose and lipid metabolism and leads to diabetic nephropathy and vascular complications [57–60].

Similar to C3a, C5a is critical for the development of insulin resistance. Increased expression of C5a is shown in adipose tissue of obese rats compared to lean ones. Deficiency of C5a and C5aR protect diabetic complications in high-fat diet fed mice by decreasing M1 polarization [61]. Acylation stimulating protein (ASP/C3adesArg) treatment increases the colocalization of C5aR and C5L2 as a homo/heterodimer structure that increases their contribution to metabolic pathways. Deletion of C5aR decreases cytokine production, triglyceride synthesis and IRS1 expressions in adipose tissue [62].

The results of the studies on the effect of C5a to macrophage polarization could show difference in different disease models. C5a is important for salt-induced hypertensive vascular pathologies. C5a plays a critical role in DOCA-salt-induced vascular injury by stimulating macrophage polarization toward a proinflammatory M1 phenotype in perivascular adipose tissue [63]. But in a tumor model, C5a acts as M2-like phenotype in tumor-associated stroma and is accepted as a potential novel marker for cancer prognosis [64].

Chronic low-level inflammation influences organ systems, and macro and microvascular complications occur in the whole body. Nephropathy, different types of neuropathies, retinopathy, peripheral and central nervous system disorders, cardiomyopathy, and immunological disorders initiated and progress during diabetes. Diabetic nephropathy is one of the most common and early complications of diabetes. Approximately 10–40% or up to 75% of patients with type 1 diabetes and almost 30–40% of type 2 diabetic patients have renal complications and diabetic nephropathy is a most common cause of the end-stage kidney diseases worldwide. Diabetic nephropathy is a microvascular disorder with glomerular and tubular basal membrane thickening, extracellular matrix disturbances, glomerular sclerosis, tubule-interstitial fibrosis and inflammation. Glomerular and tubular dysfunctions occur in diabetic patients. The degree of glomerular damage correlated with blood glucose perturbances, the level of metabolic changes. Glomerular hypertrophy, glomerulosclerosis, and tubulointerstitial inflammation and fibrosis are the most common findings in diabetic nephropathy. New biomarkers for diagnosis and prognosis, and new therapies or therapeutic agents are needed to decrease high risk of diabetic renal damage. Protein kinase A, B, C, tyrosine kinase, JAK/STAT; Nfkb are main intracellular pathways that regulate cellular disturbances and responses to hyperglycemia [65, 66]. Podocyte damage was found to be correlated with macrophage infiltration, M1 polarization, TNF-α secretion and p38MPAK activation in podocytes of streptozocin induced diabetic rats [22].

Given that diabetes is an important component of metabolic syndrome, macrophage polarization and immune response may also have a role in metabolic syndrome. In our unpublished results, when the inflammatory response in different tissues was examined in adult rats with sucrose-induced metabolic syndrome, we found that IL-6, TNF-α and IL-10 cytokines levels increased significantly in the liver, kidney and adipose tissues. In heart and pancreas tissue, IL-6 and IL-10 responses were more prominent, whereas IL-6 and TNF-α levels were significant in vascular tissue. Zinc and copper levels in the tissues were also evaluated and they were found to have different correlations with inflammatory markers in each of the liver, kidney, pancreas, heart and adipose tissues. The correlations between trace element levels and inflammatory markers examined in the study have been shown to be effective in the metabolic-inflammatory process [67]. Furthermore, we examined age-dependent changes of macrophage polarization in the age-related spontaneous metabolic syndrome. It is known that the physiopathology of the age-related development of metabolic syndrome is different from that of excessive energy intake-related metabolic syndrome. Inflammatory response with TNF-α, IL-2, IL-4, IL-6 and IL-10 levels and macrophage polarization were investigated in liver, pancreas, kidney, spleen tissues of healthy young, old MetS (−) and old MetS (+) rats in our study. We obtained tissue dependent different changes of cytokines in each group. According to our unpublished data, TNF-α, IL-6 and IL-10 levels significantly increased with aging in liver tissue while IL-4 levels decreased, especially in liver and kidney tissues of aged MetS (+) group. In the MetS (+) group, we found a correlated increased level of IL-10 with proinflammatory cytokines in liver and kidney. IL-2 levels were increased in pancreas tissue with aging compared to MetS (−) and MetS (+). Similar to sucrose induced metabolic syndrome, zinc and copper are important trace elements in control of glucose hemostasis and cytokine responses [68]. Our results showed that depression of IL-4, increased TNF-α, IL-6, IL-2 levels triggers M1 type polarization and inflammation in tissues and unbalanced distribution of zinc and copper in the body could be one of the underlying mechanisms of diabetes in elderly people.

Non-alcoholic Fatty Liver and Macrophages

Liver macrophages, mainly kupffer cells constitute 80–90% of the tissue macrophages present in the body and expected most of the macrophages located in the liver. These cells constitute approximately 20–25% of all liver cells. These cells settle along with the sinusoidal space and play a role in the regulation of physiological homeostasis [69, 70]. Liver macrophages also play a protective role against liver damage and promote regeneration and fibrosis in cholestatic liver damage. The liver is exposed to toxic substances, microbial products and structures, cell and metabolism products due to anatomical structure and physiological functions. Therefore, there must be a strong immune system organization. Specialized macrophages of the liver play an important role not only in liver diseases but also in systemic immune response.

Abnormal accumulation of lipid is the first and reversible stage of nonalcoholic fatty liver. Alterations on transport, storage and metabolism of lipids within liver lead the pathogenesis of liver disease. Fatty acid binding protein 5 (FABP5) was shown that was highly expressed in macrophages and had a role in the hepatic inflammatory response to LPS induction in FABP5 knockout mice. FABP5 deficiency enhanced the hepatic F4/80(+) macrophage levels and prevented the liver damage. Also, FABP5 knockout mice had higher mRNA levels of Il-10, arginase, YM-1 and Fizz-1 which are known as anti-inflammatory cytokines and resembles M2 type macrophages [71].

In progressive hepatosteatosis process, we could see the effects of adipose tissue macrophages, particularly in the visceral adipocytes, on disease development and insulin resistance besides the effects of polarization of liver macrophages, especially pro-inflammatory phenotype. It is known that endotoxins, bacteria, intestinal permeability and factors released from damaged heptocytes, microbiota profiles and components such as free fatty acids, cholesterol and certain metabolites have an effect on macrophage polarization and activation [72].

There were identified different immunologic pathways on disease pathogenesis and macrophage differentiation. It was mentioned the potential central switch of macrophage polarization was controlled by IL-4 receptor alpha (IL-4Rα). The role of IL-4Rα was investigated with general IL-4Rα$^{-/-}$ and macrophage-specific IL-4Rα$^{-/-}$ mice. Monocyte infiltration and hepatic fibrosis were suppressed in each study group. Macrophage-specific IL-4Ra$^{-/-}$ was silenced and delayed recovery of fibrosis. This mechanism is thought to be related to the loss of M2 macrophages, the source of MMP-12 [73]. The autophagy-specific substrate p62 as a marker of autophagic dysfunction was aggregated in approximately 88% of NAFLD specimens obtained with liver biopsy from NAFLD patients and caused increases in the number of CD11c-positive cells which identifies M1 [74]. On the other hand, CD44 role was evaluated with CD44$^{-/-}$ mice fed with a methionine-choline-deficient diet and samples from obese patients. It was shown that CD44 was significantly upregulated in obese NASH patients. However, in the animal model, CD44 deficiency stimulated the M2 polarization and significantly decreased the LPS induced macrophage activation and hepatocyte damage [75]. In the study of Krenkel et al. monocyte infiltration in NASH models was inhibited with an oral dual chemokine receptor CCR2/CCR5 antagonist and it efficiently ameliorated insulin resistance, inflammation in hepatocytes, and hepatic-fibrosis [76].

Endoplasmic reticulum (ER) stress of hepatocytes was mentioned as highly associated with fatty liver disease. Yang et al. lightened the underlying mechanism with inositol-requiring enzyme 1 α (IRE1α) [77]. ER stress-activated Kupffer cells secreted more pro-inflammatory cytokines to induce M1-phenotype in the fatty liver which resulted in severe ischemia and reperfusion injury and reactive oxygen species. Also, the reverse effect was possible with knock-down of IRE1α by regulating the STAT1 and STAT6 pathway of macrophage to decrease M1 and increase M2 types. Modifying the polarization of resident and recruited macrophage/Kupffer cells were found as new insights in NAFLD and the different micronutrients and antioxidants were seen as new therapeutic agents with their underlying molecular mechanisms in the development of NAFLD, with macrophage cell polarization in the liver, and damaged hepatocytes [78]. In atherosclerotic study model of Shibata et al., a significant elevation was seen in Cd163 levels as an M2 macrophage marker in atherosclerosis developed *Apoe*$^{-/-}$ mice fed with 0.4% mangosteen extracts compared to control group [38]. The main changes were lower total cholesterol and triglycerides, decreased hepatic HMG-CoA synthase and fatty acid transporter and an increase in IL-13 levels. The role of p38 mitogen activation on non-alcoholic steatohepatitis (NASH) was investigated through macrophage polarization with the macrophage-specific p38 knock out mice (p38$\alpha^{\Delta M\Phi}$). p38$\alpha^{\Delta M\Phi}$ mice displayed less severe steatohepatitis, inflammatory damage and insulin resistance than wild type mice in response to a high-fat high cholesterol diet or methionine-choline-deficient diet. The underlying mechanism was macrophage p38 induction of pro-inflammatory factors such as CXCL2, IL-1β, CXCL10 and IL-6. However, p38$\alpha^{\Delta M\Phi}$ mice showed M2 anti-inflammatory polarization in liver as demonstrated by increased CD45$^+$, F4/80$^+$, CD11b$^+$, CD206$^+$ M2 macrophages and increased arginase activity. Furthermore, the same study results showed that there was regulation of p38 in NASH patients' liver tissues [79]. Handa et al. indicated that iron may lead to steatohepatitis and fibrogenesis by increasing M1/M2 ratio and inflammatory markers as CCL2, CD14, iNOS, IL-1β, IL-6, TNF-α [80]. They also showed that iron overload decreased M2 markers; arginase-1, Mgl-1, and M2-specific transcriptional regulator, KLF4 which can be returned by iron chelators. In recent years, curcumin becomes one of the well-known anti-oxidant components of obesity and related diseases. The data showed that curcumin also affects macrophage polarization and infiltration. When mice with NASH were treated with tetrahydrocurcumin, inflammatory macrophage infiltration and polarization were found to be decreased in adipose tissue and there were markedly alleviation on steatosis by 28–37% due to the decreasing of lipogenesis, the increasing of AMP-activated protein kinase (AMPK), and the elevation of oxidated fatty acids [81].

References

1. WHO Diabetes
2. Murray PJ (2017) Macrophage polarization. Annu Rev Physiol 79:541–566
3. Krausgruber T, Blazek K, Smallie T et al (2011) IRF5 promotes inflammatory macrophage polarization and T H 1-T H 17 responses. Nat Immunol 12:231
4. Chinetti-Gbaguidi G, Colin S, Staels B (2014) Macrophage subsets in atherosclerosis. Nat Rev Cardiol 12:10
5. Vinchi F, Muckenthaler MU, Da Silva MC et al (2014) Atherogenesis and iron: from epidemiology to cellular level. Front Pharmacol 5:94
6. Martinez FO, Sica A, Mantovani A, Locati M (2008) Macrophage activation and polarization. Front Biosci 13:453–461
7. Colin S, Chinetti-Gbaguidi G, Staels B (2014) Macrophage phenotypes in atherosclerosis. Immunol Rev 262:153–166. https://doi.org/10.1111/imr.12218
8. Williams JW, Giannarelli C, Rahman A et al (2018) Macrophage biology, classification, and phenotype in cardiovascular disease: JACC macrophage in CVD series (part 1). J Am Coll Cardiol 72:2166–2180
9. Yap J, Cabrera-Fuentes HA, Irei J et al (2019) Role of macrophages in cardioprotection. Int J Mol Sci 20:2474
10. Bi Y, Chen J, Hu F et al (2019) M2 macrophages as a potential target for antiatherosclerosis treatment. Neural Plast. https://doi.org/10.1155/2019/6724903
11. Sodhi K, Puri N, Favero G et al (2015) Fructose mediated non-alcoholic fatty liver is attenuated by HO-1-SIRT1 module in murine hepatocytes and mice fed a high fructose diet. PLoS ONE 10:e0128648
12. Tu TH, Joe Y, Choi H-S, et al (2014) Induction of heme oxygenase-1 with hemin reduces obesity-induced adipose tissue inflammation via adipose macrophage phenotype switching. Mediators Inflamm
13. Schuch K, Wanko B, Ambroz K et al (2016) Osteopontin affects macrophage polarization promoting endocytic but not inflammatory properties. Obesity 24:1489–1498
14. Braune J, Weyer U, Hobusch C et al (2017) IL-6 regulates M2 polarization and local proliferation of adipose tissue macrophages in obesity. J Immunol 198:2927–2934. https://doi.org/10.4049/jimmunol.1600476
15. Boutens L, Hooiveld GJ, Dhingra S et al (2018) Unique metabolic activation of adipose tissue macrophages in obesity promotes inflammatory responses. Diabetologia 61:942–953. https://doi.org/10.1007/s00125-017-4526-6
16. Fujisaka S, Usui I, Ikutani M et al (2013) Adipose tissue hypoxia induces inflammatory M1 polarity of macrophages in an HIF-1α-dependent and HIF-1α-independent manner in obese mice. Diabetologia 56:1403–1412
17. Liu P-S, Wang H, Li X et al (2017) α-ketoglutarate orchestrates macrophage activation through metabolic and epigenetic reprogramming. Nat Immunol 18:985

18. Sadiku P, Walmsley SR (2019) Hypoxia and the regulation of myeloid cell metabolic imprinting: consequences for the inflammatory response. EMBO Rep 20
19. Takeda N, O'Dea EL, Doedens A et al (2010) Differential activation and antagonistic function of HIF-α isoforms in macrophages are essential for NO homeostasis. Genes Dev 24:491–501
20. Cho SH, Raybuck AL, Blagih J et al (2019) Hypoxia-inducible factors in $CD4^+$ T cells promote metabolism, switch cytokine secretion, and T cell help in humoral immunity. Proc Natl Acad Sci 116:8975–8984. https://doi.org/10.1073/pnas.1811702116
21. Choe SS, Shin KC, Ka S et al (2014) Macrophage HIF-2α ameliorates adipose tissue inflammation and insulin resistance in obesity. Diabetes 63:3359–3371. https://doi.org/10.2337/db13-1965
22. Guo Y, Song Z, Zhou M et al (2017) Infiltrating macrophages in diabetic nephropathy promote podocytes apoptosis via TNF-α-ROS-p38MAPK pathway. Oncotarget 8:53276
23. Ren W, Xia Y, Chen S et al (2019) Glutamine metabolism in macrophages: a novel target for obesity/type 2 diabetes. Adv Nutr 10:321–330
24. Calderon B, Carrero JA, Ferris ST et al (2015) The pancreas anatomy conditions the origin and properties of resident macrophages. J Exp Med 212:1497–1512. https://doi.org/10.1084/jem.20150496
25. Ferris ST, Carrero JA, Mohan JF et al (2014) A minor subset of Batf3-dependent antigen-presenting cells in islets of Langerhans is essential for the development of autoimmune diabetes. Immunity 41:657–669. https://doi.org/10.1016/j.immuni.2014.09.012
26. Carrero JA, Ferris ST, Unanue ER (2016) Macrophages and dendritic cells in islets of Langerhans in diabetic autoimmunity: a lesson on cell interactions in a mini-organ. Curr Opin Immunol 43:54–59. https://doi.org/10.1016/j.coi.2016.09.004
27. Woodland DC, Liu W, Leong J et al (2016) Short-term high-fat feeding induces islet macrophage infiltration and β cell replication independently of insulin resistance in mice. Am J Physiol Metab 311:E763–E771
28. Torres-Castro I, Arroyo-Camarena ÚD, Martínez-Reyes CP et al (2016) Human monocytes and macrophages undergo M1-type inflammatory polarization in response to high levels of glucose. Immunol Lett 176:81–89
29. Jourdan T, Szanda G, Cinar R et al (2017) Developmental role of macrophage cannabinoid-1 receptor signaling in type 2 diabetes. Diabetes 66:994–1007
30. Moganti K, Li F, Schmuttermaier C et al (2017) Hyperglycemia induces mixed M1/M2 cytokine profile in primary human monocyte-derived macrophages. Immunobiology 222:952–959
31. Cucak H, Grunnet LG, Rosendahl A (2014) Accumulation of M1-like macrophages in type 2 diabetic islets is followed by a systemic shift in macrophage polarization. J Leukoc Biol 95:149–160. https://doi.org/10.1189/jlb.0213075
32. Donath MY (2016) Multiple benefits of targeting inflammation in the treatment of type 2 diabetes. Diabetologia 59:679–682. https://doi.org/10.1007/s00125-016-3873-z
33. Imai Y, Dobrian AD, Morris MA et al (2016) Lipids and immunoinflammatory pathways of beta cell destruction. Diabetologia 59:673–678
34. Jing Y, Wu F, Li D et al (2018) Metformin improves obesity-associated inflammation by altering macrophages polarization. Mol Cell Endocrinol 461:256–264
35. Sell H, Habich C, Eckel J (2012) Adaptive immunity in obesity and insulin resistance. Nat Rev Endocrinol 8:709
36. Chung K-J, Nati M, Chavakis T, Chatzigeorgiou A (2018) Innate immune cells in the adipose tissue. Rev Endocr Metab Disord 19:283–292. https://doi.org/10.1007/s11154-018-9451-6
37. Fonseca-Alaniz MH, Takada J, Alonso-Vale MIC, Lima FB (2006) The adipose tissue as a regulatory center of the metabolism. Arq Bras Endocrinol Metabol 50:216–29
38. Shibata M-A, Harada-Shiba M, Shibata E et al (2019) Crude α-mangostin suppresses the development of atherosclerotic lesions in apoe-deficient mice by a possible M2 macrophage-mediated mechanism. Int J Mol Sci 20:1722
39. Nguyen KD, Qiu Y, Cui X et al (2011) Alternatively activated macrophages produce catecholamines to sustain adaptive thermogenesis. Nature 480:104

40. Qiu Y, Nguyen KD, Odegaard JI et al (2014) Eosinophils and type 2 cytokine signaling in macrophages orchestrate development of functional beige fat. Cell 157:1292–1308
41. Fischer K, Ruiz HH, Jhun K et al (2017) Alternatively activated macrophages do not synthesize catecholamines or contribute to adipose tissue adaptive thermogenesis. Nat Med 23:623
42. Rajasekaran M, Sul O, Choi E et al (2019) MCP-1 deficiency enhances browning of adipose tissue via increased M2 polarization. J Endocrinol 1
43. Lv Y, Zhang S-Y, Liang X et al (2016) Adrenomedullin 2 enhances beiging in white adipose tissue directly in an adipocyte-autonomous manner and indirectly through activation of M2 macrophages. J Biol Chem 291:23390–23402
44. Barrett R, Narasimhulu CA, Parthasarathy S (2018) Adrenergic hormones induce extrapituitary prolactin gene expression in leukocytes-potential implications in obesity. Sci Rep 8:1936. https://doi.org/10.1038/s41598-018-20378-1
45. Hu H, Moon J, Chung JH et al (2015) Arginase inhibition ameliorates adipose tissue inflammation in mice with diet-induced obesity. Biochem Biophys Res Commun 464:840–847
46. Jais A, Einwallner E, Sharif O et al (2014) Heme oxygenase-1 drives metaflammation and insulin resistance in mouse and man. Cell 158:25–40
47. Kralova Lesna I, Petras M, Cejkova S et al (2018) Cardiovascular disease predictors and adipose tissue macrophage polarization: Is there a link? Eur J Prev Cardiol 25:328–334
48. Tateya S, Rizzo NO, Handa P et al (2011) Endothelial NO/cGMP/VASP signaling attenuates Kupffer cell activation and hepatic insulin resistance induced by high-fat feeding. Diabetes 60:2792–2801
49. Roberts LD (2015) Does inorganic nitrate say NO to obesity by browning white adipose tissue? Adipocyte 4:311–314
50. Roberts LD, Ashmore T, Kotwica AO et al (2015) Inorganic nitrate promotes the browning of white adipose tissue through the nitrate-nitrite-nitric oxide pathway. Diabetes 64:471–484
51. Varzandi T, Abdollahifar MA, Rohani SAH et al (2018) Effect of long-term nitrite administration on browning of white adipose tissue in type 2 diabetic rats: a stereological study. Life Sci 207:219–226
52. Zhang J, Shi G-P (2012) Mast cells and metabolic syndrome. Biochim Biophys Acta (BBA)-Molecular Basis Dis 1822:14–20
53. Varricchi G, Galdiero MR, Loffredo S et al (2017) Are mast cells MASTers in cancer? Front Immunol 8:424
54. Liu J, Divoux A, Sun J et al (2009) Genetic deficiency and pharmacological stabilization of mast cells reduce diet-induced obesity and diabetes in mice. Nat Med 15:940
55. Wang J, Shi G (2011) Mast cell stabilization: novel medication for obesity and diabetes. Diabetes Metab Res Rev 27:919–924
56. Kalkman H, Feuerbach D (2017) Microglia M2A polarization as potential link between food allergy and autism spectrum disorders. Pharmaceuticals 10:95
57. Castellano-Castillo D, Moreno-Indias I, Fernandez-Garcia JC et al (2018) Complement factor C3 methylation and mRNA expression is associated to BMI and insulin resistance in obesity. Genes (Basel) 9:410. https://doi.org/10.3390/genes9080410
58. Li Y, Huang B, Jiang X et al (2018) Mucosal-associated invariant T cells improve nonalcoholic fatty liver disease through regulating macrophage polarization. Front Immunol 9:1994
59. Kremlitzka M, Nowacka A, Mohlin FC et al (2019) Interaction of serum-derived and internalized C3 with DNA in human B cells–a potential involvement in regulation of gene transcription. Front Immunol 10:493
60. Zheng J, Jiang Z, Chen D et al (2019) Pathological significance of urinary complement activation in diabetic nephropathy: a full view from the development of the disease. J Diabetes Investig 10:738–744

61. Phieler J, Chung K-J, Chatzigeorgiou A et al (2013) The complement anaphylatoxin C5a receptor contributes to obese adipose tissue inflammation and insulin resistance. J Immunol 191:4367–4374
62. Poursharifi P, Lapointe M, Fisette A et al (2014) C5aR and C5L2 act in concert to balance immunometabolism in adipose tissue. Mol Cell Endocrinol 382:325–333
63. Ruan C-C, Ge Q, Li Y et al (2015) Complement-mediated macrophage polarization in perivascular adipose tissue contributes to vascular injury in deoxycorticosterone acetate-salt mice. Arterioscler Thromb Vasc Biol 35:598–606
64. Piao C, Zhang W-M, Li T-T et al (2018) Complement 5a stimulates macrophage polarization and contributes to tumor metastases of colon cancer. Exp Cell Res 366:127–138
65. Alicic RZ, Rooney MT, Tuttle KR (2017) Diabetic kidney disease. Clin J Am Soc Nephrol 12:2032–2045. https://doi.org/10.2215/CJN.11491116
66. Tesch GH (2017) Diabetic nephropathy–is this an immune disorder? Clin Sci 131:2183–2199
67. Akdas S, Turan B, Aribal Ayral P, Yazihan N (2019) The relationship between metabolic syndrome development and tissue trace elements status and inflammatory markers. Unpublished data (Abstract was sent to ISZB 2019 congress)
68. Alemdar M, Akdas S, Biriken D, Inanc I, Billur D, Turan B, Aribal Ayral P, Yazihan N (2019) Evaluation of the effect of tissue zinc-copper level and inflammatory process in elderly and metabolic syndrome developed rats. Unpublished data (Abstract was sent to ISZB 2019 congress.)
69. Crispe IN (2009) The liver as a lymphoid organ. Annu Rev Immunol 27:147–163. https://doi.org/10.1146/annurev.immunol.021908.132629
70. Bilzer M, Roggel F, Gerbes AL (2006) Role of Kupffer cells in host defense and liver disease. Liver Int 26:1175–1186. https://doi.org/10.1111/j.1478-3231.2006.01342.x
71. Moore SM, Holt VV, Malpass LR et al (2015) Fatty acid-binding protein 5 limits the anti-inflammatory response in murine macrophages. Mol Immunol 67:265–275
72. Kazankov K, Jørgensen SMD, Thomsen KL et al (2018) The role of macrophages in nonalcoholic fatty liver disease and nonalcoholic steatohepatitis. Nat Rev Gastroenterol Hepatol 1
73. Weng S-Y, Wang X, Vijayan S et al (2018) IL-4 receptor alpha signaling through macrophages differentially regulates liver fibrosis progression and reversal. EBioMedicine 29:92–103
74. Fukushima H, Yamashina S, Arakawa A et al (2018) Formation of p62-positive inclusion body is associated with macrophage polarization in non-alcoholic fatty liver disease. Hepatol Res 48:757–767
75. Patouraux S, Rousseau D, Bonnafous S et al (2017) CD44 is a key player in non-alcoholic steatohepatitis. J Hepatol 67:328–338
76. Krenkel O, Puengel T, Govaere O et al (2018) Therapeutic inhibition of inflammatory monocyte recruitment reduces steatohepatitis and liver fibrosis. Hepatology 67:1270–1283
77. Yang F, Wang S, Liu Y et al (2018) IRE1α aggravates ischemia reperfusion injury of fatty liver by regulating phenotypic transformation of kupffer cells. Free Radic Biol Med 124:395–407
78. Kitade H, Chen G, Ni Y, Ota T (2017) Nonalcoholic fatty liver disease and insulin resistance: new insights and potential new treatments. Nutrients 9:387
79. Zhang X, Fan L, Wu J et al (2019) Macrophage p38α promotes nutritional steatohepatitis through M1 polarization. J Hepatol 71:163–174
80. Handa P, Thomas S, Morgan-Stevenson V et al (2019) Iron alters macrophage polarization status and leads to steatohepatitis and fibrogenesis. J Leukoc Biol 105:1015–1026
81. Pan M-H, Chen J-W, Kong Z-L et al (2018) Attenuation by tetrahydrocurcumin of adiposity and hepatic steatosis in mice with high-fat-diet-induced obesity. J Agric Food Chem 66:12685–12695

Part III
Neurological and Visceral Complications Due to Obesity

Chapter 9
Association Between Obesity and Poor Sleep: A Review of Epidemiological Evidence

Yaqoot Fatima, Abdullah Al Mamun and Timothy Skinner

Abstract Sleep is a multifaceted biological process linked with biochemical pathways of growth, maturation, and energy homeostasis. Due to the growing complaints of poor sleep in the general population, there has been a significant increase in sleep research aiming to identify the determinants of poor sleep. This review aims to explore the current state of the knowledge on obesity-sleep association and evaluate the role of weight loss in sleep improvement and vice versa. Current evidence suggests three directions of association between obesity and sleep: obesity leading to poor sleep, poor sleep leading to obesity and co-occurrence of obesity and poor sleep. People with obesity often report poor sleep, and obstructive sleep apnoea (OSA), and poor sleepers are often overweight or obese. Weight gain is not only associated with poor sleep and OSA prevalence but also affects the severity of OSA. Poor sleep, specifically OSA, affects metabolic hormones and influence behavioural pathways favouring unhealthy lifestyle leading to further worsening of obesity. Weight loss has shown potential in improving the quality of sleep, reducing OSA severity and decreasing metabolic abnormalities related to obesity. However, weight management is yet to be established as a clinical intervention with a long-lasting effect in improving sleep. Evidence also suggests that improvement in sleep quality lowers the odds of future obesity. Despite, the growing evidence base, the current literature has methodological limitations and fails to establish a causal link between obesity and poor sleep. Nonetheless, it is prudent to advise that weight management should be an important component of the clinical management plans for improving sleep.

Y. Fatima (✉)
Centre for Rural and Remote Health (Mount Isa),
James Cook University, Mount Isa, QLD 4825, Australia
e-mail: yaqoot.fatima@jcu.edu.au

A. A. Mamun
Institute for Social Science Research, University of Queensland,
Brisbane, QLD, Australia

T. Skinner
Department of Psychology, University of Copenhagen, Copenhagen, Denmark

P. S. Tappia et al. (eds.), *Pathophysiology of Obesity-Induced Health Complications*, Advances in Biochemistry in Health and Disease 19,
https://doi.org/10.1007/978-3-030-35358-2_9

Keywords Obesity · Poor sleep · Sleep duration · Obstructive sleep apnoea · Weight loss · Behavioural interventions · Sleep hygiene

Introduction

Sleep is a fundamental component of the biological and environmental mix that shapes the overall health and development of human beings [1, 2]. Considering that we spent a third of our lives sleeping, and essential purpose of sleep is to refresh the mind and repair the body, it is not surprising that deviations from healthy sleep lead to poor health [3, 4]. Anecdotal evidence on the needs and benefits of healthy sleep is supported by the growing epidemiological evidence highlighting a link between poor sleep and medical and psychological issues, e.g., hypertension, diabetes, depression and cardio-metabolic problems [5].

There has been a rather significant increase in the prevalence of short sleep and sleep disorders in concurrence with the obesity epidemic [6]. Recent studies indicate an upwards trend in poor sleep in the general population with nearly 30% prevalence of insomnia and around 38% prevalence of sleep disorders such as sleep apnoea [7, 8]. The concurrent rise in obesity and poor sleep led many researchers to investigate whether there is a link between obesity and poor sleep. Although there is still a lack of conclusive evidence to disentangle cause and effect, nonetheless, growing epidemiological and experimental evidence indicates entrainment of a vicious cycle of obesity and poor sleep [9]. Essentially, it means that people with obesity have further increased the risk of obesity and associated cardiovascular problems due to the lack of healthy sleep. And likewise, people with poor sleep have further aggravated the risk of impaired sleep due to consequent obesity [10]. Understandably, people with concurrent obesity and poor sleep will have further increased the risk of poor health outcomes.

Unfortunately, the bulk of the preliminary literature on obesity and sleep association is based on subjective sleep reports, and hence the evidence is marred with bias, often leading to under-reporting of poor sleep [11]. Nonetheless, the advancement in objective measures for sleep assessment, e.g., actigraphy and polysomnography, make it easier to reliably study the aetiology and subtypes of poor sleep as well as evaluate appropriate behavioural interventions for sleep improvement.

In the past decade, there have been considerable increases in the number of studies exploring the association between obesity and sleep. Due to methodological differences across the studies, there are variations in reported effect sizes and the direction of the association. That in turn, hinders the application of research evidence for clinical decision making. This review aims to evaluate the current state of knowledge and to produce high-level evidence on the bi-directional and concurrent association between obesity and sleep. We'll also highlight limitations of the existing literature so that future research is directed at the literature gaps to further extend the knowledge in this area.

Obesity and Poor Sleep: Evidence from Epidemiological Studies

Poor sleep is highly comorbid with obesity as 20–70% of obese individuals are reported to have impaired sleep [9]. Although the consequent outcomes may vary from seemingly harmless problems including trouble sleeping, to far more severe disorders, such as obstructive sleep apnoea (OSA) and restless leg syndrome, nonetheless evidence suggests that obesity and being overweight significantly influence sleep quality [12].

Amongst the various dimensions of sleep, the literature on obesity and comorbid poor sleep has primarily focused on obesity and OSA, with only a few studies exploring the impact of obesity on short sleep and insomnia. Preliminary cross-sectional studies reported a negative linear association between obesity and short sleep in adults [13]. The evidence from prospective studies further confirmed the initial findings where participants with obesity were found to have a shorter duration of sleep, higher rates of subjective sleep disturbances (47.4 vs. 25.5%; $P < 0.01$) than their counterparts [14]. Some studies also report that people with obesity have high rates of poor sleep quality and daytime sleepiness due to inadequate nighttime sleep [15].

Although the evidence is limited, the association between obesity and short sleep is reported to be mediated through sleep disturbances and emotional stress [14]. It is postulated that individuals with obesity, who are struggling with poor sleep are psychosocially stressed and have higher odds of engaging in unhealthy behaviours, including the consumption of energy-rich comfort foods that further increase the burden of obesity and associated co-morbidities [16].

Intriguingly, a recent meta-analysis on the role of obesity in insomnia and reported a non-significant difference in the odds of having insomnia in people with obesity compared with those with normal weight [9]. However, it is worth mentioning that there was significant variation in effect size distribution across the studies. The inconsistencies in the findings may be primarily due to the subjective nature of sleep assessments and the differences in defining insomnia across studies. Therefore, to confirm the impact of obesity on sleep quality, further research utilising objective or validated subjective tools is warranted.

Obesity and Obstructive Sleep Apnoea (OSA)

OSA, characterised by frequent episodes of complete (apnoea) or partial closure (hypopnea) of the pharyngeal airway, is manifested as disordered breathing during sleep [17]. The ensuing reduction in the airflow leads to gas exchange derangement, fall in the blood oxygen levels, and sleep fragmentation [18]. While the short term implications of OSA are poor sleep, fatigue and daytime dysfunction [19], chronic OSA is a strong predictor of cardio-metabolic problems, depression, and stroke [20].

In adult populations, OSA prevalence without excessive daytime sleepiness varies between 9 and 38% [7]. Whereas OSA prevalence with accompanying daytime sleepiness varies between 3 and 7% [18]. In the paediatric population, although OSA is relatively less prevalent (1–10%) [21], it significantly affects daytime learning, behaviour [22] and leads to serious complications, including neuromuscular disease, and craniofacial abnormalities [23]. The emerging research also highlights a despairingly high percentage of up to 80% percent of undiagnosed cases of moderate or severe OSA in the general population [12], underlining that the community and healthcare of burden of OSA are even higher than reported in the literature [24].

The upward increase in OSA prevalence has given rise to large scale population studies aiming to identify risk factors associated with OSA so that early screening and management could be instituted. Amongst the various risk factors, obesity emerges as one of the most influential risk factor associated with OSA [25]. Several studies have consistently demonstrated a strong link between higher body weight and increased likelihood of developing OSA [26]. The prevalence of OSA is twice as high in people with obesity than their counterparts [27]. Therefore, it is not surprising that about 60% of moderate to severe OSA is attributable to obesity [28].

In addition to cross-sectional studies, evidence from some of the prospective cohort studies also indicate a strong influence of obesity in the development of OSA. A prospective cohort study conducted by Peppard et al. reported that in patients with mild OSA, 10% gain of the baseline weight leads to a sixfold-increased risk of in the odds of developing moderate-to-severe OSA [29]. The link between obesity and OSA is also seen to be present in the paediatric population as children with obesity are found to have about 40% higher odds of OSA than children with healthy weight [30].

In addition to affecting OSA prevalence, weight gain is also seen to influence the severity of OSA strongly. It is reported that fat deposition in the tissues surrounding the upper airway leads to increased collapsibility of the upper airway, leading to further progression of OSA progression [31]. The findings of large multicentre cohort studies such as the Sleep Heart Health study provides conclusive evidence that even relatively small gains in body weight are associated with worsening of sleep apnoea [32].

Weight Loss to Improve the Quality of Sleep and OSA Symptoms

Many studies have evaluated the effectiveness of weight loss achieved by behavioural, pharmacological, and surgical approaches in improving the quality of sleep and management of OSA in obese patients [33, 34]. The effectivity of weight management in improving sleep quality and duration is not explored as extensively as in the treatment of OSA. Besides the bulk of evidence is based on behavioural interventions for weight loss [35]. Nonetheless, the existing evidence consistently suggests that weight reduction seems to help in improving overall sleep [36].

Observational studies conducted by Verhoeff et al. and Steinberg et al. suggested that about 10% loss of baseline weight loss helps in improving the quality and about 24 min increase in the duration of sleep [37, 38]. Similar improvement in sleep quality was reported by Chaput et al. with a 5% reduction in initial weight [36]. Findings of some randomised controlled trials (RCT) also confirmed the results of observational studies. An RCT of behavioural weight loss intervention (POWER-UP trial) in a primary care setting, reported significant improvement in sleep duration in participants who had lost $\geq 5\%$ of their baseline body weight [39].

Unfortunately, the positive effect of weight loss on sleep quality and sleep duration are not seen to be sustained in the long term, and many of the effects did not persist beyond six months post intervention [39]. It is also important to note that there have been studies where no significant difference in sleep quality was found between the control and intervention groups [40].

Compared with the effectivity of weight loss in improving the quality of sleep the evidence on weight loss in OSA intervention is stronger, and lifestyle modification to achieve a healthy weight are often recommended for OSA management [33]. Many studies reported that a reduction from baseline weight leads to dramatic improvements in self-reporting of frequent apnoea, snoring, and daytime sleepiness [41].

A meta-analysis found that weight reduction programs are significantly effective in reducing the severity of OSA, i.e., a decrease in apnoea-hypopnea index (AHI) [−6.04 events/h, 95% confidence interval (CI) −11.18, −0.9] [34]. The benefit of weight loss on the severity of OSA is further confirmed by reviews evaluating bariatric surgery outcomes [42]. The effectiveness of weight loss in OSA management is reported to be dependent on the severity of the underlying OSA at diagnosis. In patients with mild-to-moderate OSA, about 10% reduction of the baseline weight frequently leads to symptomatic and metabolic improvements [26].

Intriguingly, in the long-term, the positive effect of weight loss on OSA progression is found to be much less compared to the negative effect of the same amount of weight gain [32]. However, it is important to note that weight loss alone is not sufficient normalise OSA and is only 'cured' (AHI <5 events/h) in a small number of patients [43]. Nonetheless, the improvement is OSA severity, and a potential role in mitigating cardio-metabolic outcomes strongly support weight management as an adjunct to pharmacological or surgical interventions.

Poor Sleep and Consequent Obesity: Evidence from Epidemiological Studies and Experimental Studies

Evidence from cross-sectional as well as longitudinal studies has provided strong evidence on the relationship between sleep duration and obesity risk in children, adolescents and adults [44]. Several systematic reviews and meta-analyses have tried to pool the evidence on sleep duration-obesity link and consistently report an inverse link between sleep duration and obesity [45, 46]. A recent meta-analysis reported a dose-response relationship between hours of sleep and body mass index (BMI) such that for each additional hour of sleep BMI decreases by 0.35 kg/m^2 [47]. Evidence from further population studies reported a U-shaped curvilinear association between sleep duration and BMI, suggesting that long duration of sleep is also associated with higher risk of obesity (RR: 1.04; 95% CI: 1.00–1.09) [48]. Therefore, suggesting that a regular good night's sleep of optimal duration is required to maintain metabolic health with a healthy weight.

In addition to sleep duration, sleep quality was also found to be significantly associated with greater BMI and future weight gain [49]. One study found that participants with sleep problems had a significant increase in BMI over a 10-year follow-up than their counterparts [50]. Problems initiating and maintaining sleep were seen to have a dose-response association with weight gain, with a higher likelihood of weight gain in subjects with frequent sleep problems than among those with occasional sleep problems [49]. However, some studies couldn't find any association between sleep problems and BMI or change in weight [51].

Poor sleep is also seen to be linked with Ow/Ob in children and adolescents. Liu et al. reported that children with sleep problems such as snoring, restless sleep and night waking are 3.5 times more likely to be obese [52]. Another study reported a strong association between sleep disturbances and obesity in adolescents, robust to adjustment for potential covariates [53]. The results from different studies were further corroborated by pooled estimates obtained from a recent meta-analysis where inadequate sleep (including both short duration and poor quality) was found to be linked 27% higher odds of Ow/Ob [(odds ratio) OR: 1.27 95% CI: 1.05–1.53], whereas poor sleep quality (independent of duration) was associated with 46% higher odds of Ow/Ob (OR = 1.46, 95% CI: 1.24–1.72) [54].

The results of preliminary experimental sleep deprivation studies were further confirmed by experimental studies suggesting that as little as 2–3 nights of sleep deprivation can significantly affect metabolic hormones [55, 56]. It was seen that young men limited to four hours of sleep a night for six days had a 40% decrease in the rate of glucose clearance and a 30% decrease in insulin response, suggesting a possible mechanism linking sleep to obesity [57]. Taheri et al. found that subject with five hours of sleep time were found to have 15% changes in morning leptin and ghrelin levels, suggesting that the sleep duration affects leptin and ghrelin effect, that in turn influence appetite and satiety and lead to increased BMI [55]. Studies assessing sleep duration mediated leptin and ghrelin alterations are yet to be conducted in children, but it is possible that sleep loss-metabolic regulation cycles in children and adolescents work no different than the same cycle in the adult population [58].

Healthy Sleep and Weight Loss

There is ample evidence to establish a relationship between poor sleep and higher BMI, and weight gain. However, the significantly unexplored aspect of the sleep-obesity association is whether healthy sleep helps in weight loss in people with obesity. Also, there is a scarcity of evidence on whether short sleepers lose less weight than normal sleepers. Some longitudinal studies found that people who improve their night-time sleep from less than six hours to 7–8 h gained less weight than those who were persistent short sleepers [12]. These preliminary positive findings encouraged further research to evaluate the contribution of healthy sleep in weight loss interventions. A study conducted by Thomson et al. found that participants with overweight and obesity had a higher chance of continued weight-loss if they were sleeping for more than seven hours each night and reported a good quality of sleep [59].

The contribution of sleep in achieving and maintaining healthy sleep weight has implications in designing and implementing behavioural weight-loss interventions. The amount of sleep is also seen to augment the effect of dietary interventions in the maintenance of fat-free body mass. Participants of an interventional study reported that during the sleep restriction period, only 26% of the total weight loss came from body fat, while during the normal sleep period fat made up 57% of the weight loss [60].

Consistent evidence is obtained from a few other studies that examined the role of baseline sleep in the success of various weight-loss studies programs. Researchers at the Laval University in Quebec evaluated three weight-loss programs and found that even after adjusting for caloric intake, baseline BMI and other socio-demographic factors, baseline sleep duration and sleep quality was found to have a significant inverse relationship with loss of body fat [61]. Another two-phase trial comparing two alternative strategies for weight loss reported that adequate sleep duration leads to greater weight-loss success [62]. Based on their findings the authors proposed that baseline sleep assessment could help in identifying individuals who may require additional support to ensure the success of the weight loss program.

So far, only one study has evaluated the role of sleep in weight management in children with obesity and found that a 30 min increase in baseline sleep duration resulted in an additional reduction of 0.2 kg/m^2 in BMI [63]. Therefore, suggesting a potential role of sleep hygiene in paediatric weight management.

Putative Mechanisms of Association Between Poor Sleep and Obesity

Several studies have attempted to assess possible mechanisms that link inadequate sleep and obesity. It is seen that during the deeper stage of slow wave sleep (SWS) which is associated with elevation of growth hormone (GH) levels, brain glucose utilisation, and sympathetic nervous activity are decreased, and parasympathetic nervous activity is increased [64]. Hence SWS is likely to play a significant role in total body glucose regulation. Subjects with sleep deprivation or poor quality are seen to spend less time spent in SWS and have alterations in rapid eye movement (REM) sleep, inducing hormonal changes [65] mainly anorexigenic leptin and orexigenic ghrelin [66, 67]. Leptin acts on specific receptors in the hypothalamus, with high concentrations inducing satiation and reducing food intake, whereas, ghrelin is responsible for stimulating appetite, its levels increase before meals and decrease after meals [68]. Overall, these two neuropeptides significantly influence energy intake and affect the energy balance equation [68].

It is also likely that inadequate sleep leads to decreased energy expenditure through altered thermoregulation and increased fatigue [69]. Besides, insufficient sleep is also linked to alterations in the regulation of thyroid-stimulating hormones that in turn may affect the thyroid hormones, an important regulator of carbohydrate metabolism, leading to the development of glucose dysregulation and insulin resistance and predispose to diabetes in the longer term [70]. Poor sleep quality also impacts higher-order executive functions that are responsible for things like exercising or restrained eating [71]. Overall, sleep-induced alteration in metabolic hormone levels leads to increase in food intake without compensating energy expenditure, and thus play an important role in weight gain and obesity.

There are several pathways through which obesity leads to OSA, and most of the times these work synergistically. It is hypothesised that increased fat deposition in the peri-pharyngeal region results in mechanical loading that counterbalances airway patency maintenance by the dilator muscles, leading to increased collapsibility particularly during sleep [33]. Some studies also report the role of central obesity in having detrimental effects on neuromuscular activity in the upper airway that further contributes to pharyngeal collapsibility [72].

Once OSA is established a self-perpetuating cycle of increase appetite due to hormonal dysregulation, decreased physical activity and further weight ensues, that aggravates both obesity and worsens OSA symptoms [27]. Figure 9.1 is a diagrammatic representation of the intertwined association between obesity, sleep deprivation, and OSA sharing the common pathways linked with the energy balance equation.

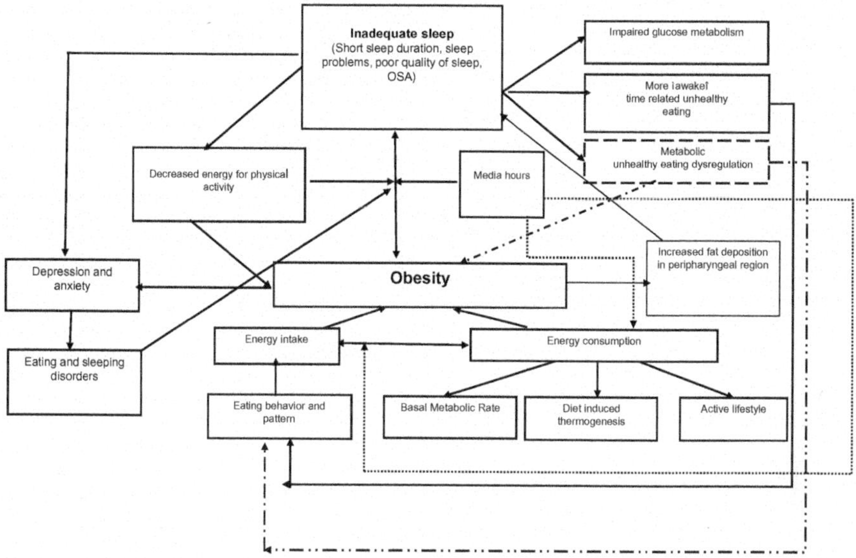

Fig. 9.1 Diagrammatic representation of possible pathways linking obesity and sleep

Limitations of the Literature

Overall, there is strong evidence linking obesity and sleep, but there are some methodological limitations to these studies that should be noted. The common bias in most of the studies is selection bias where the participants were not representative of the general population. Prevalence estimations based on subjects referred to sleep clinics lead to inaccurate estimation of the burden of sleep disorders. Although some studies have used two stages of design in which large screening study is followed by a random selection of participants for diagnostic testing [73], overall there are limited and unreliable estimates of the burden of sleep disorders in the general population.

There is also significant variation in the approach to defining sleep problems and the criteria to recognise respiratory events. Different methodologies were utilised for the confirmation of OSA such as polysomnography, self-reports, and respiratory polygraphs. Reliance on subjective non-validated tools is likely to have affected the reliability of the data collected through such sources. Also, there was significant variation across the studies for assessing and defining the severity of obesity and OSA. That, in turn, would have affected the estimation of effect size and effectiveness of the interventions in controlled trials. Moreover, the small number of participants and the uncontrolled design of interventional trials could have affected the power of the study to detect the true effect size.

The missing information on potential covariates such the concurrent psychological and medical problems, diet habits and exercise have also confounded the reported association between obesity and sleep. Most of the studies did not account for subjects lost to follow-up that might have introduced attrition bias and incorrect estimation of the improvement associated with the intervention. Also, there is limited information on whether short term positive gains in OSA management after weight management or reduction in weight after sleep improvement was sustained in the long run.

Although, the existing literature is difficult to compare across the studies and has limitations in extrapolation to the broader population. Nonetheless, the evidence base is strong enough to warrant a large-scale population and clinical research to explore the role of weight management in sleep disorders and vice versa.

Future Research and Conclusion

The association between obesity and sleep is complex, and the current evidence base lacks unbiased information on the temporality of the association. Therefore, a definitive causal role of obesity in OSA or vice versa is yet to be established. Nonetheless, the co-occurrence of obesity and poor sleep, specifically OSA, is a clinically significant phenotype warranting further research into the pathways linking obesity and poor sleep. There is substantial evidence to issue prudent advice that in the coming years in addition to managing the direct burden of the obesity epidemic, health services should be ready to handle a significant increase in the burden of co-morbid sleep problems.

Although, pharmacological and surgical interventions have shown success in reducing the burden of disease and associated co-morbidities. However, considering the complexity of obesity—sleep association, the treatment plan should not be limited to any single intervention. A multidisciplinary strategy is needed to achieve positive and long-lasting results. Current evidence suggests a promising role of behavioural interventions in improving sleep and reducing future weight gain, but further evidence from RCTs is needed to conclusively establish the role of healthy sleep in obesity prevention and management. Findings from multidisciplinary weight loss strategies will offer critical insight into mechanisms linking obesity and sleep, and potentially facilitate the exploration of the pathways crucial for the success of weight loss programs.

Considering the health and financial burden associated with the problem, there is a need for more awareness and research to reduce the burden of poor sleep. It is suggested that future studies should consider the role of sleep compensation on weekends and sleep timing to provide clarity to the role of poor sleep in obesity. It will also be interesting to see the impact of poor sleep on work stress, family functioning and relationships. Overall, it is recommended that in addition to diet and physical activity, sleep hygiene should also be an essential component of lifestyle package for achieving and maintaining healthy weight.

References

1. Nathaniel K (1957) Sleep, wakefulness, and consciousness. Psychol Bull 54(4):354–359
2. Pieron H (1913) Le Probleme Physiologique du Sommeil. Masson, Paris
3. Gregory AM et al (2009) The direction of longitudinal associations between sleep problems and depression symptoms: a study of twins aged 8 and 10 years. Sleep 32(2):189–199
4. Liu J, Hay J, Faught BE (2013) The association of sleep disorder, obesity status, and diabetes mellitus among US adults-the NHANES 2009–2010 survey results. Int J Endocrinol 2013:234129
5. Colten HR, Altevogt B (2006) Sleep disorders and sleep deprivation: an unmet public health problem. In: Colten HR, Altevogt BM (eds). National Academy of Sciences, Washington DC
6. Marshall NS, Bin YS, Glozier N (2013) Sleeping at the limits: the changing prevalence of short and long sleep durations in 10 countries. Am J Epidemiol 177(8):826–833
7. Senaratna CV et al (2017) Prevalence of obstructive sleep apnea in the general population: a systematic review. Sleep Med Rev 34:70–81
8. Singh-Manoux A et al (2011) Sleep epidemiology—a rapidly growing field. Int J Epidemiol 40(6):1431–1437
9. Chan WS, Levsen MP, McCrae CS (2018) A meta-analysis of associations between obesity and insomnia diagnosis and symptoms. Sleep Med Rev 40:170–182
10. Mavanji V et al (2012) Sleep and obesity: a focus on animal models. Neurosci Biobehav Rev 36(3):1015–1029
11. Ritter PS et al (2016) Comparison of subjective and objective sleep estimations in patients with bipolar disorder and healthy control subjects. Sleep Disorders 2016:5
12. Lee W et al (2008) Epidemiology of obstructive sleep apnea: a population-based perspective. Expert Rev Respir Med 2(3):349–364
13. Vorona RD et al (2005) Overweight and obese patients in a primary care population report less sleep than patients with a normal body mass index. Arch Intern Med 165(1):25–30
14. Vgontzas AN et al (2008) Short sleep duration and obesity: the role of emotional stress and sleep disturbances. Int J Obes (Lond) 32(5):801–809
15. Akinnusi ME et al (2012) Sleep disorders in morbid obesity. Eur J Intern Med 23(3):219–226
16. Vgontzas AN, Bixler EO (2008) Short sleep and obesity: are poor sleep, chronic stress, and unhealthy behaviors the link? Sleep 31(9):1203
17. Leinum CJ, Dopp JM, Morgan BJ (2009) Sleep-disordered breathing and obesity: pathophysiology, complications, and treatment. Nutr Clin Pract 24(6):675–687
18. Punjabi NM (2008) The epidemiology of adult obstructive sleep apnea. Proc Am Thorac Soc 5(2):136–143
19. Ong JC, Crawford MR (2013) Insomnia and obstructive sleep apnea. Sleep Med Clin 8 (3):389–398
20. Schulz R et al (2006) Obstructive sleep apnea-related cardiovascular disease. Med Klin (Munich) 101(4):321–327
21. Capdevila OS et al (2008) Pediatric obstructive sleep apnea: complications, management, and long-term outcomes. Proc Am Thorac Soc 5(2):274–282
22. Yoon SY, Jain U, Shapiro C (2012) Sleep in attention-deficit/hyperactivity disorder in children and adults: past, present, and future. Sleep Med Rev 16(4):371–388
23. Chan J, Edman JC, Koltai PJ (2004) Obstructive sleep apnea in children. Am Fam Phys 69(5):1147–1154
24. Heinzer R et al (2015) Prevalence of sleep-disordered breathing in the general population: the HypnoLaus study. Lancet Respir Med 3(4):310–318
25. Tuomilehto H, Seppa J, Uusitupa M (2013) Obesity and obstructive sleep apnea–clinical significance of weight loss. Sleep Med Rev 17(5):321–329
26. Hamilton GS, Joosten SA (2017) Obstructive sleep apnoea and obesity. Aust Fam Physician 46(7):460–463

27. Romero-Corral A et al (2010) Interactions between obesity and obstructive sleep apnea: implications for treatment. Chest 137(3):711–719
28. Resta O et al (2001) Sleep-related breathing disorders, loud snoring and excessive daytime sleepiness in obese subjects. Int J Obes Relat Metab Disord 25(5):669–675
29. Peppard PE et al (2000) Longitudinal study of moderate weight change and sleep-disordered breathing. JAMA 284(23):3015–3021
30. Rudnick EF et al (2007) Prevalence and ethnicity of sleep-disordered breathing and obesity in children. Otolaryngol Head Neck Surg 137(6):878–882
31. Shelton KE et al (1993) Pharyngeal fat in obstructive sleep apnea. Am Rev Respir Dis 148(2): 462–466
32. Newman AB et al (2005) Progression and regression of sleep-disordered breathing with changes in weight: the Sleep Heart Health Study. Arch Intern Med 165(20):2408–2413
33. Cowan DC, Livingston E (2012) Obstructive sleep apnoea syndrome and weight loss: review. Sleep Disord 2012:163296
34. Araghi MH et al (2013) Effectiveness of lifestyle interventions on obstructive sleep apnea (OSA): systematic review and meta-analysis. Sleep 36(10):1553–1562
35. St-Onge M-P, Shechter A (2014) Sleep disturbances, body fat distribution, food intake and/or energy expenditure: pathophysiological aspects. Horm Mol Biol Clin Invest 17(1):29–37
36. Chaput JP et al (2005) Psychobiological impact of a progressive weight loss program in obese men. Physiol Behav 86(1–2):224–232
37. Gonnissen HK et al (2013) Concomitant changes in sleep duration and body weight and body composition during weight loss and 3-mo weight maintenance. Am J Clin Nutr 98(1):25–31
38. Steinberg DM et al (2017) Preventing weight gain improves sleep quality among black women: results from a RCT. Ann Behav Med 51(4):555–566
39. Alfaris N et al (2015) Effects of a 2-year behavioral weight loss intervention on sleep and mood in obese individuals treated in primary care practice. Obesity (Silver Spring, Md.) 23(3):558–564
40. Rubin RR et al (2013) Patient-reported outcomes in the practice-based opportunities for weight reduction (POWER) trial. Qual Life Res 22(9):2389–2398
41. Nerfeldt P et al (2010) A two-year weight reduction program in obese sleep apnea patients. J Clin Sleep Med 6(5):479–486
42. Buchwald H et al (2004) Bariatric surgery: a systematic review and meta-analysis. JAMA 292(14):1724–1737
43. Mitchell LJ et al (2014) Weight loss from lifestyle interventions and severity of sleep apnoea: a systematic review and meta-analysis. Sleep Med 15(10):1173–1183
44. Miller MA et al (2018) Sleep duration and incidence of obesity in infants, children, and adolescents: a systematic review and meta-analysis of prospective studies. Sleep 41(4)
45. Wu Y, Zhai L, Zhang D (2014) Sleep duration and obesity among adults: a meta-analysis of prospective studies. Sleep Med 15(12):1456–1462
46. Fatima Y, Doi SA, Mamun AA (2015) Longitudinal impact of sleep on overweight and obesity in children and adolescents: a systematic review and bias-adjusted meta-analysis. Obes Rev 16(2):137–149
47. Cappuccio F, Miller MA (2010) Sleep, health and society: from aetiology to public health in The epidemiology of sleep and cardiovascular risk and disease. In: Cappuccio F, Miller M, Lockley S (eds). Oxford University Press, Oxford, pp 83–110
48. Liu W et al (2018) Long sleep duration predicts a higher risk of obesity in adults: a meta-analysis of prospective cohort studies. J Public Health (Oxf)
49. Lyytikainen P et al (2011) Sleep problems and major weight gain: a follow-up study. Int J Obes (Lond) 35(1):109–114
50. Janson C et al (2001) Insomnia in men-a 10-year prospective population based study. Sleep 24(4):425–430
51. Owens JF, Matthews KA (1998) Sleep disturbance in healthy middle-aged women. Maturitas 30(1):41–50

52. Liu J et al (2011) Sleep difficulties and obesity among preadolescents. Can J Public Health 102(2):139–143
53. Jarrin DC, McGrath JJ, Drake CL (2013) Beyond sleep duration: distinct sleep dimensions are associated with obesity in children and adolescents. Int J Obes (Lond) 37(4):552–558
54. Fatima Y, Doi SA, Mamun AA (2016) Sleep quality and obesity in young subjects: a meta-analysis. Obes Rev 17(11):1154–1166
55. Taheri S et al (2004) Short sleep duration is associated with reduced leptin, elevated ghrelin, and increased body mass index. PLoS Med 1(3):e62–e62
56. Simpson NS, Banks S, Dinges DF (2010) Sleep restriction is associated with increased morning plasma leptin concentrations, especially in women. Biol Res Nurs 12(1):47–53
57. Spiegel K, Leproult R, Van Cauter E (1999) Impact of sleep debt on metabolic and endocrine function. Lancet 354(9188):1435–1439
58. Beccuti G, Pannain S (2011) Sleep and obesity. Curr Opin Clin Nutr Metab Care 14(4): 402–412
59. Thomson CA et al (2012) Relationship between sleep quality and quantity and weight loss in women participating in a weight-loss intervention trial. Obesity (Silver Spring, Md.) 20(7): 1419–1425
60. Nedeltcheva AV et al (2010) Insufficient sleep undermines dietary efforts to reduce adiposity. Ann Intern Med 153(7):435–441
61. Chaput JP, Tremblay A (2012) Sleeping habits predict the magnitude of fat loss in adults exposed to moderate caloric restriction. Obes Facts 5(4):561–566
62. Elder CR et al (2012) Impact of sleep, screen time, depression and stress on weight change in the intensive weight loss phase of the LIFE study. Int J Obes (Lond) 36(1):86–92
63. Sallinen BJ et al (2013) Longer weekly sleep duration predicts greater 3-month BMI reduction among obese adolescents attending a clinical multidisciplinary weight management program. Obes Facts 6(3):239–246
64. Van Cauter E (2011) Sleep disturbances, obesity and diabetes: Interacting epidemics. Sleep Biol Rhythms 9(4):195
65. Rutters F et al (2012) Distinct associations between energy balance and the sleep characteristics slow wave sleep and rapid eye movement sleep. Int J Obes (Lond) 36(10): 1346–1352
66. Schmid SM et al (2008) A single night of sleep deprivation increases ghrelin levels and feelings of hunger in normal-weight healthy men. J Sleep Res 17(3):331–334
67. Spiegel K et al (2004) Leptin levels are dependent on sleep duration: relationships with sympathovagal balance, carbohydrate regulation, cortisol, and thyrotropin. J Clin Endocrinol Metab 89(11):5762–5771
68. Horne J (2008) Short sleep is a questionable risk factor for obesity and related disorders: statistical versus clinical significance. Biol Psychol 77(3):266–276
69. Patel SR, Hu FB (2008) Short sleep duration and weight gain: a systematic review. Obesity (Silver Spring) 16(3):643–653
70. Potenza M, Via MA, Yanagisawa RT (2009) Excess thyroid hormone and carbohydrate metabolism. Endocr Pract 15(3):254–262
71. Auger RR (2006) Sleep-related eating disorders. Psychiatry (Edgmont (Pa.: Township)) 3(11):64–70
72. Schwartz AR et al (2008) Obesity and obstructive sleep apnea: pathogenic mechanisms and therapeutic approaches. Proc Am Thorac Soc 5(2):185–192
73. Popko K et al (2007) Frequency of distribution of leptin receptor gene polymorphism in obstructive sleep apnea patients. J Physiol Pharmacol 58(Suppl 5):551–561

Chapter 10
Exploration of the Bidirectionality of Obesity and Depression by Means of the Neuropsychological Model of Obesity Genesis

Matthew Ramjiawan and Paramjit S. Tappia

Abstract Obesity and depression are co-morbid disorders that place pervasive and significant burdens on individuals, their families and society. The neuropsychological model of obesity offers a framework for understanding their etiology, bidirectionality and impact on neurological, gastrointestinal, endocrine, cardiovascular and psychosocial systems. In recent years, research has focused on the role of stress as a perpetuator of this disabling cycle. Though complex and difficult to treat, the multi-systemic nature of these combined disorders offer several entry points for intervention and management. Several evidence-based treatment options are offered that, particularly when attempted concurrently, relieve suffering and promote health and wellbeing.

Keywords Obesity · Depression · Neuropsychological model · Hypothalamus pituitary adrenal axis · Stress-induced obesity · Corticotropin releasing factor · Adiposity · Body mass index

Introduction

Historically, Hippocrates was the first to make note that obesity negatively affected health outcomes, as his obese patients were unable to conceive and tended to die prematurely [1]. We now know that obesity is a pandemic that has numerous comorbid diseases including cardiovascular disease (CVD), diabetes mellitus, neurodegenerative diseases such as Alzheimer's disease, chronic obstructive pulmonary disease (COPD), metabolic syndrome, gout, reproductive disorders, stroke, certain cancers, depression, anxiety and many other chronic and inflammatory illnesses. [2–4]. Obesity is described as an excessive accumulation of white adipose

M. Ramjiawan · P. S. Tappia (✉)
Asper Clinical Research Institute & Office of Clinical Research, St. Boniface Hospital, Winnipeg, MB R2H 2A6, Canada
e-mail: ptappia@sbrc.ca

P. S. Tappia et al. (eds.), *Pathophysiology of Obesity-Induced Health Complications*, Advances in Biochemistry in Health and Disease 19,
https://doi.org/10.1007/978-3-030-35358-2_10

tissue (WAT) that causes an increase in body weight leading to a body mass index above 30 (BMI > 30 kg/m^2). According to the World Health Organization (WHO), over 650 million adults worldwide are obese, a number which has tripled over the past three decades [2]. Although global rates of obesity are an obvious concern, developed countries such as Canada show similar trends. Statistics Canada states that nearly 17.3 million Canadians (26%) self-report height and weight values that indicate a BMI over 30 kg/m^2, classifying themselves as obese. Combining this statistic with the population catalogued as overweight, a grand total of 63.1% of Canadians suffer from weight-related complications [5–7]. Obesity is a leading cause of morbidity among Canadians, with over 61% of type 2 diabetes, 14% of colorectal cancers, 17% of osteoarthritis, 14% of depression and 20% of premature deaths reported to be directly related to obesity [3].

In a similar fashion, the prevalence of mental health disorders has been increasing over the last several decades. The WHO, in the "World Health Ministers Call to Action" estimates that mental illness, including stress-related disorders, will be the second leading cause of disability by 2020 [8]. In her 2019 Health Report, Canada's Chief Medical Officer suggests that over 30% of Canadians will experience mental illness at some point during their lifetime [9]. Mental illness, particularly anxiety and depression, places significant financial, emotional and psychological burdens on individuals, their caregivers, and society as a whole, with treatment costs accounting for only a small proportion of the total burden [10]. Mental health disorders, along with their resultant complications and co-morbidities, have the capacity to compromise all areas of an individual's life in pervasive and debilitating ways.

Quite often, there is a cyclical co-existence of obesity and mental illness. Not only does obesity negatively impact physiological health in a variety of ways, it also leads to increased social withdrawal causing subsequent deterioration in psychological health. Concurrently, depression also results in further withdrawal, and since the diseases are comorbid, an individual suffering from both may be forced into a state of isolation. Social isolation may also further reduce employability, which may result in lessened work experience, social skills, or professional abilities to find employment. Mental health diseases may also further reduce an individual's ability to work. Not only does this result in further social isolation and inactivity, but also poses serious financial concerns. This is what some call the "double burden" of disease—the heavy economic drain combined with the physical encumbrance [2]. Mental health disorders and obesity are also both major sources of financial stress within the healthcare system. The Public Health Agency of Canada projected the annual healthcare cost for the treatment of obese individuals to reach $8.8 billion by 2021 [7]. This figure is expected to rise as comorbid diseases hinder treatment and may even mask crucial underlying casual factors [11].

This review will focus on the bidirectionality of obesity and depression by examining the Neuropsychological Model of Obesity Genesis, prevailing factors that perpetuate the cycle of obesity and depression, and evidence-based interventions that interrupt it in ways that relieve suffering and promote wellbeing.

Neuropsychological Model of Obesity Genesis

The most exhaustive model to explain the etiology of both obesity and depression, and their resultant maladaptive overeating patterns, is the neuropsychological model of obesity genesis. Jauch-Chara and Oltmanns describe obesity in this model as the end product of a vicious cycle of chronic stress activation and the body's physiological and psychological coping response. Further, it purports that obesity evolves from impaired neuropsychological functioning, specifically with regards to dysfunctional stress-response and reward systems [12]. This model suggests that chronic stress causes neurochemical imbalances and brain dysfunction that drive individuals to overconsume high caloric diets, adopt sedentary isolated lifestyles, develop poor motivation and negative mood. Similarly, depression has also been linked to dysfunctional stress-response and reward systems. Herein lies the connection of obesity genesis and depressive mood states. The negative consequences of chronic stress, as mentioned above, results in both obesity and depression. Both of these states cause further increases in stress, and a cycle of overindulgence in comfort food, inactivity and social withdrawal begins. Ultimately, the end-result is pervasive, excessive weight gain, along with negative mood states and potentially clinical depression. It is this recurrent and chronic nature of stress, along with its affiliated neuropsychological responses, that both induces and perpetuates obesity and depression.

While short-term stress serves as an adaptive response, prolonged stress can lead to long-term damage and disease. When chronically stressed, the brain and body are in a state of threatened homeostasis and initiate a wide range of neuroendocrine, immune, cardiovascular and metabolic responses. Allostatic load refers to the cost of chronic stress exposure that leads to atherosclerosis, impaired immunity, neural cell atrophy, obesity, depression and anxiety [13, 14]. As the control centre of the body, the brain is responsible for the body's coping responses to stress. The structures that are particularly involved in these processes include the hippocampus, pre-frontal cortex and amygdala. Combined, these regions are responsible for learning, memory, executive functioning and managing intense emotions such as fear and anger. While the hippocampus and pre-frontal cortex are targets of stress hormones and vulnerable to atrophy, the amygdala is prone to growth and hyperarousal. The result is dysregulation and dysfunction, which will be evaluated further [14].

Neurological Feedback Loop of Stress and Food-Seeking Compulsion

Stress is a state of threatened homeostasis within the body that motivates a wide range of responses in order to restore balance. The primary pathways of the stress system in the body are the hypothalamic-pituitary-adrenal (HPA) axis and the sympathetic nervous system [15–17]. The HPA axis is a major neuroendocrine system that regulates many of the body's physiological processes, including digestion, sexuality, energy storage and outflow, mood and emotions [15, 18–21]. When the HPA axis is activated, the hypothalamus releases corticotropin releasing factor (CRF), which then stimulates the release of adrenocorticotropic hormone (ACTH) from the anterior lobe of the pituitary which then stimulates the release of cortisol from the adrenal gland [15, 17, 19, 21]. Cortisol has a wide variety of effects on the body, including increasing blood glucose levels through gluconeogenesis as well as suppression of the immune system. Although the resultant action of CRF is immunosuppression through cortisol release, CRF overexpression can cause inflammation and increase physiological arousal, producing common symptoms of anxiety that leads to a vicious cycle of stress-induced stress [17, 22, 23]. Elevated CRF levels are consistently found in patients with depression and anxiety disorders, as well as those who suffer from obesity [17, 23–25].

The negative consequences of chronic stress create a variety of physiological and psychological effects in the body that, through similar pathways, result in both obesity and depression. And one inextricably perpetuates the other. Several studies have demonstrated that individuals who underwent early life trauma have an enhanced HPA axis with elevated salivary cortisol levels and potentially altered glucocorticoid feedback regulation [26–28]. Ultimately, it is this stress-induced state that may serve as an instigating factor for the cycle of obesity and depression. As seen in Fig. 10.1, the overexpression of HPA axis activity will perpetuate a state of dysregulation and preserve the body's distress.

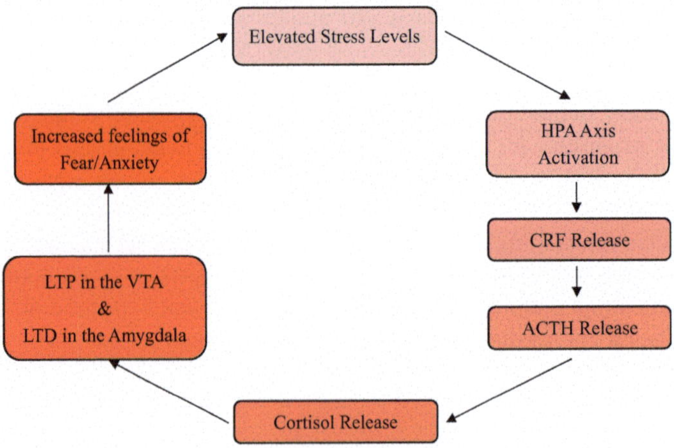

Fig. 10.1 Cyclical nature of HPA axis activation and reactivation due to chronic stress. Chronic (or traumatic acute) stressors leads to activation of the Hypothalamic Pituitary Adrenal (HPA) Axis, which causes a chain reaction of Corticotropin Releasing Factor (CRF), Adrenocorticotropic Hormone (ACTH), and Cortisol release. This chain of events leads to Long-Term Potentiation (LTP) in the Ventral Tegmental Area (VTA), creating an environment where increased reward is needed to stimulate the reward pathway. Long-Term Depression (LTD) in the amygdala also occurs, which results in increased feelings of fear and anxiety. This causes further activations of stress which reactivates the cycle

Chronic hyperactivity of the HPA axis increases the degree and length of CRF expression in the hypothalamus. As CRF expression is increased, the HPA axis reaches a maximum stimulation point where it is unable to be stimulated further for any perceivable effect. Once this occurs, the neurons involved become persistently activated, which leads to impaired neurotransmission [12, 29]. Ultimately, this leads to long-term potentiation (LTP) in the Ventral Tegmental Area (VTA), which is a crucial component of the mesolimbic reward pathway. Chronic HPA axis activation also induces long-term depression (LTD) in the amygdala, which is involved in anxiety and fear encoding [12, 30–32]. While LTP is the strengthening of synaptic connectivity caused by repetitive increases in activity, the opposite is LTD, which is simply the reduction of signal strength between two neurons due to a relative period of inactivity [33, 34]. As a result of these changes in synaptic connectivity, chronic stress causes an individual to feel increased reward, encode fear more often, and encode safety less [31]. It is important to note that the VTA is a dopamine-enriched area of the mesolimbic pathway that is directly involved and responsible for the learning of pleasurable activities. Chronic stress, and the subsequent LTP of the VTA, results in sensitization towards palatable food reinforcement [32, 35–38]. Consequently, palatable high calorie food results in greater pleasure and reward during a stressed state compared to a non-stressed state.

Behavioral Response of Stress-Reduction and Palatable Food Intake

The use of food as a means of stress-reduction is well documented and easily observable in everyday life. Individuals often turn to hyperpalatable (high fat/sugar content) foods during periods of stress, which activates the VTA and creates a rewarding stress-reduction mechanism that is reinforced over decades [23, 39]. This forms a strong association and an addictive-like compulsion to seek out hyper-palatable foods when stressed, further strengthening the association between stress, stress-reduction and food intake as demonstrated in Fig. 10.2 [36, 39–43]. Food is inherently dopaminergic and highly palatable foods that are high in sugar have an addictive pleasurable quality that mimics some drugs [35, 36, 44, 45].

In a study conducted on social defeat as a stressor in mice, it was found that immediately after suffering social defeat, mice displayed an acute increase in food intake, but in quantities that are insignificant for weight gain. However, *chronic* exposure to social defeat caused a cumulative increase of food intake significantly beyond that of controls [46]. A similar study was conducted among Syrian hamsters, which found that chronic social defeat significantly increased food intake, body mass and adiposity, proving that stress-induced obesity is a cross-species phenomenon [47]. Studies with monkeys have also demonstrated that increased adiposity is connected to the time in which feeding occurs. That is, food intake that is limited during daylight hours is less likely to result in overindulgence, whereas

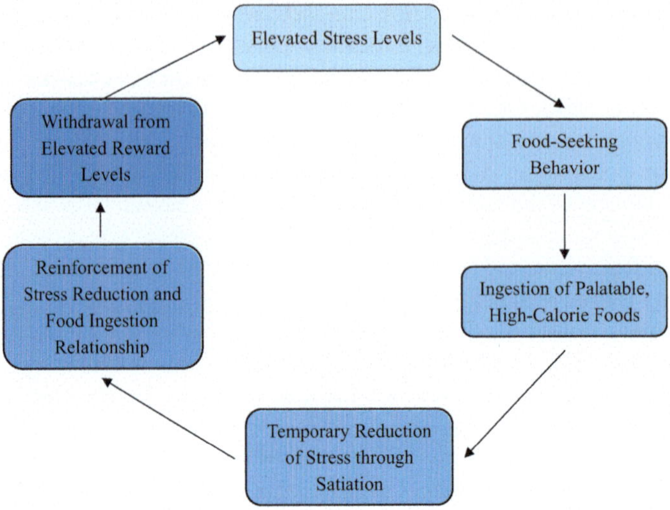

Fig. 10.2 Behavioral relationship between stress and food-seeking compulsions. Increased stress levels cause an individual to seek measures that return the body back to homeostasis, with food serving as a primary stress-reduction medium. Once the individual engages in reducing their stress by eating food that has a high calorie content, their stress is abated for a short while. The stress-reducing nature of this satiation causes both a positive and negative reinforcing relationship where the induction of food is rewarding, but also the removal of stress is also rewarding. This is significant to note as the ingestion of food is not the primary reinforcer in this case, but instead it is the removal of stress that has the greatest salience upon the individual. This negative reinforcement of stress reduction through food consumption highly strengthens the associations formed between elevated stress levels and palatable high-calorie food. When the highly-rewarding stimulus is removed from circumstance (i.e. the food has been eaten), the individual goes into acute, albeit minor, withdrawal and temporarily spikes in cortisol levels. Overindulgence may lead to addictive-like behavior

intake that occurred at night was salient among monkeys with high anxiety-like behaviors and resulted in greater adiposity [48]. Additionally, social status and gender differences were observed among mice, hamsters and monkeys in terms of eating behaviors, suggesting that nondominant females expressed the greatest amount of adiposity following chronic social stress [46–48].

It is interesting to note that chronic adaptation of stress-eating behaviors results in a feedback loop where the consumption of palatable foods reduces stress, but the body's withdrawal from the reward gained from the palatable food causes a sharp increase of stress as the body experiences a sudden termination of dopamine release —leading to the compulsion of seeking more palatable food [49].

Cycle of Cerebral Energy and Metabolic Functions

It is well documented that both depression and obesity present with decreased activity in certain regions in the brain [50–52]. Observations were made that while stressed, human subjects displayed significantly decreased activity in the dorsolateral prefrontal cortex (DLPFC) [53]. Studies were also conducted that found reduced grey matter among obese patients in the same brain regions as depression, even with several other comorbidities held constant [54, 55]. Hypothalamic neurons are chiefly involved with food intake regulation by transmissions between the brain stem and spinal cord with regards to stimulation or inhibition of food intake, ingestion behavior and neurochemical energy balances [56]. The hypothalamus, with regards to food intake, is regulated by a number of central nervous system functions, which include determining the current rate of adenosine triphosphate (ATP) production and use. During a state of reduced energy, such as hypoglycemia, the hypothalamus is activated to stimulate glucose production and hunger induction [57–61]. Additionally, a glucose-deficient state also reduces insulin secretion to allow for an increased rate of gluconeogenesis [57, 60]. It is conceivable then, that chronic overactivity of the HPA axis will result in impairment of ingestion behavior and energy levels. In a study conducted by Schmoller et al., it was determined that an inverse correlation exists between ATP and BMI [62]. These results are consistent with the idea that the brain's supply of ATP is involved with the regulation of body weight and energy usage [63, 64]. In addition to obesity, depression is characterized by reduced levels of ATP as well. The consequence of an overactive HPA axis is that the brain is induced into a state of chronic starvation, with food intake being the primary drive that is stimulated. The starved brain is a depressed brain in this case as the energy required to function is depressed to a point of inactivity. As a direct result, overconsumption for the sake of satiation leads to obesity [12, 63–65].

Therapeutic Interventions

A significant number of therapeutic options exist for both obesity and depression, so many that it is unlikely that an exhaustive list exists. This author will focus on some of the therapeutic interventions that target both obesity and depression concurrently, are of current interest and may be clinically relevant for individuals who show resistance to current therapeutic models. Many other treatment options exist for either obesity or depression including bariatric surgery and multiple pharmacological therapies such as selective serotonin reuptake inhibitors, selective norepinephrine reuptake inhibitors, tricyclic antidepressants, etc. When pursuing therapeutic options in a clinical setting, a combination of treatments is best considered for the individual.

It has been known for centuries that a healthy diet and exercise promote positive physical and psychological health. It has even shown to be as, if not more effective than traditional pharmacological treatments for mild-to-moderate cases of depression [66–69]. With the countless and repetitive research conducted on both a healthy diet and exercise, it has become evident that this method should and must be used as a concurrent treatment to supplement all additional therapeutic options.

In addition to energy intake and output, new evidence is emerging that engaging in mindfulness-based interventions that target stress, negative thoughts and eating behaviors also aid in positive life outcomes [70–72]. As a program designed specifically for stress reduction, it was found that nearly half of patients (n = 145) demonstrated a reduced relapse rate [73]. Mindfulness has also been shown to have positive effects on obesity as a subset of the practice involves mindful eating—where participants are actively engaged in the ingestion process and are conscious in their decisions in when and how often to partake in indulgent consumption [71, 74]. In a methodical review (n = 529), Carriere et al. [75] examined the effects of mindfulness as the sole intervention for weight loss and found that participants lost an average of 3.3% of their baseline bodyweight. This figure is noteworthy when compared to the lifestyle (diet/exercise) control group, who lost 4.7% [75]. One of the primary reasons as to why mindfulness works for both obesity and depression is that it directly intervenes within the stress and stress-response cycle, allowing for new homeostatic functions to be engaged.

A technique that is gaining traction and is of significant interest is Repetitive Transcranial Magnetic Stimulation (rTMS) within the area of neuromodulation and neuropsychiatry. This involves placing a magnetic coil over the target area that directly stimulates the surface of the brain. As noted above, significant reduction of activity in the DLPFC is noted during periods of elevated stress. rTMS has been used in a randomized controlled study (n = 60), which examined its effects on body weight in obese patients [76]. It was found that when administered to the DLPFC, rTMS was effective in reducing food intake and aiding weight loss among obese individuals [76, 77]. It is interesting to note that current depression-treatment protocols using rTMS include directly stimulating the DLPFC as well [78–80].

Conclusions

As both depression and obesity are leading illnesses in the modern world, the examination of their comorbidities and common physiological pathways is crucial in the discussion of holistic healthcare. The Neuropsychological Model of Obesity purports that obesity may be caused by a combination of factors which are psychogenic in nature with a physiological expression. Both obesity and depression share many of the same physiological pathways that lead to overeating and one of the potential causes includes chronic stress. Breaking the cyclical nature of stress and dysfunctional stress-response systems offers opportunities for a healthier life.

References

1. Haslam D (2007) Obesity: a medical history. Obes Rev 8:31–36. https://doi.org/10.1111/j.1467-789X.2007.00314.x
2. World Health Organization (2018) Obesity and overweight
3. Janssen I (2013) The public health burden of obesity in Canada. Can J Diabetes 37:90–96. https://doi.org/10.1016/j.jcjd.2013.02.059
4. Low S, Chin MC, Deurenberg-Yap M (2009) Review on epidemic of obesity. Ann Acad Med Singapore 38:57–59
5. Kwan A, Corscadden L (2012) Obesity in Canada. Can J Diabetes 35:152. https://doi.org/10.1016/s1499-2671(11)52056-8
6. (2018) Overweight and obese adults, 2018. Ottawa
7. Bancej C, Jayabalasingham B, Wall RW et al (2015) Evidence brief—trends and projections of obesity among Canadians. Heal Promot Chronic Dis Prev Canada 35:109–112. https://doi.org/10.24095/hpcdp.35.7.02
8. Reddy MS (2010) Depression: the disorder and the burden. Indian J Psychol Med 32:1–2. https://doi.org/10.4103/0253-7176.70510
9. Tam T (2019) The chief public health officer's report on the state of public health in Canada 2018: preventing problematic substance use in youth. Ottawa
10. Trautmann S, Rehm J, Wittchen H (2016) The economic costs of mental disorders. EMBO Rep 17:1245–1249. https://doi.org/10.15252/embr.201642951
11. Sagar R, Gupta T (2018) Psychological aspects of obesity in children and adolescents. Indian J Pediatr 85:554–559
12. Jauch-chara K, Öltmanns KM (2014) Obesity: a neuropsychological disease? Systematic review and neuropsychological model. Prog Neurobiol 114:84–101. https://doi.org/10.1016/j.pneurobio.2013.12.001
13. McEwen BS (2008) Central effects of stress hormones in health and disease: understanding the protective and damaging effects of stress and stress mediators. Eur J Pharmacol 583:174–185. https://doi.org/10.1016/j.ejphar.2007.11.071
14. Mcewen BS (2009) Protection and damage from acute and chronic stress: allostasis and allostatic overload and relevance to the pathophysiology of psychiatric disorders. Ann N Y Acad Sci 1032:1–7. https://doi.org/10.1196/annals.1314.001
15. Kyrou I, Tsigos C (2009) Stress hormones: physiological stress and regulation of metabolism. Curr Opin Pharmacol 9:787–793. https://doi.org/10.1016/j.coph.2009.08.007
16. Tomiyama AJ (2019) Stress and Obesity. Annu Rev Psychol 70:703–718
17. Warne JP (2009) Shaping the stress response: interplay of palatable food choices, glucocorticoids, insulin and abdominal obesity. Mol Cell Endocrinol 300:137–146. https://doi.org/10.1016/j.mce.2008.09.036
18. Pariante CM, Lightman SL (2008) The HPA axis in major depression: classical theories and new developments. Trends Neurosci 31:464–468. https://doi.org/10.1016/j.tins.2008.06.006
19. Tsigos C, Chrousos GP (2002) Hypothalamic–pituitary–adrenal axis, neuroendocrine factors and stress. J Psychosom Res 53:865–871. https://doi.org/10.1016/S0022-3999(02)00429-4
20. Pariante CM (2003) Depression, stress and the adrenal axis. J Neuroendocrinol 15:811–812. https://doi.org/10.1046/j.1365-2826.2003.01058.x
21. Stephens MAC, Wand G (2012) Stress and the HPA axis: role of glucocorticoids in alcohol dependence. Alcohol Res 34:468–483
22. Risbrough VB, Stein MB (2006) Role of corticotropin releasing factor in anxiety disorders: a translational research perspective. Horm Behav 50:550–561. https://doi.org/10.1016/j.yhbeh.2006.06.019
23. Sominsky L, Spencer SJ (2014) Eating behavior and stress: a pathway to obesity. Front Psychol 5:1–8. https://doi.org/10.3389/fpsyg.2014.00434

24. Wang L, Goebel-Stengel M, Yuan P-Q et al (2017) Corticotropin-releasing factor overexpression in mice abrogates sex differences in body weight, visceral fat, and food intake response to a fast and alters levels of feeding regulatory hormones. Biol Sex Differ 8:2. https://doi.org/10.1186/s13293-016-0122-6
25. Sharma R, Banerji MA (2012) Corticotropin releasing factor (CRF) and obesity. Maturitas 72:1–3. https://doi.org/10.1016/j.maturitas.2012.01.015
26. Kempke S, Luyten P, De Coninck S et al (2015) Effects of early childhood trauma on hypothalamic–pituitary–adrenal (HPA) axis function in patients with Chronic Fatigue Syndrome. Psychoneuroendocrinology 52:14–21. https://doi.org/10.1016/j.psyneuen.2014.10.027
27. von Baes CW, Martins CMS, Tofoli SM de, Juruena MF (2014) Early life stress in depressive patients: HPA axis response to GR and MR agonist. Front Psychiatry 5:2. https://doi.org/10.3389/fpsyt.2014.00002
28. Shea A, Walsh C, MacMillan H, Steiner M (2005) Child maltreatment and HPA axis dysregulation: relationship to major depressive disorder and post traumatic stress disorder in females. Psychoneuroendocrinology 30:162–178. https://doi.org/10.1016/j.psyneuen.2004.07.001
29. Gallagher JP, Orozco-Cabal LF, Liu J, Shinnick-Gallagher P (2008) Synaptic physiology of central CRH system. Eur J Pharmacol 583:215–225. https://doi.org/10.1016/J.EJPHAR.2007.11.075
30. Saal D, Dong Y, Bonci A, Malenka RC (2003) Drugs of abuse and stress trigger a common synaptic adaptation in dopamine neurons. Neuron 37:577–582. https://doi.org/10.1016/S0896-6273(03)00021-7
31. Maroun M (2006) Stress reverses plasticity in the pathway projecting from the ventromedial prefrontal cortex to the basolateral amygdala. Eur J Neurosci 24:2917–2922. https://doi.org/10.1111/j.1460-9568.2006.05169.x
32. Vucetic Z, Reyes TM (2010) Central dopaminergic circuitry controlling food intake and reward: implications for the regulation of obesity. Wiley Interdiscip Rev Syst Biol Med 2:577–593. https://doi.org/10.1002/wsbm.77
33. Massey PV, Bashir ZI (2007) Long-term depression: multiple forms and implications for brain function. Trends Neurosci 30:176–184. https://doi.org/10.1016/j.tins.2007.02.005
34. Cooke SF (2006) Plasticity in the human central nervous system. Brain 129:1659–1673. https://doi.org/10.1093/brain/awl082
35. Wise RA (2004) Dopamine and food reward: back to the elements. Am J Physiol Integr Comp Physiol 286:R13–R13. https://doi.org/10.1152/ajpregu.00590.2003
36. Baik J-H (2013) Dopamine signaling in food addiction: role of dopamine D2 receptors. BMB Rep 46:519–526. https://doi.org/10.5483/BMBRep.2013.46.11.207
37. Wise RA (2004) Dopamine, learning and motivation. Nat Rev Neurosci 5:483–494. https://doi.org/10.1038/nrn1406
38. Wise RA (2006) Role of brain dopamine in food reward and reinforcement. Philos Trans R Soc B Biol Sci 361:1149–1158. https://doi.org/10.1098/rstb.2006.1854
39. Yau YHC, Potenza MN (2013) Stress and eating behaviors. Minerva Endocrinol 38:255–267
40. Isabel A, Rudá G, Braz F et al (2019) Body composition, biochemical, behavioral and molecular alterations in overfed rats after chronic exposure to SSRI. Behav Brain Res 356: 62–70. https://doi.org/10.1016/j.bbr.2018.08.007
41. Tomycz N, Oh MY, Whiting D (2015) Future targets. 38:1–9. https://doi.org/10.3171/2015.3.FOCUS1542.Disclosure
42. Ulrich-Lai YM, Fulton S, Wilson M et al (2015) Stress exposure, food intake and emotional state. Stress 18:381–399. https://doi.org/10.3109/10253890.2015.1062981
43. Osdoba KE, Mann T, Redden JP, Vickers Z (2015) Using food to reduce stress: effects of choosing meal components and preparing a meal. Food Qual Prefer 39:241–250. https://doi.org/10.1016/j.foodqual.2014.08.001
44. Lenoir M, Serre F, Cantin L, Ahmed SH (2007) Intense sweetness surpasses cocaine reward. PLoS ONE 2:e698. https://doi.org/10.1371/journal.pone.0000698

45. Volkow ND, Wang G-J, Baler RD (2011) Reward, dopamine and the control of food intake: implications for obesity. Trends Cogn Sci 15:37–46. https://doi.org/10.1016/j.tics.2010.11.001
46. Bhatnagar S, Vining C, Iyer V, Kinni V (2006) Changes in hypothalamic-pituitary-adrenal function, body temperature, body weight and food intake with repeated social stress exposure in rats. J Neuroendocrinol 18:13–24. https://doi.org/10.1111/j.1365-2826.2005.01375.x
47. Foster MT, Solomon MB, Huhman KL, Bartness TJ (2006) Social defeat increases food intake, body mass, and adiposity in Syrian hamsters. Am J Physiol Integr Comp Physiol 290: R1284–R1293. https://doi.org/10.1152/ajpregu.00437.2005
48. Wilson ME, Fisher J, Fischer A et al (2008) Quantifying food intake in socially housed monkeys: social status effects on caloric consumption. Physiol Behav 94:586–594. https://doi.org/10.1016/j.physbeh.2008.03.019
49. Pecoraro N, Reyes F, Gomez F et al (2004) Chronic stress Promotes palatable feeding, which reduces signs of stress: feed forward and feedback effects of chronic stress. Endocrinology 145:3754–3762. https://doi.org/10.1210/en.2004-0305
50. Pandya M, Altinay M, Malone DA Jr, Anand A (2012) Where in the brain is depression? Curr Psychiatry Rep 14:634–642. https://doi.org/10.1007/s11920-012-0322-7
51. Gluck ME, Alonso-Alonso M, Piaggi P et al (2015) Neuromodulation targeted to the prefrontal cortex induces changes in energy intake and weight loss in obesity. Obesity 23:2149–2156. https://doi.org/10.1002/oby.21313
52. Le DSN, Pannacciulli N, Chen K et al (2006) Less activation of the left dorsolateral prefrontal cortex in response to a meal: a feature of obesity. Am J Clin Nutr 84:725–731. https://doi.org/10.1093/ajcn/84.4.725
53. Qin S, Hermans EJ, van Marle HJF et al (2009) Acute psychological stress reduces working memory-related activity in the dorsolateral prefrontal cortex. Biol Psychiatry 66:25–32. https://doi.org/10.1016/j.biopsych.2009.03.006
54. Jagust W, Harvey D, Mungas D, Haan M (2005) Central obesity and the aging brain. JAMA Neurol 62:1545–1548. https://doi.org/10.1001/archneur.62.10.1545
55. Stillman CM, Weinstein AM, Marsland AL et al (2017) Body-brain connections: the effects of obesity and behavioral interventions on neurocognitive aging. Front Aging Neurosci 9:115. https://doi.org/10.3389/fnagi.2017.00115
56. Palkovits M (2003) Hypothalamic regulation of food intake. Ideggyogy Sz 56:288–302
57. Epstein FH, Cryer PE, Gerich JE (1985) Glucose counter regulation, hypoglycemia, and intensive insulin therapy in diabetes mellitus. N Engl J Med 313:232–241. https://doi.org/10.1056/NEJM198507253130405
58. Tesfaye N, Seaquist ER (2010) Neuroendocrine responses to hypoglycemia. Ann N Y Acad Sci 1212:12–28. https://doi.org/10.1111/j.1749-6632.2010.05820.x
59. McCrimmon RJ (2012) Update in the CNS response to hypoglycemia. J Clin Endocrinol Metab 97:1–8. https://doi.org/10.1210/jc.2011-1927
60. Sprague JE, Arbeláez AM (2011) Glucose counterregulatory responses to hypoglycemia. Pediatr Endocrinol Rev 9:463–73; quiz 474–475
61. Diggs-Andrews KA, Zhang X, Song Z et al (2010) Brain insulin action regulates hypothalamic glucose sensing and the counterregulatory response to hypoglycemia. Diabetes 59:2271–2280. https://doi.org/10.2337/db10-0401
62. Schmoller A, Hass T, Strugovshchikova O et al (2010) Evidence for a relationship between body mass and energy metabolism in the human brain. J Cereb Blood Flow Metab 30:1403–1410. https://doi.org/10.1038/jcbfm.2010.48
63. Peters A, Schweiger U, Pellerin L et al (2004) The selfish brain: competition for energy resources. Neurosci Biobehav Rev 28:143–180. https://doi.org/10.1016/j.neubiorev.2004.03.002
64. Peters A, Pellerin L, Dallman M et al (2007) Causes of obesity: Looking beyond the hypothalamus. Prog Neurobiol 81:61–88. https://doi.org/10.1016/j.pneurobio.2006.12.004
65. Miki T, Liss B, Minami K et al (2001) ATP-sensitive K^+ channels in the hypothalamus are essential for the maintenance of glucose homeostasis. Nat Neurosci 4:507–512. https://doi.org/10.1038/87455

66. Knapen J, Vancampfort D, Moriën Y, Marchal Y (2015) Exercise therapy improves both mental and physical health in patients with major depression. Disabil Rehabil 37:1490–1495. https://doi.org/10.3109/09638288.2014.972579
67. Russo-Neustadt A, Beard RC, Cotman CW (1999) Exercise, antidepressant medications, and enhanced brain derived neurotrophic factor expression. Neuropsychopharmacology 21:679–682. https://doi.org/10.1016/S0893-133X(99)00059-7
68. Duman CH, Schlesinger L, Russell DS, Duman RS (2008) Voluntary exercise produces antidepressant and anxiolytic behavioral effects in mice. Brain Res 1199:148–158. https://doi.org/10.1016/j.brainres.2007.12.047
69. Ransford CP (1982) A role for amines in the antidepressant effect of exercise: a review. Med Sci Sports Exerc 14:1–10
70. O'Reilly GA, Cook L, Spruijt-Metz D, Black DS (2014) Mindfulness-based interventions for obesity-related eating behaviours: a literature review. Obes Rev 15:453–461. https://doi.org/10.1111/obr.12156
71. Dalen J, Smith BW, Shelley BM et al (2010) Pilot study: mindful eating and living (MEAL): weight, eating behavior, and psychological outcomes associated with a mindfulness-based intervention for people with obesity. Complement Ther Med 18:260–264. https://doi.org/10.1016/j.ctim.2010.09.008
72. Grossman P, Niemann L, Schmidt S, Walach H (2004) Mindfulness-based stress reduction and health benefits. J Psychosom Res 57:35–43. https://doi.org/10.1016/S0022-3999(03)00573-7
73. Teasdale JD, Segal ZV, Williams JMG et al (2000) Prevention of relapse/recurrence in major depression by mindfulness-based cognitive therapy. J Consult Clin Psychol 68:615–623. https://doi.org/10.1037/0022-006X.68.4.615
74. Dunn C, Haubenreiser M, Johnson M et al (2018) Mindfulness approaches and weight loss, weight maintenance, and weight regain. Curr Obes Rep 7:37–49. https://doi.org/10.1007/s13679-018-0299-6
75. Carrière K, Khoury B, Günak MM, Knäuper B (2018) Mindfulness-based interventions for weight loss: a systematic review and meta-analysis. Obes Rev 19:164–177. https://doi.org/10.1111/obr.12623
76. Kim S-H, Chung J-H, Kim T-H et al (2018) The effects of repetitive transcranial magnetic stimulation on eating behaviors and body weight in obesity: a randomized controlled study. Brain Stimul 11:528–535. https://doi.org/10.1016/j.brs.2017.11.020
77. Bou Khalil R, El Hachem C (2014) Potential role of repetitive transcranial magnetic stimulation in obesity. Eat Weight Disord Stud Anorexia, Bulim Obes 19:403–407. https://doi.org/10.1007/s40519-013-0088-x
78. Teng S, Guo Z, Peng H et al (2017) High-frequency repetitive transcranial magnetic stimulation over the left DLPFC for major depression: Session-dependent efficacy: a meta-analysis. Eur Psychiatry 41:75–84. https://doi.org/10.1016/j.eurpsy.2016.11.002
79. Du L, Liu H, Du W et al (2018) Stimulated left DLPFC-nucleus accumbens functional connectivity predicts the anti-depression and anti-anxiety effects of rTMS for depression. Transl Psychiatry 7:3. https://doi.org/10.1038/s41398-017-0005-6
80. Yadollahpour A, Hosseini SA, Shakeri A (2016) rTMS for the treatment of depression: a comprehensive review of effective protocols on right DLPFC. Int J Ment Health Addict 14:539–549. https://doi.org/10.1007/s11469-016-9669-z

Chapter 11
Obesity-Induced Non-alcoholic Fatty Liver Disease (NAFLD): Role of Hyperhomocysteinemia

Santosh Kumar, Sreyoshi F. Alam and Paul K. Ganguly

Abstract Non-alcoholic fatty liver disease (NAFLD) is one of the most common causes of chronic liver disease globally. NAFLD encompasses broader spectrum of liver pathology from fatty liver to non-alcoholic steatohepatitis (NASH) and cirrhosis. Its burden is expected to rise due to the increasing prevalence of obesity and diabetes mellitus worldwide. However, pathogenic mechanisms underlying NAFLD are not clearly understood. Hyperhomocysteinemia is increasingly being associated with NAFLD. Pathogenesis of NAFLD involves dysregulation/derangements of homocysteine, folate, and vitamin B_6 and B_{12} metabolism. Obesity is also shown to be associated with these vitamin deficiencies and homocysteine elevation. Recently, Cystathionine-β-synthase (CBS) and Cystathionine-γ-lyase (CSE), key enzymes that regulate the levels of homocysteine and cysteine and production of endogenous hydrogen sulfide (H_2S), have been implicated in NAFLD. In this review, we have examined known associations between NAFLD, hyperhomocysteinemia and obesity.

Keywords Non-alcoholic fatty liver disease · NAFLD · NASH · Homocysteine · Hyperhomocysteinemia · Obesity · Metabolic syndrome

Pathogenic Link Between Non-alcoholic Fatty Liver Disease and Obesity—An Overview

Non-alcoholic fatty liver disease (NAFLD) is one of the most common causes of chronic liver disease globally [1–3]. The diagnosis of NAFLD is made based on histological and radiological evidence of steatosis in the liver [4]. Histological diagnosis requires macrovesicular steatosis in $\geq 5\%$ of hepatocytes. This accumulation of fat should not be due to high alcohol intake, hereditary disorders or use of steatotic medications [4, 5]. NAFLD is histopathologically characterized into two

S. Kumar · S. F. Alam · P. K. Ganguly (✉)
College of Medicine, Alfaisal University, Box 50927, Riyadh 11533, Saudi Arabia
e-mail: pganguly@alfaisal.edu

P. S. Tappia et al. (eds.), *Pathophysiology of Obesity-Induced Health Complications*, Advances in Biochemistry in Health and Disease 19,
https://doi.org/10.1007/978-3-030-35358-2_11

forms, (1) non-alcoholic fatty liver (NAFL) and (2) non-alcoholic steatohepatitis (NASH) [4]. NAFLD leads to an incremental liver metabolic dysfunction beginning with hepatic steatosis and steatohepatitis and possibly progressing to cirrhosis and/or hepatocellular carcinoma [6]. In about 30% of patients, fatty liver can progress to non-alcoholic steatohepatitis (NASH) [7]. NASH falls under the broad umbrella of the NAFLD spectrum, but unlike fatty liver, it has a more complex histopathology involving steatosis, necrosis, inflammation and fibrosis [8] and has a predilection towards progression to more severe liver injury. NAFLD has shown to be associated with various metabolic disorders including insulin resistance, diabetes mellitus, hyperlipidemia and obesity [4]. However, there are regional differences in NAFLD's association with these comorbidities, especially with body mass index (BMI) [5]. NAFLD is also considered as the liver component of the metabolic syndrome [4, 5] and therefore, remains an indirect cardiovascular risk factor.

Correlations Between Metabolic Syndrome, NAFLD and BMI

As indicated above NAFLD is one of the most common liver diseases [9]. It has been shown to be most prevalent in the Middle East and South America [10]. With the improvement of viral hepatitis treatment, NAFLD has replaced viral hepatitis as the forefront of clinical hepatology [11]. Since NAFLD is associated with multiple comorbidities such as obesity, type 2 diabetes mellitus, hypertension and cardiovascular disease [11, 12], the interaction between these metabolic syndrome components and NAFLD is believed to be bidirectional, meaning that NAFLD may either precede or promote metabolic comorbidities [13]. One of the major causes of hepatic steatosis is obesity and hepatic steatosis is one of the manifestations affecting the liver in metabolic syndrome [14, 15]. About 90% of NAFLD patients develop at least one feature of metabolic syndrome and 33% possess the characteristics in its entirety [16].

A classic measure of obesity has been the body mass index (BMI). Until better tools are standardized, this provides a sufficiently usable tool for assessment. Risk factor assessment for fatty liver disease has indicated its correlation to a higher BMI as an independent risk factor [17–20]. The risk of fatty liver was about 4.1 to 14-fold higher in individuals with a higher BMI [18, 19]. Most studies had defined BMI as a categorical variable, and this meant that a dose-response relationship was not easily discernible [18–20]. Most studies calculated the strength of its association with fatty liver disease using logistic or cox model, and this blurred the continuity of BMI and fatty liver risk [21]. This artificial subdivision of BMI into segments of underweight, normal, overweight and obese, leads to both loss of information as well as a somewhat inaccurate result [21]. Fan et al. presented a dose response analysis of 3202 subjects, in a cross-sectional study, which helped associate fatty liver risk and BMI in a nonlinear fashion with a significantly increased trend of odds ratio as per 1 kg/m^2 increase in BMI [21]. The study adjusted for age, gender, hypertension, total cholesterol, triglycerides, glucose, HDL, LDL, uric acid, homocysteine, creatinine, AST and ALT. When BMI <23 kg/m^2, it did not affect the risk of fatty liver but a BMI >23 kg/m^2 increased the risk of fatty liver significantly with a 1 kg/m^2 increase in BMI. Fan et al. showed that an overweight and obese BMI was significantly associated with the risk of fatty liver, with odds ratio of 3.55 and 7.59 compared to the normal weight population, respectively [21]. In the dose-response analysis, BMI also showed a nonlinear, statistically significant relationship with fatty liver both in the total population as well as in the subgroups that were stratified by gender, age (older/younger than 50), and presence of hypertension (yes/no). This indicates that a higher BMI is an independent, dose-dependent risk factor for fatty liver [21].

Patients with a higher BMI usually have greater quantities of adipose tissue, but this point is highly debatable and therefore is one of the drawbacks of the possible mechanism of the link between BMI and fatty liver disease. BMI does not account for more specific analysis of composition (i.e. lean and fat masses), therefore, subjects with high BMI may not be obese, and the subjects with normal BMI may counterintuitively obese [21]. The mechanism as described by Musso et al. and Stefan et al. indicates that patients with a higher BMI have more adipose tissue and thus an increase in fatty acid flow to the liver [22, 23]. Higher BMI patients are presumed to consume a high-fat diet over long periods of time, which might increase the absorption of exogenous fat, leading to increase of fatty acids and their lipid deposition in liver [21]. This study has certain drawbacks other than its use of BMI as a variable. Firstly, the population studied is derived from China, where BMI cut-off points are lower than the WHO classification, leading to an increase in higher BMI values (overweight/obesity). Thus the association between BMI and fatty liver risk will vary and may be different from the rest of the world [21]. The diagnosis of fatty liver via ultrasonography is valid, but the user dependent nature of the technique may lead to variation. The definitive method would have been a liver biopsy. The sample size was large but stratification for subgroup analysis may lead to loss of power [21].

Hyperhomocysteinemia and Obesity

Obesity has also been correlated with vitamin B_6, B_{12} and folate deficiencies [24–28]. Guven et al. reported that vitamin B_{12} level was significantly lower in patients with metabolic syndrome than those without metabolic syndrome [28]. Although the study by Baltaci et al. did not find statistically significant difference between vitamin B_{12} and metabolic syndrome, the levels were lower in patients with metabolic syndrome [24]. All the patients with metabolic syndrome were obese [24]. Therefore, vitamin B_{12} deficiencies are correlated to obesity and since this deficiency is also related to hyperhomocysteinemia, a tertiary association is possible. Unfortunately, this study had limitations since it is cross-sectional and serum homocysteine and methylmalonic acid level were not assayed [24]. Obesity is a consequence of inappropriate nutrition in majority of the cases [27]. Hence its correlation with vitamin deficiencies is unavoidable. Folate deficiency has also been directly correlated with NAFLD [27].

Over the years, homocysteine has garnered attention due to its notoriety in causing multiple comorbidities. Homocysteine, a sulfhydryl-containing amino acid, is an intermediate product in the normal biosynthesis of the amino acids, methionine and cysteine [29]. Homocysteine has been seen to be able to alter intracellular lipid metabolism and this might enable hepatic fat accumulation [30, 31] and provides a possible management route for prevention of NAFLD progression and its related cardiovascular complications [9]. Obesity has been reported to be associated with two key liver enzymes involved in homocysteine metabolism which may influence the catabolism of homocysteine in liver damage [32].

Hyperhomocysteinemia and fatty liver are quite often observed simultaneously though the mechanism is still unclear [33]. Wang et al. showed that a rich, fatty diet not only increased cholesterol, but also doubled homocysteine levels in animals [34]. This might indicate that high intake of dietary cholesterol and fat might increase homocysteine values resulting in progression of atherosclerosis. Another proposed theory says that homocysteine-induced endoplasmic reticulum stress can lead to increased biosynthesis and uptake of cholesterol and triglycerides by the liver [30]. Another proposed possibility is hypomethylation associated with hyperhomocysteinemia that results in lipid accumulation, decreased synthesis of phosphatidylcholine required for very low-density lipoprotein assembly and homeostasis [33]. Boushey et al. carried out a meta-analysis that concluded that homocysteine is an independent, graded risk factor for atherosclerotic disease in the coronary, cerebral and peripheral vessels [35]. Homocysteine may also alter the endothelial cell surface by changing their phenotype from anticoagulant to procoagulant, via expression modulation of the enzymes glutathione peroxidase and nitric oxide synthase [33].

Hyperhomocysteinemia is also associated with adipose tissue dysfunction via inhibition of lipolysis, and some reports indicate that exogenous dietary homocysteine for two weeks lowered circulating glycerol and free fatty acids levels [33]. Homocysteine supplementation was associated with increased leptin and decreased adiponectin levels in plasma [33]. Sreckovic et al. found higher homocysteine levels in metabolic syndrome patients along with an exaggeration of components of metabolic syndrome such as abdominal obesity, glycemia, blood pressure and hypertriglyceridemia with lower HDL-C [33].

Furthermore, an establishment of a link between hyperhomocysteinemia and NAFLD, opens up avenues towards treatment options. The cystathione-β-synthase (CBS) and cystathione-γ-lyase (CSE) regulates homocysteine and cysteine metabolism and contributes to endogenous hydrogen sulfide (H2S) biosynthesis [8].

The CBS/CSE system has recently been positioned as a potential therapeutic target in NAFLD since disturbances in circulating homocysteine and H2S levels have been reported in NAFLD patients and associated co-morbidities. Homocysteine could possibly be a non-invasive marker of NAFLD as well as a predictor of NASH. On the contrary, there are other conflicting findings regarding the relative change in homocysteine levels across the various NAFLD spectrums that leaves this open to debate [8].

Some evidence indicates the CBS/CSE system may be regulated by lipids [8]. Polyunsaturated n-3 fatty acids docosahexaenoicacid (DHA) and eicosapentaenoicacid (EPA) showed an increase in CSE gene expression in human hepatocytes. Similar findings were also reported for dietary supplementation of DHA or conjugated linoleic acid (CLA), which also increased hepatic CBS gene expression and reduced plasma homocysteine levels in rats [8]. Endoplasmic reticulum (ER) stress has also been reported as a contributing factor in NAFLD pathogenesis [8]. There is already an established interdependent relationship between lipids and hepatic ER stress. Lately, a relationship between ER stress and the CBS/CSE system is emerging [8]. The product of CBS-mediated transsulfuration reaction, known as cystathione has shown to prevent ER stress induced steatosis liver injury. Oxidative stress is responsible for the progression of NAFLD to NASH [8]. The CBS/CSE system is emerging as a potent regulator of oxidative stress due to the transulfuration and desulfuration capacities of the enzymes [8].

Hyperhomocysteinemia can lead to NAFLD by disrupting hepatic lipid metabolism as well as increasing oxidative injury [8]. Bravo et al., Hwang et al., and Peh et al. demonstrated that high fat diets alter sulfur amino acid metabolism in rodents [36–38]. Hwang et al. reported that CBS and CSE expression was increased in mice that were put on a high-fat diet for 5 weeks, although Peh et al. showed that an 8-week regimen of high fat diet led to reduction in CSE protein levels and despite a compensatory elevation in CBS expression, hepatic H2S biosynthesis was impaired [38]. An 18-week high fat diet in mice showed reduction in the transulfuration activities of CBS and CSE and caused hyperhomocysteinemia in these animals [36]. Hwang et al. also reiterated that CBS and CSE expression activated the transsulfuration pathway, increased hepatic H2S biosynthesis, and reduced circulating homocysteine levels [37]. Dahloff et al. investigated whether NAFLD caused by a high-fat (HF) diet over 8 weeks in mice, can be reversed by additional 4 weeks of a dietary methyl-donor supplementation (MDS) [39]. His study failed to reverse NAFLD, but halted the progression of hepatic steatosis associated with major changes in key hepatic C1-metabolites, including *S*-adenosyl-methionine and *S*-adenosylhomocysteine. A possible mechanism involves that dietary methyl-donors activate AMPK, a key enzyme in fatty acid β-oxidation control, mediates increased fatty acid utilization and thereby prevents further hepatic lipid accumulation [39].

Role of Homocysteine Metabolism in Non-alcoholic Fatty Liver Disease

Despite increasing morbidity and mortality associated with this disease, the mechanisms involved in development and progression are not clearly understood. However, the dysregulation/derangements of homocysteine, folate, vitamin B_6, vitamin B_{12} and CBS/CSE system have been implicated in the pathogenesis of NAFLD.

A cross sectional study ($N = 7203$) individuals was carried out for a period of 1 year in subjects aged 18 years and older with available abdominal ultrasound and undergoing plasma homocysteine measurements simultaneously. Subjects with factors that can affect data, such as alcohol consumption, viral hepatitis, schistosomiasis or other chronic liver diseases were excluded. Hepatic steatosis was diagnosed via abdominal ultrasonography by experienced radiologists who were blinded to the subjects' diagnoses and tests. Ultrasound was considered positive in the presence of diffusely increased liver near field ultrasound echo ('bright liver'), liver echo greater than kidney, vascular blurring and the gradual attenuation of far field ultrasound echo. Subjects with at least two of the abnormal findings listed above were diagnosed with hepatic steatosis [40, 41]. Age, gender, BMI (categorized as normal: <24 kg/m^2, overweight: 24 to <28 kg/m^2, and obesity: $\geq$28 kg/m^2, (Working Group on Obesity in China guidelines), current smoking, hypertension, and diabetes were also evaluated to assess whether there was any significant interaction between these variables and the relationship between homocysteine levels and the prevalence of NAFLD [9]. 2370 (32.9%) subjects were diagnosed with NAFLD, and the mean homocysteine level was 7.0 μmol/L. The prevalence of NAFLD progressively increased in the higher quartiles of homocysteine (19.8, 28.3, 37.7, and 46.6%, respectively). Adjustment for gender, BMI, current smoker, physical activity, education, drinking, hypertension, diabetes, uric acid, ALT, TBIL, ALB, PLT, TG, TC, HDL-C, hs-CRP, and creatinine, the risk for NAFLD increased across the homocysteine quartiles, and the OR in the highest quartile compared with the lowest quartile was 2.08 (95% CI 1.61, 2.67). Similar results were also observed when homocysteine was considered as a continuous exposure variable. To determine the effect of potential confounding factors, the associations between quartiles of homocysteine and NAFLD were further investigated among subgroups. As a result, the association differed significantly according to gender, BMI category and smoking status (p value for interaction: 0.001, 0.002 and <0.001, respectively) [9]. Homocysteine was strongly associated with NAFLD in females but was weaker in males. Obesity and non-smoking status were also related to an enhanced association between homocysteine and the prevalence of NAFLD [9]. No effect modifications by age, hypertension, or diabetes on the association between homocysteine and NAFLD were observed (all p for interaction >0.05) [9]. Therefore, elevated homocysteine levels were positively associated with the prevalence of NAFLD in Chinese adults. On subgroup analyses, effect modification by gender, BMI and smoking on the association were found. A stronger association of homocysteine with the prevalence of NAFLD was observed in female, obese and non-smoking adults than in male, normal weight and smoking subjects [9]. Cross sectional studies are limited by their lack of establishment of a causal relationship and the use of ultrasound although the acceptable norm, but a liver biopsy leads to definitive diagnosis [9]. In contrast to this cross-sectional study, interestingly another study found that homocysteine was closely associated with the prevalence of NAFLD in males but not in females [42].

It has been shown that mice fed on high fat diet develop fatty liver due to activation of transsulfuration pathway [37]. Due to this activation, H_2S levels increase, and plasma homocysteine levels increase [37]. In the liver, cysteine is produced as a result of transsulfuration of homocysteine by two key enzymes, Cystathionine-β-synthase (CBS) and Cystathionine-γ-lyase (CSE) [43, 44]. Glutathione synthesis and its storage in the liver cells is regulated by the production of cysteine by the transsulfuration pathway and catabolism of homocysteine [45, 46]. The evidence of CBS/CSE system in NAFLD has been gathered through knock out studies on mice which have demonstrated that homozygous deletion of CBS results in decreased survival of the animal with damage to the liver [47]. Further, it has been shown that this deletion can increase the homocysteine levels by 40 to 50-fold [47, 48]. Increase in homocysteine levels is accompanied by oxidative stress in hepatocytes, mononuclear infiltration and fibrosis [48, 49]. On the contrary, CSE deletions will not have similar harmful effects on growth and survival if dietary supplementation of cysteine is provided [50, 51]. However, in case of dietary restriction of cysteine, mice showed 90% mortality rate [52]. CSE deletion also leads to 15 times increase in plasma homocysteine levels [52].

NAFLD is presumed to be benign disease however its progression to NASH has been seen in 30% of patients [7]. It has been seen that NAFLD is associated with numerous metabolic disorders including diabetes, and obesity [53–55]. Serum ALT levels have been seen to be elevated in NAFLD while folate and vitamin B12 levels are seen to be reduced [56, 57]. Homocysteine and H_2S levels in plasma have been shown to be regulated through their metabolism in the liver [58–60]. It has been reported previously that changes in homocysteine and H_2S levels have been associated with NAFLD and other metabolic disorders. However, there have been conflicting reports; one study showed a positive relationship between NASH and homocysteine levels [61], while another showed an inverse relationship between homocysteine levels and fibrotic & inflammatory changes in the liver [62]. The systemic homocysteine levels depend upon its release from the liver [58–60] and its clearance via kidney [43]. Thus, the inconsistency among these results between the two studies could be multifactorial. As folate and vitamin B_{12} levels are inversely related with homocysteine [61], one reason for dysregulation of homocysteine could be a defective re-methylation cycle (in which homocysteine is converted to methionine). However, it has been shown that NAFLD is not always associated with perturbations in these vitamins [62–64]. Current evidence suggests that perturbations in CBS/CSE pathway are partly responsible in the pathogenesis of NAFLD. In addition, studies have shown dysregulation of endogenous H_2S in the co-morbidities associated with NAFLD such as diabetes mellitus [65, 66].

There are some genetic mutations which affect the homocysteine metabolism [4, 61]. DNA methylation is inversely related with levels of homocysteine and positively related to folate levels [67]. The methylenetetrahydrofolate reductase (MTHFR) polymorphisms including C677T and A1298C are common gene

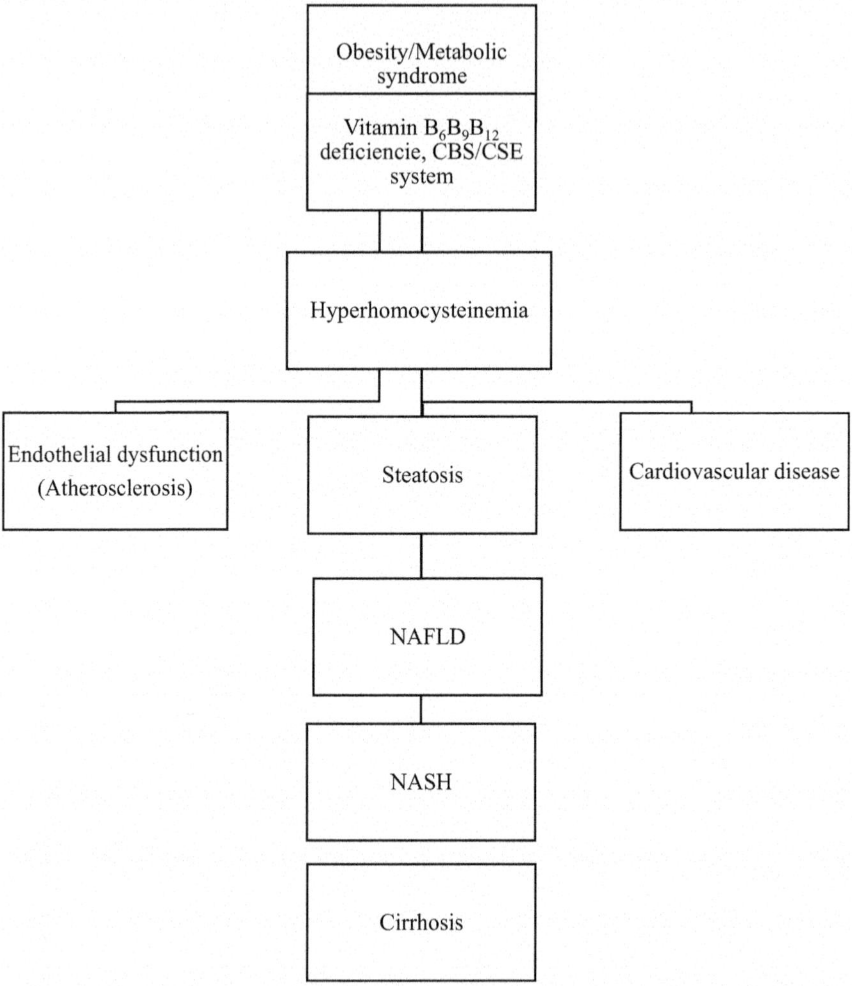

Fig. 11.1 Flowchart on the association of obesity, hyperhomocystenemia and non-alcoholic fatty liver disease. The mechanism of hyperhomocystenemia in obesity is still unclear. The above mentioned are some examples of possible association shown in literature. Both vitamin deficiencies as well as malfunction of the cystathione β synthase, cystathione γ lyase (CBS/CSE) pathway contribute to hyperhomocysteinemia and NAFLD in obesity. There are various other possibilities that remain to be explored. Hyperhomocystenemia is also associated with endothelial dysfunction and cardiovascular disease as discussed earlier in literature [29]. CBS/CSE system: Cystathione β synthase, Cystathione γ lyase. NAFLD: non-alcoholic fatty liver disease. NASH: non alcoholic steatohepatitis

polymorphisms found in different ethnic populations [68]. These genetic changes result in decrease MTHFR activity and increase in homocysteine levels due to dysfunction of remethylation pathway [69, 70].

Conclusion

NAFLD is a growing concern over time owing to the changes in lifestyle and diet in recent times. Physicians face a dilemma in the management of fatty liver. Obesity is one of the key players in NAFLD. Homocysteine has revealed its notoriety in multiple comorbidities, including NAFLD. This provides us with a myriad of opportunities to not only better understand the pathogenic basis of NAFLD, but also to progress towards further establishment of blood-based biomarkers of NAFLD. Association of homocysteine with NAFLD also opens the discussion over the therapeutic approach to NAFLD via modulation of the enzymatic pathways involved. There is still a long way to go down this path of intensive study and research, but it is a promising avenue to proceed upon. A proposed hypothesis by which obesity can lead to non-alcoholic liver disease and cirrhosis can be seen in Fig. 11.1.

References

1. Rector RS, Thyfault JP, Wei Y, Ibdah JA (2008) Non-alcoholic fatty liver disease and the metabolic syndrome: an update. World J Gastroenterol 14(2):185–192
2. Bellentani S, Tiribelli C (2001) The spectrum of liver disease in the general population: lesson from Dionysus study. J Hepatol 35:531–537
3. Shen L, Fan JG, Shao Y et al (2003) Prevalence of nonalcoholic fatty liver among administrative officers in Shanghai: an epidemiological survey. World J Gastroenterol 9:1106–1110
4. Chalasani N, Younossi Z, Lavine JE et al (2012) The diagnosis and management of non-alcoholic fatty liver disease: practice Guideline by the American Association for the Study of Liver Diseases, American College of Gastroenterology, and the American Gastroenterological Association. Hepatology 55(6):2005–2023
5. Loomba R, Sanyal AJ (2013) The global NAFLD epidemic. Nat Rev Gastroenterol Hepatol 10:686–690
6. Chen L, Corrales FJ, Avila MA et al (2002) Methionine adenosyltransferase 1A knockout mice are predisposed to liver injury and exhibit increased expression of genes involved in proliferation. Proc Natl Acad Sci. https://doi.org/10.1073/pnas.091016398
7. Pais R, Charlotte F, Fedchuk L et al (2013) A systematic review of follow-up biopsies reveals disease progression in patients with non-alcoholic fatty liver. J Hepatol 59:550–556
8. Sarna LK, Siow YL, Karmin O (2014) The CBS/CSE system: a potential therapeutic target in NAFLD? Can J Physiol Pharmacol 93(1):1–11. https://doi.org/10.1139/cjpp-2014-0394
9. Dai H, Wang W, Tang X et al (2016) Association between homocysteine and non-alcoholic fatty liver disease in Chinese adults: a cross-sectional study. Nutr J. https://doi.org/10.1186/s12937-016-0221-6

10. Younossi ZM, Koenig AB, Abdelatif D et al (2016) Global epidemiology of nonalcoholic fatty liver disease—meta-analytic assessment of prevalence, incidence, and outcomes. Hepatology. https://doi.org/10.1002/hep.28431
11. George J, Anstee Q, Ratziu V, Sanyal A (2018) NAFLD: the evolving landscape. J Hepatol 68(2):227–229. https://doi.org/10.1016/j.jhep.2017.11.016
12. Angulo P (2002) Nonalcoholic fatty liver disease. N Engl J Med 346(16):1221–1231. https://doi.org/10.1056/NEJMra011775
13. Lonardo A, Nascimbeni F, Mantovani A, Targher G (2018) Hypertension, diabetes, atherosclerosis and NASH: cause or consequence? J Hepatol. https://doi.org/10.1016/j.jhep.2017.09.021
14. Lazo M, Clark JM (2008) The epidemiology of nonalcoholic fatty liver disease: a global perspective. Semin Liver Dis. https://doi.org/10.1055/s-0028-1091978
15. Wree A, Kahraman A, Gerken G, Canbay A (2010) Obesity affects the liver—the link between adipocytes and hepatocytes. Digestion. https://doi.org/10.1159/000318741
16. Varela-Rey M, Embade N, Ariz U et al (2009) Non-alcoholic steatohepatitis and animal models: understanding the human disease. Int J Biochem Cell Biol. https://doi.org/10.1016/j.biocel.2008.10.027
17. Miyake T, Kumagi T, Hirooka M et al (2013) Body mass index is the most useful predictive factor for the onset of nonalcoholic fatty liver disease: a community-based retrospective longitudinal cohort study. J Gastroenterol. https://doi.org/10.1007/s00535-012-0650-8
18. Saida T, Fukushima W, Ohfuji S et al (2014) Effect modification of body mass index and body fat percentage on fatty liver disease in a Japanese population. J Gastroenterol Hepatol. https://doi.org/10.1111/jgh.12377
19. Katrina Loomis A, Kabadi S, Preiss D et al (2016) Body mass index and risk of nonalcoholic fatty liver disease: two electronic health record prospective studies. J Clin Endocrinol Metab. https://doi.org/10.1210/jc.2015-3444
20. Wang L, Guo J, Lu J (2016) Risk factor compositions of nonalcoholic fatty liver disease change with body mass index in males and females. Oncotarget. https://doi.org/10.18632/oncotarget.9691
21. Fan R, Wang J, Du J (2018) Association between body mass index and fatty liver risk: a dose-response analysis. Sci Rep. https://doi.org/10.1038/s41598-018-33419-6
22. Musso G, Gambino R, Cassader M (2009) Recent insights into hepatic lipid metabolism in non-alcoholic fatty liver disease (NAFLD). Prog Lipid Res. https://doi.org/10.1016/j.plipres.2008.08.001
23. Stefan N, Kantartzis K, Häring HU (2008) Causes and metabolic consequences of fatty liver. Endocr Rev. https://doi.org/10.1210/er.2008-0009
24. Baltaci D, Kutlucan A, Turker Y et al (2013) Association of vitamin B12 with obesity, overweight, insulin resistance and metabolic syndrome, and body fat composition; primary care-based study. Med Glas 10(2):203–210
25. Aasheim ET, Hofsø D, Hjelmesœth J et al (2008) Vitamin status in morbidly obese patients: a cross-sectional study. Am J Clin Nutr 87(2):362–369
26. Thomas-Valdés S, Tostes MDGV, Anunciação PC et al (2017) Association between vitamin deficiency and metabolic disorders related to obesity. Crit Rev Food Sci Nutr 57(15):3332–3343. https://doi.org/10.1080/10408398.2015.1117413
27. Sid V, Siow YL, Karmin O (2017) Role of folate in nonalcoholic fatty liver disease. Can J Physiol Pharmacol 95(10):1141–1148. https://doi.org/10.1139/cjpp-2016-0681
28. Guven A, Inanc F, Kilinc M, Ekerbicer H (2005) Plasma homocysteine and lipoprotein (a) levels in Turkish patients with metabolic syndrome. Heart Vessels 20(6):290. https://doi.org/10.1007/s00380-004-0822-4
29. Ganguly P, Alam SF (2015) Role of homocysteine in the development of cardiovascular disease. Nutr J 14(1):6. https://doi.org/10.1186/1475–2891-14-6
30. Werstuck GH, Lentz SR, Dayal S et al (2001) Homocysteine-induced endoplasmic reticulum stress causes dysregulation of the cholesterol and triglyceride biosynthetic pathways. J Clin Invest. https://doi.org/10.1172/JCI11596

31. Obeid R, Herrmann W (2009) Homocysteine and lipids: *S*-Adenosyl methionine as a key intermediate. FEBS Lett. https://doi.org/10.1016/j.febslet.2009.03.038
32. Fonseca V, Dicker-Brown A, Ranganathan S et al (2000) Effects of a high-fat-sucrose diet on enzymes in homocysteine metabolism in the rat. Metabolism. https://doi.org/10.1053/meta.2000.6256
33. Sreckovic B, Sreckovic VD, Soldatovic I et al (2017) Homocysteine is a marker for metabolic syndrome and atherosclerosis. Diabetes Metab Syndr Clin Res Rev. https://doi.org/10.1016/j.dsx.2016.08.026
34. Wang H, Jiang XH, Yang F et al (2003) Hyperhomocysteinemia accelerates atherosclerosis in cystathionine β-synthase and apolipoprotein E double knock-out mice with and without dietary perturbation. Blood. https://doi.org/10.1182/blood-2002-08-2606
35. Boushey CJ, Beresford SAA, Omenn GS, Motulsky AG (1995) A quantitative assessment of plasma homocysteine as a risk factor for vascular disease: probable benefits of increasing folic acid intakes. JAMA J Am Med Assoc. https://doi.org/10.1001/jama.1995.03530130055028
36. Bravo E, Palleschi S, Aspichueta P et al (2011) High fat diet-induced non alcoholic fatty liver disease in rats is associated with hyperhomocysteinemia caused by down regulation of the transsulphuration pathway. Lipids Health Dis. https://doi.org/10.1186/1476-511X-10-60
37. Hwang S-Y, Sarna LK, Siow YL, Karmin O (2013) High-fat diet stimulates hepatic cystathionine β-synthase and cystathionine γ-lyase expression. Can J Physiol Pharmacol 91 (11):913–919. https://doi.org/10.1139/cjpp-2013-0106
38. Peh MT, Anwar AB, Ng DS et al (2014) Effect of feeding a high fat diet on hydrogen sulfide (H2S) metabolism in the mouse. Nitric Oxide Biol Chem. https://doi.org/10.1016/j.niox.2014.03.002
39. Dahlhoff C, Worsch S, Sailer M et al (2014) Methyl-donor supplementation in obese mice prevents the progression of NAFLD, activates AMPK and decreases acyl-carnitine levels. Mol Metab. https://doi.org/10.1016/j.molmet.2014.04.010
40. Chitturi S, Farrell GC, Hashimoto E et al (2007) Non-alcoholic fatty liver disease in the Asia-Pacific region: definitions and overview of proposed guidelines. J Gastroenterol Hepatol. https://doi.org/10.1111/j.1440-1746.2007.05001.x
41. Fan JG, Jia JD, Li YM et al (2011) Guidelines for the diagnosis and management of nonalcoholic fatty liver disease: Update 2010: (Published in Chinese on Chinese Journal of Hepatology 2010; 18:163–166) JG Fan et al. Diagnosis and management of NAFLD. J Dig Dis. https://doi.org/10.1111/j.1751-2980.2010.00476.x
42. Won BY, Park KC, Lee SH et al (2016) Sex difference in the association between serum homocysteine level and non-alcoholic fatty liver disease. Korean J Fam Med. https://doi.org/10.4082/kjfm.2016.37.4.242
43. Finkelstein JD (1998) The metabolism of homocysteine: pathways and regulation. Eur J Pediatr 157(Suppl 2):S40–S44
44. Stipanuk MH (2004) Sulfur amino acid metabolism: pathways for production and removal of homocysteine and cysteine. Annu Rev Nutr 24:539–577
45. Mosharov E, Cranford MR, Banerjee R (2000) The quantitatively important relationship between homocysteine metabolism and glutathione synthesis by the transsulfuration pathway and its regulation by redox changes. Biochemistry 39:13005–13011
46. Vitvitsky V, Mosharov E, Tritt M et al (2003) Redox regulation of homocysteine-dependent glutathione synthesis. Redox Rep 8:57–63
47. Watanabe M, Osada J, Aratani Y et al (1995) Mice deficient in cystathionine beta-synthase: animal models for mild and severe homocyst(e)inemia. Proc Natl Acad Sci USA 92(5):1585–1589
48. Robert K, Nehme J, Bourdon E et al (2005) Cystathionine beta synthase deficiency promotes oxidative stress, fibrosis, and steatosis in mice liver. Gastroenterology 128(5):1405–1415
49. Woo CW, Prathapasinghe GA, Siow YL, Karmin O (2006) Hyperhomocysteinemia induces liver injury in rat: protective effect of folic acid supplementation. Biochim Biophys Acta 1762 (7): 656–665

50. Yang G, Wu L, Jiang B et al (2008) H2S as a physiologic vasorelaxant: hypertension in mice with deletion of cystathionine gammalyase. Science 322(5901):587–590
51. Sturman JA, Gaull G, Raiha NC (1970) Absence of cystathionase in human fetal liver: is cystine essential? Science 169(3940):74–76
52. Mani S, Yang G, Wang RA (2011) Critical life-supporting role for cystathionine gamma-lyase in the absence of dietary cysteine supply. Free Radic Biol Med 50(10):1280–1287
53. Adams LA, Angulo P (2005) Recent concepts in non-alcoholic fatty liver disease. Diabet Med 22:1129–1133
54. Sartorio A, Del Col A, Agosti F et al (2007) Predictors of non-alcoholic fatty liver disease in obese children. Eur J Clin Nutr 61:877–883
55. Anstee QM, Targher G, Day CP (2013) Progression of NAFLD to diabetes mellitus, cardiovascular disease or cirrhosis. Nat Rev Gastroenterol Hepatol 10:330–344
56. Rodríguez-Hernández H, Gonzalez JL, Márquez-Ramirez MD et al (2008) Risk factors associated with nonalcoholic fatty liver disease and its relationship with the hepatic histological changes. Eur J Gastroenterol Hepatol 20(5):399–403
57. Koplay M, Gulcan E, Ozkan F (2011) Association between serum vitamin B12 levels and the degree of steatosis in patients with nonalcoholic fatty liver disease. J Investig Med 59 (7):1137–1140
58. Jensen KK, Geoghagen NS, Jin L et al (2011) Pharmacological activation and genetic manipulation of cystathionine betasynthase alter circulating levels of homocysteine and hydrogen sulfide in mice. Eur J Pharmacol 650(1):86–93
59. Norris EJ, Culberson CR, Narasimhan S, Clemens MG (2011) The liver as a central regulator of hydrogen sulfide. Shock 36(3):242–250
60. Stead LM, Brosnan ME, Brosnan JT (2000) Characterization of homocysteine metabolism in the rat liver. Biochem J 350(3):685–692
61. Gulsen M, Yesilova Z, Bagci S et al (2005) Elevated plasma homocysteine concentrations as a predictor of steatohepatitis in patients with non-alcoholic fatty liver disease. J Gastroenterol Hepatol 20(9):1448–1455
62. Polyzos SA, Kountouras J, Patsiaoura K et al (2012) Serum homocysteine levels in patients with nonalcoholic fatty liver disease. Ann Hepatol 11(1):68–76
63. Bosy-Westphal A, Petersen S, Hinrichsen H et al (2001) Increased plasma homocysteine in liver cirrhosis. Hepatol Res 20(1):28–38
64. Hirsch S, Poniachick J, Avendaño M et al (2005) Serum folate and homocysteine levels in obese females with non-alcoholic fatty liver. Nutrition 21(2):137–141
65. Jain SK, Bull R, Rains JL et al (2010) Low levels of hydrogen sulfide in the blood of diabetes patients and streptozotocin-treated rats causes vascular inflammation? Antioxid Redox Signal 12(11):1333–1337
66. Whiteman M, Gooding KM, Whatmore JL et al (2010) Adiposity is a major determinant of plasma levels of the novel vasodilator hydrogen sulphide. Diabetologia 53(8):1722–1726
67. de Carvalho SC, Muniz MT, Siqueira MD et al (2013) Plasmatic higher levels of homocysteine in non-alcoholic fatty liver disease (NAFLD). Nutr J 12:37. https://doi.org/10.1186/1475-2891-12-37
68. Fukuda N, Hamajima N, Wakai K, Suzuki K (2014) A cross-sectional study to find out the relationship of methylenetetrahydrofolate reductase (MTHFR) C677T genotype with plasma levels of folate and total homocysteine by daily folate intake in Japanese. J Nutr Sci Vitaminol (Tokyo) 60(4):231–238
69. Zhu Y, Zhu RX, He ZY et al (2015) Association of MTHFR C677T with total homocysteine plasma levels and susceptibility to Parkinson's disease: a meta-analysis. Neurol Sci 36 (6):945–951
70. García-Minguillán CJ, Fernandez-Ballart JD, Ceruelo S et al (2014) Riboflavin status modifies the effects of methylenetetrahydrofolate reductase (MTHFR) and methionine synthase reductase (MTRR) polymorphisms on homocysteine. Genes Nutr. 9(6):435

Chapter 12
Mechanisms for Obesity Related Kidney Disease

Praveen Murlidharan, Sreelekshmi Kamaladevan, Satish Balan and Chandrasekharan C. Kartha

Abstract Obesity is a major cause for the initiation and progression of kidney injury resulting in chronic kidney disease (CKD) and end stage renal disease. High body mass index is a major risk factor for new-onset CKD as well. The well recognized kidney disease secondary to obesity is glomerulopathy. Typical histological features of obesity related glomerulopathy include glomerulomegaly and focal segmental glomerulosclerosis. In obese individuals, excess excretory load induces hyperperfusion and hyperfiltration by the kidneys leading to glomerulomegaly. Lipid accumulation in the kidney which accompanies excessive fat deposition in the body is implicated in the development of CKD. Obesity is associated with metabolic abnormalities in the adipose tissue such as increased free fatty acids, hyperinsulinemia, insulin resistance, pro inflammatory conditions, reduced adiponectin, leptin resistance and activation of the renin-angiotensin-aldosterone system, all of which mediate injury to the cells of glomeruli and tubules leading to CKD. Despite much progress in our understanding of the mechanisms of obesity related kidney disease, several questions about the pathogenesis of nephropathy associated with obesity remain to be answered. Delineating obesity linked factors, which lead to adaptive and maladaptive changes in the kidney and predispose patients to renal disease could lead to identification of molecular targets and reno-protective and therapeutic strategies to improve outcomes for obese patients with CKD.

P. Murlidharan · S. Balan
Department of Nephrology, Kerala Institute of Medical Sciences, Anayara 695029, Trivandrum, India

P. Murlidharan · S. Kamaladevan
Division of Clinical Research, Kerala Institute of Medical Sciences, Anayara 695029, Trivandrum, India

C. C. Kartha (✉)
Society for Continuing Medical Education and Research, Kerala Institute of Medical Sciences, Anayara 695029, Trivandrum, India
e-mail: drkartha.cc@kimsglobal.com

P. S. Tappia et al. (eds.), *Pathophysiology of Obesity-Induced Health Complications*, Advances in Biochemistry in Health and Disease 19,
https://doi.org/10.1007/978-3-030-35358-2_12

Keywords Obesity · Chronic kidney disease · Glomerulopathy · Glomerulosclerosis · Adipose tissue · Adipokines · Insulin resistance · Cytokines

Introduction

Obesity is currently an epidemic in developed countries and a global health challenge. A significant association has been noted between obesity and initiation and progression of chronic kidney disease (CKD) in population-based studies [1–6]. The incidence of obesity-related renal disease has increased ten times in recent years. Prevalence of obesity which is estimated to rise by 40% in the next decade is expected to result in a parallel escalation in the incidence of obesity related CKD [7, 8].

High body mass index (BMI) is a major risk factor for new-onset CKD [1, 2, 9]. In a study of 75,000 Norwegians, who were followed up for 21 years, increased BMI was found to correlate with initiation of kidney disease or CKD related mortality [5]. A 600% increase in end stage renal disease (ESRD) was observed in those with a BMI more than 40 kg/m^2 in a survey of more than 3 lakhs individuals enrolled in the Kaiser Permanent Health System between 1964 and 1985 [8, 10]. Obese individuals are at a greater risk for acute kidney injury, for nephrolithiasis and kidney cancer as well [7].

Until recently, obesity related kidney disease was thought to be linked to obesity associated hypertension, diabetes and cardiovascular disease, which are all common causes of renal disease. There is increasing evidence that even without other risk factors, obesity can initiate kidney disease and also accelerate progression of pre-existing renal disease [5, 11, 12]. Several obesity-induced disorders are nephrotoxic [13].

Obesity Related Glomerulopathy

The first record on the link between obesity and massive proteinuria was by Weisinger et al. [14]. Later, many reports confirmed the association of obesity with proteinuria, glomerulomegaly and frequently, focal segmental glomerulosclerosis (FSGS) [15–18]. FSGS is not observed in all cases of obesity related kidney disease; its presence may depend on the degree of obesity or renal impairment. Among various sub-types of FSGS, the perihilar variant is more common [17].

In 2001, Kambham and colleagues proposed the term 'obesity-related glomerulopathy' (ORG) for the microscopic lesions observed in kidney biopsies from obese individuals [17]. The diagnostic criteria for ORG are: BMI values of 30 kg/m^2 or greater and absence of clinical as well as biopsy evidence of other renal diseases [17]. The concept of ORG as a nephropathy does not for its diagnosis, depend on the manifestation of proteinuria. Patients with ORG may also not have edema [17]. Kambham et al. noted that ORG is less likely than idiopathic FSGS to present with edema or the degree of proteinuria as seen in nephrotic syndrome. ORG is however frequently associated with hypertension and dyslipidemia [17].

Isolated proteinuria of unknown onset with or without renal impairment, is the initial symptom in most cases. Even with relatively high excretion of protein in the urine, hypoalbuminemia may not be present [17]. The mechanism for this remains obscure. Typically, the clinical condition is stable. Alternately, the patient may have slowly progressive proteinuria. Significantly, weight loss can strikingly reduce proteinuria associated with obesity [18, 19]. In 10–33% of patients diagnosed with ORG, long-term outcomes include progression to ESRD [17].

Renal Changes in Obesity

Structural Changes

Much is currently known about the structural and hemodynamic changes in the kidneys of the obese (Table 12.1) [20–23].

Autopsies have revealed an increase in kidney weight in those with a high BMI [22]. Increase in kidney weight may result from hypertrophy of individual nephrons secondary to increased tubular and glomerular functions related to increase in BMI.

Table 12.1 Structural and functional changes in kidney in obesity

A. Structural changes	
1.	Increase in kidney weight
2.	Intra cellular (mesangial and tubular cells) and extracellular lipid accumulation
3.	Glomerulomegaly
4.	Thickening of glomerular basement membrane
4.	Focal segmental glomerulosclerosis
5.	Enlargement of podocyte foot processes
6.	Podocyte detachment
7.	Decreased podocyte density
8.	Increased cross sectional area of tubular epithelial cells
9	Tubulointerstitial damage
10.	Increased microvessels
B. Hemodynamic changes	
1.	Increased glomerular filtration rate
2.	Increased plasma flow
3.	Increased filtration fraction
4.	Afferent arteriolar vasodilatation
5.	Efferent arteriolar vasoconstriction
6.	Disrupted tubuloglomerular feedback

Intracellular or extracellular accumulation of fluid and lipid components may also contribute to a larger kidney.

Glomerular sizes are larger in the obese, even without obvious renal disease [23–25]. There is no unanimity on the definition of glomerulomegaly in ORG. A morphometric study found that in patients with ORG and preserved renal function, when compared to control subjects, the mean glomerular volume is increased about 3-fold [26]. The glomerular capillaries are increased in number. The increase in vascular endothelial growth factor (VEGF) expression in the glomeruli of patients with ORG may contribute to the formation of new micro vessels [27]. Cell proliferation and matrix synthesis could also contribute to glomerular enlargement. Glomerulomegaly is accompanied by a 45% reduction in podocyte density [24]. Thickening of the glomerular basement membrane (GBM) considered as an early manifestation of diabetic nephropathy, is seen with obesity as well. Both glycemic and lipid abnormalities in obesity may contribute to GBM thickening, which may not be as severe as seen in patients with type 2 diabetes.

Obesity-induced glomerular hypertrophy and glomerulomegaly may cause glomerular podocytes to enlarge their foot processes to cover the expanded glomerular surface area. Consistent with this, a relative reduction in the coating area of glomerular podocytes on the glomerular surface is found in patients with ORG [28]. This may cause changes in podocyte function and a consequent loss in protein selectivity, podocyte detachment, and replacement by matrix deposition, leading to FSGS. Incidence of foot process fusion among glomerular podocytes is significantly lower in ORG than in idiopathic FSGS [17].

Focal lipid vacuoles are occasionally seen in the cytoplasm of glomerular mesangial cells and tubular epithelial cells [29]. A study comparing the kidney biopsies from obese patients with proteinuria and biopsies from non-obese patients with proteinuria found a 33% increase in the cross-sectional area of proximal tubular epithelial cells and 54% increase in the lumen of the proximal tubules [30].

Changes in Hemodynamics

Hemodynamics in the kidney is markedly altered by obesity. In obese individuals, there is compensatory hyperfiltration to meet the high metabolic needs of the larger fat mass.

A recent study found that glomerular filtration rate (GFR) is higher in obese adults than in those with normal body weight [30, 31]. Renal plasma flow (RPF) is also increased, though not to the same degree. As a result, the filtration fraction (FF) increases, a hemodynamic adjustment that parallel the degree of BMI and adipose tissue mass. Studies in obese individuals suggest that afferent arteriolar vasodilatation, together with efferent arteriolar vasoconstriction, contribute to the increase in FF [31–34]. By lowering tubular sodium chloride relative to GFR, obesity-dependent mechanisms disrupt the tubuloglomerular feedback (TGF) response, preventing suppression of GFR [35]. Given the high rate of association of hypertension with obesity, inadequate TGF may result in the transmission of systemic blood pressure to the glomerulus contributing not only to increased GFR but also to structural changes in the kidney [35]. Increase in obesity-induced GFR is not however permanent [36, 37].

The changes in renal hemodynamics found in obesity are closely linked to increased salt sensitivity [38]. Activation of the renin–angiotensin–aldosterone system (RAAS) in the kidney is an important mechanism by which salt sensitivity is increased in obesity [37, 39]. Activation of renal sympathetic nerves may also be involved in the increased salt reabsorption seen with obesity [40, 41]. The glomerulus enlarges in response to increases in GFR, RPF, FF and tubular sodium reabsorption.

Potential Mechanisms of Renal Injury

Kidney disease is initiated with increases in GFR and pressure in glomerular capillaries (P_{GC}), followed by hypertrophy of the glomerulus as well as podocytes [42, 43]. Micro albuminuria ensues and progresses to proteinuria. Mesangial cell proliferation, mesangial matrix expansion, nodular glomerulosclerosis and tubulointerstitial injury contribute to a decrease in GFR culminating in ESRD [44, 45]. The exact mechanisms for obesity associated CKD in humans are not yet completely clear. Current understanding is based upon association studies, cell and animal studies and pharmacological manipulations.

Several obesity-related factors have been implicated in the progression of CKD (Fig. 1). These may act singly or together to cause renal injury. Potential mechanisms are: (i) adverse effects of adaptations to increase in body mass and excretory load, (ii) adverse effects of adaptations to obesity-induced sodium retention [44], (iii) insulin resistance (IR) and progressive hyperglycemia (iv) direct or indirect effects of hyperinsulinemia (v) renal lipotoxicity [46–48] (vi) rise in free fatty acids (FFAs) which may aggravate insulin resistance and (vii) fatty acid/triglyceride accumulation within tissues resulting in cellular dysfunction [49].

The effects of IR, dyslipidemia and oxidative stress are likely to be intensified by dysregulation of adipocytokines such as leptin and adiponectin and proinflammatory cytokines [49]. Resistin, corticosteroids, nutritional status and genetic factors are the other factors implicated.

Excess Excretory Load

Obesity is associated with an excess excretory load resulting from increased body mass and the increased energy intake and tissue turnover required to maintain it. Fat-free body mass also increases in obesity harmonious with functional overload [50]. Organomegaly in obesity includes the kidneys [15]. Chagnac et al. confirmed

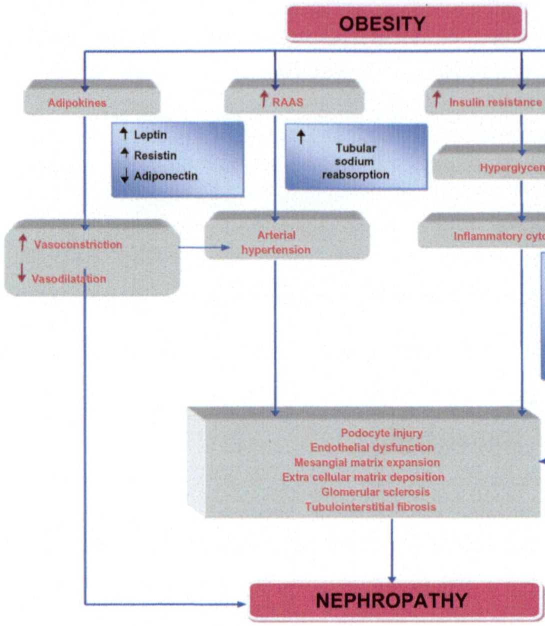

Fig. 1 Mechanisms of renal disease in Obesity. RAAS—Renin-angiotensin-aldosterone system, FFAs—free fatty acids, TNF—tumor necrosis factor, IL—interleukin, CRP—C reactive protein, MCP—monocyte chemo attractive protein, TGF—transforming growth factor, PAI—plasminogen activator inhibitor

a 51% increase in renal perperfusion and 31% increase in filtration in severe obesity [33]. The reduced renal resistance with increased FF is compatible with glomerular capillary hypertension [33], a perfect setting for future glomerulosclerosis. Obesity induces single-nephron adaptations typical of the reduced nephron number accompanying CKD.

Adverse Adaptations to Excess Retention of Sodium

Hall et al. proposed that in obesity, there is reduced capacity for sodium excretion, acting at sites proximal to the macula densa through Ang II and sympathetic activation. The reduced sodium chloride delivery to the macula densa site induces afferent vasodilation and renin release to produce compensatory glomerular hyperfiltration, thus restoring normal distal delivery [46]. Thomson et al. considers a mechanism similar for the increase in proximal tubular sodium reabsorption induced by hyperglycemia seen in diabetes [42]. The result is intraglomerular hypertension and proteinuria which form the final common pathway for chronic glomerular and tubular injury, as with excess excretory load induced hyperfiltration.

Hypertension

Hypertension is common in obesity and may be due to a range of factors including sympathetic nervous system activation and angiotensinogen release from adipose tissue [36].

The rise in blood pressure may damage microvasculature within the kidney through intensified RAAS activity. Ang II excess in the kidney can enhance renal injury through proinflammatory mechanisms and can also promote proteinuria-related renal damage [49].

The Role of Adipose Tissue

Adipose tissue is a highly active endocrine organ involved in the clearance and storage of fatty acids, regulation of energy homeostasis and metabolism, insulin function and inflammatory processes. Adipose tissue is involved in progression of diseases associated with metabolic dysfunction [51–56]. Increased fat deposition in obesity also correlates with renal injury [29].

Obesity is a chronic low-grade inflammatory condition, in which adipose tissue serves as the source of inflammatory cytokines [57]. Visceral adipose tissue produces less adiponectin and more of pro-inflammatory cytokines, including tumor necrosis factora (TNF-α) and interleukin-6 (IL-6), which can induce IR and promote endothelial dysfunction [58].

There is in obesity, lipid accumulation in the kidney suggesting a role for fat accumulation in the organ for the development of CKD [59–62]. Visceral fat elaborates bioactive substances which contribute to the abnormal hemodynamic and structural changes leading to obesity-related nephropathy. Visceral adipose tissue mediates obesity-related disease through production of pro-inflammatory cytokines (adipokines) and causing IR [63]. Adipocytes secretes angiogenic factors as well and thus can facilitate vasculogenesis locally and in distal organs [64]. In a study using obese Zucker rats, in parallel with intrarenal inflammation, significant increases in the density of cortical and medullary microvascular was found [65].

Adipocytes contain adipocyte-specific metabolites such as free fatty acids, leptin, and adiponectin and all the components of the RAAS, plasminogen activator inhibitor (PAI), all of which affect renal structure and function. In addition, fat is infiltrated by macrophages that can alter their phenotype and foster a proinflammatory environment which in the kidney, advances pathophysiologic changes associated with obesity. Adipose tissue of patients with ESRD has elevated levels of the pro-inflammatory cytokines TNF-α and monocyte chemoattractant protein-1 (MCP-1) and more infiltration of macrophages [66]. Elevated levels of pro-inflammatory cytokines or chemokines (TNF-α, IL-6, IL-1β, MCP-1) and infiltrated macrophages in the adipose tissue correlate with renal inflammation in obese rodents [67].

Metabolic Factors

Altered fatty acid and cholesterol metabolism is responsible for ipid accumulation, inflammation, oxidative stress and fibrosis in the kidney [68]. BMI seems to determine the degree of accumulation of triglycerides in the human renal cortex [69]. Triglyceride accumulation is seen in both glomerular and tubular cells, more in proximal tubular cells [69].

Lipid disturbances in obesity appears to directly involve in renal damage. Young C57BL/6 mice fed a high fat diet (HFD) have increased body weight, and elevated blood levels of glucose, insulin, triglycerides and cholesterol and lower circulating adiponectin. Proteinuria, glomerulomegaly, expanded mesangial matrix, thickened glomerular basement membrane and podocyte effacement have been observed in these mice [59]. There is evidence that lipid moieties can injure mesangial cells. Low density lipoprotein (LDL), oxidized LDL, and glycated LDL, at levels observed in blood significantly increase synthesis of mesangial matrix components, fibronectin and laminin [70]. Lipid moieties also stimulate production of macrophage migration inhibitory factor in mesangial cells and expression as well as release of inflammatory activators, CD40 and IL-6 [70]. Renal toxicity and proteinuria in mice with hyperlipidemia can be attenuated by treatment with anti-IL-6 monoclonal antibody [71].

Lipids also damage podocytes [28]. Oxidized LDL decreases phosphorylation of AKT, involved in cell survival and can thus contribute to nephrin loss and podocyte apoptosis [72, 73]. In podocytes cultured with palmitate, there is increase in the synthesis of ceramide resulting in reduced insulin-stimulated glucose uptake [74]. Thus, lipid abnormalities may interact with changes in glucose metabolism to actuate nephropathy.

Normal insulin/phosphatidylinositol 3-kinase/AKT and mTOR signalling are critical for podocyte hypertrophy and adaptation. Adipokines and lipid stores in the kidney result in IR in podocytes and maladaptive responses, to cope with the mechanical forces of hyperfiltration [68].

Sterol regulatory element binding protein-1 (SREBP-1) also seem to play a role in the damage to the kidney from lipid accumulation and ensuing inflammatory and fibrotic responses [75, 76]. The effects of HFD on the kidney were not seen in SREBP-1c −/− mutant mice, while SREBP-1a transgenic mice had lipid deposition in the glomeruli, glomerulosclerosis and albuminuria. In patients with ORG, expression of SREBP-1 in the glomeruli is up-regulated [27].

Adipokines

Visceral fat releases into the circulation a large number of adipocytokines with autocrine, paracrine and endocrine activities, which contribute to the pathogenesis of renal injury [77].

The secreted peptides include leptin, adiponectin, resistin, TNF-α, IL-6 and components of the RAAS such as angiotensinogen, ACE and ATII-1R as well as

VEGF, MCP-1, RBP-4 and TIMP-1. There is a positive correlation between increase in adipose tissue and elevated inflammatory markers [78]. Axelsson et al. 2004 observed a positive relationship between truncal fat mass and inflammation in patients with ESRD [79]. Adipose tissue also has macrophages and immune cells which secrete pro-inflammatory cytokines. The link between dysfunctional adipose tissue in obesity and raised proinflammatory adipocytokine patterns, systemic inflammation, IR and cellular dysfunction, is well recognized [49]. Be that as it may, role of this link in the pathogenesis of CKD is not well elucidated.

Leptin

Leptin levels rise in response to increase in fat mass. Elevated leptin levels in obese individuals correspond to the fat stores. Central hypothalamic resistance to leptin may also be present in the obese. Leptin has immuno-regulatory and proinflammatory actions in the obese [80].

In patients with obesity, leptin levels rise and there is leptin resistance during the initial stages of development of CKD [81]. When CKD progresses, reduced renal clearance of leptin contributes to hyperleptinemia and is associated with concomitant inflammation [82, 83].

Mice overexpressing leptin have more renal disease than leptin deficient mice [84]. Long-term infusion of recombinant leptin in rats is associated with proteinuria, increased expression of extracellular matrix proteins (collagen type IV), TGF-β and other pro-inflammatory cytokines, macrophage infiltration and glomerulosclerosis [85]. These observations indicate that leptin promotes renal injury in obese subjects. Despite severe obesity, renal dysfunction is not seen in the absence of leptin or mutation in the leptin receptor gene [13].

Adiponectin

Adiponectin (also called Acrp30), one of the most abundant adipokines produced by the adipocytes is down-regulated in obesity and type 2 diabetes [86]. Low adiponectin levels is linked to inflammation, atherosclerosis, IR, and raised blood pressure [87]. Endothelial cell dysfunction, impaired endothelium-dependent vasodilation, enhanced leukocyte-endothelium adhesion and activation of RAAS have been found in both humans and experimental animals with decreased levels of adiponectin.

Adiponectin is also known to aid the normal function of the podocyte [88]. In hypoadiponectinemia, function of podocytes in maintaining glomerular filtration breaks down causing glomerular damage and sclerosis. Glomerulomegaly, collagen deposition in glomeruli, loss of podocyte foot processes, increased TGF-β, and albuminuria have all been observed in adiponectin null mutant mice [89]. Adiponectin restitution reverts podocyte effacement and albuminuria. This benefit is attributed to reduction in oxidant stress [88, 90]. Adiponectin deficiency can also lead to increased NADPH oxidase activity and a rise in the levels of reactive oxygen species.

Adiponectin is an insulin-sensitizing factor as well and has also anti-inflammatory effects. Reduced plasma adiponectin level is inversely correlated with IR in obese patients [88, 89]. Though obese subjects have consistently low circulating adiponectin levels, in patients with CKD due to obesity, adiponectin levels are increased, possibly because of renal dysfunction [90, 91].

The role of adiponectin in obesity-related disease has been extensively investigated using transgenic mice or pharmacological globular Acrp30 compound [92–96]. Adiponectin is an important regulator of lipid and glucose metabolism and a key link among TNF-α, MCP-1, and IR.

Effects of adiponectin are tightly linked to the activation of AMP-activated protein kinase (AMPK) [88, 97]. Obesity is known to modulate the activity of AMPK. Steinberg et al. demonstrated that TNF-α could suppress AMPK activation through the TNF receptor 1 (TNFR1), suppress fatty acid oxidation and promote IR in skeletal muscle [98]. How TNF-α inhibits AMPK activation is unclear. Steinberg et al. showed that this process might involve the upregulation of protein phosphatase 2C (PP2C) by TNF-α. Treatment with TNF-α increases PP2C activity and decreases AMPK activation in WT mice but not in the transgenic ob/ob TNFR$^{-/-}$ mice. This change is also accompanied with a reduction of fatty acid oxidation and an increase of diacylglycerol (DAG) and triacylglycerol (TAG) in skeletal muscle [103]. AMPK activation reduces TNF-α and increases adiponectin levels in human adipose tissue, improving insulin sensitivity [99]. A decrease of adiponectin and AMPK activation is also associated with increase in the levels of MCP-1 in human adipocytes [100].

Recently, a role for AMPK in regulating macrophage infiltration and activation has been proposed [101]. AMPK activation completely reverses HFD induced infiltration of macrophages in the kidney [62]. AMPK activation is also a key regulator of lipid accumulation in vacuolated proximal tubular cells, maintenance of integrity of the brush border, as well as nitrotyrosine and NOX4 levels. These findings suggest that AMPK activation may have a role in tubular dysfunction [62].

Macrophage Infiltration of Adipose Tissue and Phenotype Switch in Macrophages

There are evidences for macrophage influx in adipose tissue of obese humans and animal models of obesity [102–104]. Macrophage infiltration of adipose tissue results in inflammation and IR [102, 103]. Obesity induces a phenotype switch in macrophages [105–108]. In lean rodents, M2 phenotype, involved in the resolution of inflammation and tissue repair are predominant; proinflammatory M1 macrophage population is dominant in obese animals [106]. Macrophages that infiltrate the adipose tissue are a source of a large number of proinflammatory mediators such as TNF-α, IL-6, C-reactive protein (CRP), MCP-1 and macrophage migration inhibitory factor [109]. Inhibition of proinflammatory macrophages reduces kidney injury [110–112]. There is indication from animal experiments that macrophage AT1 receptor may mediate macrophage polarization [113].

Adiposity-Driven Proinflammatory Cytokines

Inflammation markers are inversely associated with measures of kidney function and positively with albuminuria [114]. Thus, there is a strong evidence for the contribution of inflammation in obesity associated renal disease.

Rapid expansion of adipose tissue results in an altered synthesis of pro-inflammatory adipokines which leads to a state of low-grade inflammation [115]. Among the large number of pro-inflammatory adipokines, TNF-α is one of the most critical mediators of adipose tissue inflammation and development of IR [116, 117].

Fatty acids released by adipocytes promote TNF-α release by macrophages which, in turn, increases IL-6 production in fat cells and thus an inflammatory milieu in both adipose tissue and kidney [82]. TNF-α is an important mediator of progressive renal fibrosis. Gene expressions of both TNF-α and its receptors, as well as IL-6, a signal transducer are seen increased in glomeruli of patients with ORG and indicates the importance of TNF-α and IL-6 in development of ORG [27].

A rise in TNF-α levels is generally associated with increased production of MCP-1 by both adipocytes and macrophages. MCP-1 is also a key mediator of both adipose tissue inflammation and development of IR [117–119]. The mediatory role of MCP-1 and its receptor CCR2 in chronic kidney disease has recently received much attention [120]. In human podocytes, MCP-1 regulates nephrin expression via CCR2 [121]. Studies in mesangial cells have revealed that palmitate stimulates marked secretion of MCP-1 indicating that in obesity, circulating saturated fatty acids, such as palmitate may trigger production of MCP-1 [61].

The major source of pro-inflammatory cytokines that directly contribute to renal injury in obese subjects are the infiltrated macrophages [84]. In addition, renal parenchyma also releases proinflammatory cytokines in response to hyperglycemia or locally active vasoactive peptides such as Ang II [122]. These mediators produce low grade chronic inflammation and participate in the pathogenesis of ORG. TNF-α has been shown to reduce the expression of key components of the slit diaphragm, nephrin and podocin, thus contributing to podocytopathy [78]. IL-6 promotes expression of adhesion molecules and oxidative stress [123].

Insulin Resistance

Insulin resistance (IR) is a common accompaniment of obesity and related metabolic syndrome [124, 125]. In obese rodents and humans, inflamed adipose tissue is known to contribute to development of IR [124, 126]. IR is a salient metabolic risk for CKD [127]. Many studies indicate the association between IR and CKD. This association is seen even before the onset of diabetes [128, 129].

The direct link between adipose tissue dysfunction and associated IR with obesity related kidney disease is presently more evident. The low adiponectin levels in obesity is associated with reduced insulin sensitivity, which leads to a pro-inflammatory state in the kidney [130].

In the captive rhesus monkey with spontaneous obesity, glomerular hypertrophy appears in the prediabetic hyperinsulinemic phase; hyperglycemia, hypertension, renal dysfunction, and increase in mesangial matrix deposition are absent at that stage [131].

Insulin, although a weak vasodilator, augments endothelial-dependent vasodilation. Hyperinsulinemia can contribute to preglomerular vasodilation, glomerular hypertension and increase in glomerular capillary permeability [132, 133]. Structural damage is not manifest during this period. Hyperinsulinemia may induce glomerular hypertrophy either directly or by stimulating the IGF-1 receptor [134]. Hyperinsulinemia can also augment Ang II contraction of glomerular mesangial cells [135]. Abrass et al. demonstrated that high-dose insulin stimulates expression of inflammatory collagens in renal mesangial cells in culture [136]. They also found that the altered gene expression after exposure to high insulin is not reversible by later withdrawal of insulin [137, 138].

Walsh et al. through a study in transgenic mice missing insulin receptors in their podocytes, have elucidated the critical role of insulin signaling in normal podocytes [139]. They showed that these mice have normal glomeruli when they are at their early age (three weeks old), but later starting at the age of 5 weeks, have loss of podocyte foot processes, increased glomerular matrix and albuminuria [139].

Free Fatty Acids (FFA) might also contribute to IR [74]. Increased FFA flux from excess adipose tissue to non-adipose organs results in lipid accumulation in ectopic organs including the kidney. Lipid stores later advances impairment of glucose metabolism and insulin sensitivity in these organs. This adverse effect is associated with an increase of ceramide, a highly lipotoxic molecule, that is related to IR [140]. In addition, a dysregulation of the insulin receptor and the impairment of recruitment of the glucose transporter GLUT4 to the cell surface have been noted with FFA increase [74].

Though there are evidences for considering IR as a driver of the renal disease in obesity, how critical is IR for progression of the disease is still unclear.

There are many similarities between the 'obese' and 'diabetic' kidney; there are also features unique to obesity sans diabetes. Kidneys of obese individuals frequently have lipid deposits (foam cells) in glomeruli and mesangium. This is an evidence for the concept that renal injury is caused by lipotoxicity. Lipid accumulation in the glomerulus may result in the upregulation of SREBP-1 and 2 and promote podocyte apoptosis, mesangial cell proliferation and cytokine synthesis [140].

Vasoactive Peptides

Several vasoactive peptides have been implicated in the pathogenesis of OGR. There is upregulation in the intrarenal RAAS in obesity [141, 142]. Activation of the RAAS leads to both hemodynamic and cellular effects. Ang II leads to increases in efferent arteriolar vasoconstriction and glomerular pressure, sodium retention and cell proliferation [143–145]. In cells, Ang II activates protein kinase C (PKC), MAP kinase (MAPK) and transcription factors such as nuclear factor-κB. Their activation leads to alteration in the expression of genes of a number of growth factors and cytokines. Increase in TGF-β promotes podocyte apoptosis, mesangial cell proliferation and extracellular matrix synthesis, cellular events that are important in the development of obesity- and diabetes-associated glomerulopathy [146].

Renin-Angiotensin-Aldosterone System (RAAS)

Adipose tissue has all the components of the RAAS system. RAAS is activated in the adipose tissue of the obese [147, 148]. In obese adipocytes, production of angiotensinogen, aldosterone, and aldosterone-stimulating factor is increased [149–151]. In obese individuals, the RAAS is activated in renal tissue as well, resulting in increase in sodium reabsorption through many mechanisms [152–154]. Activation of the RAAS in the kidney, especially of aldosterone and or its receptor, is likely to play a major role in the development of kidney injury and proteinuria associated with obesity. Hyperglycemia and angiotensin II are known to upregulate the expression of sodium glucose co-transporter (SGLT-2) [155, 156]. Thus, in obesity, in which both hyperglycemia and RAAS activation occur, renal tubular reabsorption of glucose may be increased via upregulation of the expression of SGLT-2.

RAAS is a major regulator of vasomotor tone and cellular proliferation and thus regulates renal function and structure. Adipocytes and adipose-infiltrating macrophages are important sources of RAAS [157]. Circulating levels of angiotensinogen (Aog) increase with increasing BMI [158]. Adipose-derived increase in circulating RAAS ligands together with adipose-driven increase in renal AT1 is a potent alliance for efferent arteriolar vasoconstriction, to increase glomerular pressure and FF, as well as cellular proliferation and thus culminate in renal damage [159, 160]. There are also evidences to suggest that AT2 may mediate the substantial adipose inflammatory response associated with increased Aog [161].

Aldosterone blockade reduces renal injury. This benefit is independent of its antihypertensive effects and possibly relates to the blocking effects of aldosterone on PAI-1 and TGF-β, reactive oxygen intermediates, inflammatory mediators, and podocyte function [162–164]. Adipose tissue produces aldosterone independent of AngII as well. At least one oxidized derivative of linoleic acid is known to stimulate aldosterone synthesis [165]. Complement-C1q TNF-related protein 1 (CTRP1), prominently expressed by adipose tissue may also mediate Ang II-independent aldosterone production [166]. Thus, elevated aldosterone in obesity could be injurious to glomeruli through its indirect action which increases GFR and also through its direct effects on the podocyte.

Plasminogen Activator Inhibitor-1 (PAI-1)

Obesity induces PAI-1 in adipose tissue and glomerular cells. PAI-1 is an independent risk factor for renal damage. PAI-1 can decrease protease-dependent matrix degradation and cellular migration [167]. In a model of podocyte injury-associated glomerulosclerosis, renoprotection by PPAR-γ agonist is partially through reduced PAI-1 [168]. PAI-1 modulates podocyte injury as well. Ablation of the kidney in PAI-1 deficient mice caused less of proteinuria, podocyte damage and glomerular sclerosis [169].

Several mechanisms involved in obesity-related organ dysfunction are concomitant. IR is linked to increased levels of Ang II, which is associated with the progression of renal damage in obesity. Indeed, Angiotensin II is an important mediator in the progression of obesity related kidney disease [170–172]. Angiotensin II contributes to hyperfiltration and glomerulosclerosis through both hemodynamic and non-hemodynamic effects [18, 173–175]. ANG II produces vasoconstriction while insulin induces vascular relaxation by promoting NO production through the phosphatidylinositol 3-kinase (PIK3-Akt) signaling pathway [176]. The inhibitory effects of ANG II on the insulin action is mediated by production of reactive oxygen species (ROS) [177, 178]. In turn, ROS induces inflammatory cytokines such as MCP-1 or TNFα which can then impair the PI3K-Akt pathway of insulin signaling, resulting in IR [116, 179, 180].

Factors Which Increase Susceptibility for Renal Injury in the Obese

Severity of renal impairment does not always correlate with the severity of obesity. Considerable differences in susceptibility to renal injury have been noted among obese individuals and are possibly related to other predisposing factors (Table 2).

Obesity is a risk factor for type 2 diabetes, hypertension and other components of the metabolic syndrome as well as cardiovascular disease. All of them add to the risk for CKD in obese individuals. Each component of MetS (impaired glucose tolerance, hypertension, and dyslipidemia) can induce kidney injury and may also exacerbate pre-existing kidney disease [181]. Combinations of components synergistically increase the risk for CKD, and the risk of progression of pre-existing CKD [182]. The odds ratio for kidney disease is 1.89 when only one component of MetS is present, leaping to 5.85 when all five components of MetS are present together in an individual [183, 184]. A meta-analysis of eleven studies [185] involving 30,146 subjects revealed that MetS is associated with development of

Table 2 Predisposing factors for renal injury in obesity

1. Type of obesity type: Visceral fat obesity
2. Components of metabolic syndrome: impaired glucose tolerance, hypertension, dyslipidemia
3. Associated conditions: sleep apnea syndrome, pulmonary hypertension and right ventricular overload, nonalcoholic fatty liver disease
4. Decrease in nephron number: low birth weight, intra-uterine growth retardation, preterm birth
5. Reduced nephron mass: congenital anomalies of the kidney and urinary tract, nephrectomy
6. Progressive loss of functioning nephron: chronic kidney disease of any cause, aging

Adapted from Tsuboi N, Okabayashi Y, Shimizu A and Yokoo T. The renal pathology of obesity. *Kidney International Reports* 2:251–260, 2017

Stage III CKD with an odds ratio (OR) of 1.55 (95% CI: 1.34–1.80) [185]. Global as well as segmental glomerulosclerosis are seen in those with MetS. There is also a higher prevalence of tubular atrophy, interstitial fibrosis, and arterial sclerosis [186].

Interestingly, in comparison to normal subjects, patients with obesity related kidney disease have a significantly lower glomerular density as seen in kidney biopsies [26]. Aging related decreases in the number of glomerular podocytes can also greatly influence susceptibility to renal injury in obese individuals [187].

In obesity, nephron overwork and risk of intraglomerular hypertension would be exaggerated in those with intra uterine growth retardation and those who have had unilateral nephrectomy, as they have already a lower nephron number [188, 189].

Pulmonary hypertension secondary to sleep apnea and non-alcoholic fatty liver disease which commonly accompany obesity may also catalyze renal injury associated with obesity [190, 191].

Conclusions

Obesity is a major cause for the initiation and progression of renal injury resulting in CKD and ESRD. Many factors together cause renal vasodilation, glomerular hyperfiltration and albuminuria, leading to glomerulopathy. In obese individuals, excess excretory load induces hyperperfusion and hyperfiltration by the kidneys leading to glomerulomegaly. Obesity associated metabolic abnormalities such as increased FFA, hyperinsulinemia, IR, pro inflammatory conditions, reduced adiponectin, and leptin resistance mediate injury to glomerular and tubular cells and lead to CKD.

To a large extent, both obesity and related CKD can be prevented by following a healthy lifestyle. Renin-angiotensin-aldosterone blockade is effective in the short-term. SREBP antagonists, PPARα agonists, FXR and TGR5 agonists and LXR agonists directly target lipid metabolism and hence are considered to have therapeutic value. Delineation of obesity linked factors, which lead to adaptive and maladaptive changes in the kidney and predispose patients to renal disease could lead to identification of molecular targets and reno protective and therapeutic strategies to improve outcomes for obese patients with CKD.

Several questions about the pathogenesis of nephropathy associated with obesity remain to be answered. Whether glomerulomegaly is a cause or only an associated feature of proteinuria in ORG is unclear. Whether glomerulomegaly is a precursor of obesity-related FSGS lesion is also obscure. Whether genetic factors have any role in determining the time of onset and the rate of progression of kidney disease in obese subjects is also unknown.

References

1. Foster MC, Hwang SJ, Larson MG et al (2008) Overweight, obesity, and the development of stage 3 CKD: the Framingham Heart Study. Am J Kidney Dis 52:39–48
2. Gelber RP, Kurth T, Kausz AT et al (2005) Association between body mass index and CKD in apparently healthy men. Am J Kidney Dis 46:871–880
3. Kramer H, Luke A, Bidani A et al (2005) Obesity and prevalent and incident CKD: the hypertension detection and follow-up program. Am J Kidney Dis 46:587–594
4. Lu JL, Kalantar-Zadeh K, Ma JZ et al (2014) Association of body mass index with outcomes in patients with CKD. J Am Soc Nephrol 25:2088–2096
5. Munkhaugen J, Lydersen S, Wideroe TE, Hallan S (2009) Prehypertension, obesity, and risk of kidney disease: 20-year follow-up of the HUNT I study in Norway. Am J Kidney Dis 54:638–646
6. Vivante A, Golan E, Tzur D et al (2012) Body mass index in 1.2 million adolescents and risk for end-stage renal disease. Arch Intern Med 172:1644–1650
7. Kovesdy CP, Furth SL, Zoccali C, On Behalf of the World Kidney Day Steering Committee (2017) Obesity and kidney disease: hidden consequences of the epidemic. Am J Nephrol 45:283–291
8. Chang A, Kramer H (2012) CKD progression: a risky business. Nephrol Dial Transplant 27:2607–2609
9. Pinto-Sietsma SJ, Navis G, Janssen WM et al (2003) A central body fat distribution is related to renal function impairment, even in lean subjects. Am J Kidney Dis 41:733–741
10. Olsen NSC, Iseki K, Kramer H, Liu Z, Sharma K (2017) Kidney disease and obesity: epidemiology, mechanisms and treatment. Nat Rev Nephrol 13:181–190
11. Goncalves Torres MR, Cardoso LG, de Abreu VG et al (2009) Temporal relation between body mass index and renal function in individuals with hypertension and excess body weight. Nutrition 25:914–919
12. Nomura I, Kato J, Kitamura K (2009) Association between body mass index and chronic kidney disease: a population-based, cross-sectional study of a Japanese community. Vasc Health Risk Manag 5:315–320
13. Hunley TE, Ma LJ, Kon V (2010) Scope and mechanisms of obesity-related renal disease. Curr Opin Nephrol Hypertens 19:227–234
14. Weisinger JR, Kempson RL, Eldridge FL, Swenson RS (1974) The nephrotic syndrome: a complication of massive obesity. Ann Intern Med 81:440–447

15. Verani RR (1992) Obesity-associated focal segmental glomerulosclerosis: Pathological features of the lesion and relationship with cardiomegaly and hyperlipidemia. Am J Kidney Dis 20:629–634
16. Adelman RD, Restaino IG, Alon US, Blowey DL (2001) Proteinuria and focal segmental glomerulosclerosis in severely obese adolescents. J Pediatr 138:481–485
17. Kambham N, Markowitz GS, Valeri AM, Lin J, D'Agati VD (2001) Obesity-related glomerulopathy: an emerging epidemic. Kidney Int 59:1498–1509
18. Praga M, Hernandez E, Morales E et al (2001) Clinical features and long-term outcome of obesity-associated focal segmental glomerulosclerosis. Nephrol Dial Transplant 16:1790–1798
19. Morales E, Vlaero A, Leon M et al (2003) Benefical effects of weight loss in overweight patients with chronic proteinuric nephropathies. Am J Kidney Dis 41:319–327
20. Chen HM, Li SJ, Chen HP et al (2008) Obesity-related glomerulopathy in China: a case series of 90 patients. Am J Kidney Dis 52:58–65
21. Tsuboi N, Koike K, Hirano K et al (2013) Clinical features and long-term renal outcomes of Japanese patients with obesity-related glomerulopathy. Clin Exp Nephrol 17:379–385
22. Mandal R, Loeffler AG, Salamat S, Fritsch MK (2012) Organ weight changes associated with body mass index determined from a medical autopsy population. Am J Forensic Med Pathol 33:382–389
23. Samuel T, Hoy WE, Douglas-Denton R et al (2005) Determinants of glomerular volume in different cortical zones of the human kidney. J Am Soc Nephrol 16:3102–3109
24. Hoy WE, Hughson MD, Zimanyi M et al (2010) Distribution of volumes of individual glomeruli in kidneys at autopsy: association with age, nephron number, birth weight and body mass index. Clin Nephrol 74(suppl 1):S105–S112
25. Puelles VG, Zimanyi MA, Samuel T et al (2012) Estimating individual glomerular volume in the human kidney: clinical perspectives. Nephrol Dial Transplant 27:1880–1888
26. Tsuboi N, Utsunomiya Y, Kanzaki G et al (2012) Low glomerular density with glomerulomegaly in obesity-related glomerulopathy. Clin J Am Soc Nephrol 7:735–741
27. Wu Y, Liu Z, Xiang Z et al (2006) Obesity-related glomerulopathy: insights from gene expression profiles of the glomeruli derived from renal biopsy samples. Endocrinology 147:44–50
28. Chen HM, Liu ZH, Zeng CH et al (2006) Podocyte lesions in patients with obesity-related glomerulopathy. Am J Kidney Dis 48:772–779
29. de Vries AP, Ruggenenti P, Ruan XZ, ERA-EDTA Working Group Diabesity et al (2014) Fatty kidney: emerging role of ectopic lipid in obesity-related renal disease. Lancet Diab Endocrinol 2:417–426
30. Tobar A, Ori Y, Benchetrit S et al (2013) Proximal tubular hypertrophy and enlarged glomerular and proximal tubular urinary space in obese subjects with proteinuria. PLoS ONE 8:e75547
31. Henegar JR, Bigler SA, Henegar LK et al (2001) Functional and structural changes in the kidney in the early stages of obesity. J Am Soc Nephrol 12:1211–1217
32. Reisin E, Messerli FG, Ventura HO, Frohlich ED (1987) Renal haemodynamic studies in obesity hypertension. J Hypertens 5:397–400
33. Chagnac A, Weinstein T, Korzets A et al (2000) Glomerular hemodynamics in severe obesity. Am J Physiol Renal Physiol 278:F817–F822
34. Chagnac A, Weinstein T, Herman M et al (2003) The effects of weight loss on renal function in patients with severe obesity. J Am Soc Nephrol 14:1480–1486
35. Vallon V, Richter K, Blantz RC et al (1999) Glomerular hyperfiltration in experimental diabetes mellitus: potential role of tubular reabsorption. J Am Soc Nephrol 10:2569–2576
36. Hall JE, do Carmo JM, Silva AA et al (2015) Obesity-induced hypertension: interaction of neurohumoral and renal mechanisms. Circ Res 13:991–1006
37. Kim S, Soltani-Bejnood M, Quignard-Boulange A, et al (2006) The adipose renin-angiotensin system modulates systemic markers of insulin sensitivity and activates

the intrarenal renin-angiotensin system. J Biomed Biotechnol 27012. doi:10.1155/JBB/2006/27012
38. Strazzullo P, Barba G, Cappuccio FP et al (2001) Altered renal sodium handling in men with abdominal obesity: a link to hypertension. J Hypertens 19:2157–2164
39. Li C, Lin Y, Luo R et al (2016) Intrarenal renin-angiotensin system mediates fatty acid-induced ER stress in the kidney. Am J Physiol Renal Physiol 310:F351–F363
40. Moss NG (1982) Renal function and renal afferent and efferent nerve activity. Am J Physiol 243:F425–F433
41. Kassab S, Kato T, Wilkins FC et al (1995) Renal denervation attenuates the sodium retention and hypertension associated with obesity. Hypertension 25:893–897
42. Thomson SC, Vallon V, Blantz RC (2004) Kidney function in early diabetes: the tubular hypothesis of glomerular filtration. Am J Physiol Renal Physiol 286:F8–F15
43. Hostetter TH (2003) Hyperfiltration and glomerulosclerosis. Semin Nephrol 23:194–199
44. Caramori ML, Mauer M (2003) Diabetes and nephropathy. Curr Opin Nephrol Hypertens 12:273–282
45. Leon CA, Raij L (2005) Interaction of haemodynamic and metabolic pathways in the genesis of diabetic nephropathy. J Hypertens 23:1931–1937
46. Hall JE, Henegar JR, Dwyer TM et al (2004) Is obesity a major cause of chronic kidney disease? Adv Ren Replace Ther 11:41–54
47. Adelman RD (2002) Obesity and renal disease. Curr Opin Nephrol Hypertens 11:331–335
48. Praga M (2002) Obesity—a neglected culprit in renal disease. Nephrol Dial Transplant 17:1157–1159
49. Slee AD (2012) Exploring metabolic dysfunction in chronic kidney disease. Nutr Metab 9:36
50. Colman RJ, Roecker EB, Ramsey JJ, Kemnitz JW (1998) The effect of dietary restriction on body composition in adult male and female rhesus macaques. Aging (Milano) 10:83–92
51. Rajala MW, Scherer PE (2003) Minireview: the adipocyte-at the crossroads of energy homeostasis, inflammation, and atherosclerosis. Endocrinology 144:3765–3773
52. Unger RH (2003) Minireview: weapons of lean body mass destruction: the role of ectopic lipids in the metabolic syndrome. Endocrinology 144:5159–5165
53. Meier U, Gressner AM (2004) Endocrine regulation of energy metabolism: review of pathobiochemical and clinical chemical aspects of leptin, ghrelin, adiponectin, and resistin. Clin Chem 50:1511–1525
54. Schäffler A, Müller-Ladner U, Schölmerich J, Büchler C (2006) Role of adipose tissue as an inflammatory organ in human diseases. Endocr Rev 27:449–467
55. Katagiri H, Yamada T, Oka Y (2007) Adiposity and cardiovascular disorders. Circ Res 101:27–39
56. Musso C, Javor E, Cochran E, Balow JE, Gorden P (2006) Spectrum of renal diseases associated with extreme forms of insulin resistance. Clin J Am Soc Nephrol 1:616–622
57. Wisse BE (2004) The inflammatory syndrome: the role of adipose tissue cytokines in metabolic disorders linked to obesity. J Am Soc Nephrol 15:2792–2800
58. Hamdy O, Porramatikul S, Al-Ozairi E (2006) Metabolic obesity: the paradox between visceral and subcutaneous fat. Curr Diabetes Rev 2:367–373
59. Deji N, Kume S, Araki S et al (2009) Structural and functional changes in the kidneys of high-fat diet-induced obese mice. Am J Physiol Renal Physiol 296:F118–F126
60. Kume S, Uzu T, Araki S et al (2007) Role of altered renal lipid metabolism in the development of renal injury induced by a high-fat diet. J Am Soc Nephrol 18:2715–2723
61. Decleves AE, Mathew AV, Cunard R, Sharma K (2011) AMPK mediates the initiation of kidney disease induced by a high-fat diet. J Am Soc Nephrol 22:1846–1855
62. Decleves AE, Zolkipil Z, Satriano J et al (2014) Regulation of lipid accumulation by AMP-activated kinase [corrected] in high fat diet-induced kidney injury. Kidney Int 85:611–623
63. Jensen MD (2006) Adipose tissue as an endocrine organ: implications of its distribution on free fatty acid metabolism. Eur Heart J Suppl 8 (Supplement B):B13–B19

64. Cao Y (2010) Adipose tissue angiogenesis as a therapeutic target for obesity and metabolic diseases. Nat Rev Drug Discov 9:107–115
65. Iliescu RI, Chade AR (2010) Progressive renal vascular proliferation and injury in obese Zucker rats. Microcirculation 17:250–258
66. Roubicek T, BArtlova M, Krajickova J et al (2009) Increased production of proinflammatory cytokines in adipose tissue of patients with end-stage renal disease. Nutrition 25:762–768
67. Stemmer K, Pervez-Tilve D, Ananthakrishnan G et al (2012) High-fat-diet-induced obesity causes an inflammatory and tumor-promoting microenvironment in the rat kidney. Dis Model Mech 5:627–635
68. D'Agati VD, Chagnac A, De Vries AP et al (2016) Obesity-related glomerulopathy: clinical and pathologic characteristics and pathogenesis. Nat Rev Nephrol 12:453–471
69. Bobulescu IA, Lotan Y, Zhang J et al (2014) Triglycerides in the human kidney cortex: relationship with body size. PLoS ONE 9(e101285):44–50
70. Santini E, Lupi R, Baldi S et al (2008) Effects of different LDL particles on inflammatory molecules in human mesangial cells. Diabetologia 51:2117–2125
71. Tomiyama-Hanayama M, Rakugi H, Kohara M et al (2009) Effect of interleukin-6 receptor blockage on renal injury in apolipoprotein E-deficient mice. Am J Physiol Renal Physiol 297:F679–F684
72. Bussolati B, Deregibus MC, Fonsato V et al (2005) Statins prevent oxidized LDL-induced injury of glomerular podocytes by activating the phosphatidyl inositol 3-kinase/AKT signaling pathway. J Am Soc Nephrol 16:1936–1947
73. Cormack-Aboud FC, Brinkkoetter PT, Pippin JW et al (2009) Rosuvastatin protects against podocyte apoptosis in vitro. Nephrol Dial Transplant 24:404–412
74. Lennon R, Pons D, Sabin MA et al (2009) Saturated fatty acids induce insulin resistance in human podocytes: implications for diabetic nephropathy. Nephrol Dial Transplant 24:3288–3296
75. Sun L, Halaihel N, Zhang W et al (2002) Role of sterol regulatory element-binding protein 1 in regulation of renal lipid metabolism and glomerulosclerosis in diabetes mellitus. J Biol Chem 277:18919–18927
76. Jiang T, Wang Z, Proctor G et al (2005) Diet-induced obesity in C57BL/6J mice causes increased renal lipid accumulation and glomerulosclerosis via a sterol regulatory element-binding protein-1c-dependent pathway. J Biol Chem 280:32317–32325
77. Tang J, Yan H, Zhuang S (2012) Inflammation and oxidative stress in obesity-related glomerulopathy. Int J Nephrol 608397
78. Ikezumi Y, Suzuki T, Karasawa T et al (2008) Activated macrophages down-regulate podocyte nephrin and podocin expression via stress-activated protein kinases. Biochem Biophys Res Commun 376:706–711
79. Axelsson J, Qureshi AR, Suliman ME et al (2004) Truncal fat mass as a contributor to inflammation in end-stage renal disease. Am J Clin Nutr 80:1222–1229
80. Fernández-Riejos P, Najib S, Santos-Alvarez J et al (2010) Role of leptin in the activation of immune cells. Mediators Inflamm 568343
81. Kastarinen H, Kesaniemi YA, Ukkola O (2009) Leptin and lipid metabolism in chronic kidney failure. Scand J Clin Lab Invest 69:401–408
82. Sharma K, Considine RV, Michael B et al (1997) Plasma leptin is partly cleared by the kidney and is elevated in hemodialysis patients. Kidney Int 51:1980–1985
83. Mak RH, Cheung W (2007) Adipokines and gut hormones in end-stage renal disease. Perit Dial Int 27(Suppl 2):S298–S302
84. Mathew AV, Okada S, Sharma K (2011) Obesity related kidney disease. Curr Diabetes Rev 7:41–49
85. Wolf G, Ziyadeh FN (2006) Leptin and renal fibrosis. Contrib Nephrol 151:175–183
86. Hotta K, Funahashi T, Arita Y et al (2000) Plasma concentrations of a novel, adipose-specific protein, adiponectin, in type 2 diabetic patients. Arterioscler Thromb Vasc Biol 20:1595–1599

87. Kadowaki T, Yamauchi T, Kubota N et al (2006) Adiponectin and adiponectin receptors in insulin resistance, diabetes, and the metabolic syndrome. J Clin Invest 116:1784–1792
88. Sharma K, Ramachandrarao S, Qiu G et al (2008) Adiponectin regulates albuminuria and podocyte function in mice. J Clin Invest 118:1645–1656
89. Ohashi K, Iwatani H, Kihara S et al (2007) Exacerbation of albuminuria and renal fibrosis in subtotal renal ablation model of adiponectin-knockout mice. Arterioscler Thromb Vasc Biol 27:1910–1917
90. Cammisotto PG, Bendayan M (2008) Adiponectin stimulates phosphorylation of AMP-activated protein kinase alpha in renal glomeruli. J Mol Histol 39:579–584
91. Guebre-Egziabher F, Bernhard J, Funahashi T et al (2005) Adiponectin in chronic kidney disease is related more to metabolic disturbances than to decline in renal function. Nephrol Dial Transplant 20:129–134
92. Saraheimo M, Forsblom C, Thorn L et al (2008) Serum adiponectin and progression of diabetic nephropathy in patients with type 1 diabetes. Diabetes Care 31:1165–1169
93. Berg AH, Combs TP, Du X et al (2001) The adipocyte-secreted protein Acrp30 enhances hepatic insulin action. Nat Med 7:947–953
94. Yamauchi T, Kamon J, Waki H et al (2003) Globular adiponectin protected ob/ob mice from diabetes and ApoE-deficient mice from atherosclerosis. J Biol Chem 278:2461–2468
95. Combs TP, Pajvani UB, Berg AH et al (2004) A transgenic mouse with a deletion in the collagenous domain of adiponectin displays elevated circulating adiponectin and improved insulin sensitivity. Endocrinology 145:367–383
96. Maeda N, Shimomura I, Kishida K et al (2002) Diet-induced insulin resistance in mice lacking adiponectin/ACRP30. Nat Med 8:731–737
97. Kubota N, Terauchi Y, Yamauchi T et al (2002) Disruption of adiponectin causes insulin resistance and neointimal formation. J Biol Chem 277(29):25863–25866
98. Yamauchi T, Kamon J, Minokoshi Y et al (2002) Adiponectin stimulates glucose utilization and fatty-acid oxidation by activating AMP-activated protein kinase. Nat Med 8:1288–1295
99. Steinberg GR, Michell BJ, van Denderen BJ et al (2006) Tumor necrosis factor alpha-induced skeletal muscle insulin resistance involves suppression of AMP-kinase signaling. Cell Metab 4:465–474
100. Lihn AS, Jessen N, Pedersen SB et al (2004) AICAR stimulates adiponectin and inhibits cytokines in adipose tissue. Biochem Biophys Res Commun 316:853–858
101. Sell H, Dietze-Shroeder D, Eckardt K, Eckel J (2006) Cytokine secretion by human adipocytes is differentially regulated by adiponectin, AICAR, and troglitazone. Biochem Biophys Res Commun 343:700–706
102. O'Neill LA, Hardie DG (2013) Metabolism of inflammation limited by AMPK and pseudo-starvation. Nature 493:346–355
103. Weisberg SP, Mc Cann D, Desai M et al (2003) Obesity is associated with macrophage accumulation in adipose tissue. J Clin Invest 112:1796–808
104. Xu H, Barnes GT, Yang Q et al (2003) Chronic inflammation in fat plays a crucial role in the development of obesity-related insulin resistance. J Clin Invest 112:1821–1830
105. Stienstra R, Van Diepen JA, Tack CJ et al (2011) Inflammasome is a central player in the induction of obesity and insulin resistance. Proc Natl Acad Sci U S A 108:15324–15329
106. Lumeng CN, Bodzin JL, Saltiel AR (2007) Obesity induces a phenotypic switch in adipose tissue macrophage polarization. J Clin Invest 117:175–184
107. Gordon S (2003) Alternative activation of macrophages. Nat Rev Immunol 3:23–35
108. Lumeng CN, DelProposto JB, Westcott DJ, Saltiel AR (2008) Phenotypic switching of adipose tissue macrophages with obesity is generated by spatiotemporal differences in macrophage subtypes. Diabetes 57:3239–3246
109. King GL (2008) The role of inflammatory cytokines in diabetes and its complications. J Periodontol 79(8 Suppl):1527–1534
110. Duffield JS, Tipping PG, Kipari T et al (2005) Conditional ablation of macrophages halts progression of crescentic glomerulonephritis. Am J Pathol 167:1207–1219

111. Lim AK, Ma FY, Nikolic-Paterson DJ et al (2009) Antibody blockade of c-fms suppresses the progression of inflammation and injury in early diabetic nephropathy in obese db/db mice. Diabetologia 52:1669–1679
112. Wang Y, Wang YP, Zheng G et al (2007) Ex vivo programmed macrophages ameliorate experimental chronic inflammatory renal disease. Kidney Int 72:290–299
113. Mal J, Corsa BA, Zhou J et al (2011) Angiotensin type 1 receptor modulates macrophage polarization and renal injury in obesity. Am J Physiol Renal Physiol 300:F1203–F1213
114. Gupta J, Mitra N, Kanetsky PA et al (2012) Association between albuminuria, kidney function, and inflammatory biomarker profile in CKD in CRIC. Clin J Am Soc Nephrol 7:1938–1946
115. Berg AH, Scherer PE (2005) Adipose tissue, inflammation, and cardiovascular disease. Circ Res 96:939–949
116. Uysal KT, Wiesbrock SM, Hotamisligil GS (1998) Functional analysis of tumor necrosis factor (TNF) receptors in TNF-alpha-mediated insulin resistance in genetic obesity. Endocrinology 139(12):4832–4838
117. Hivert MF, Sullivan LM, Fox CS et al (2008) Associations of adiponectin, resistin, and tumor necrosis factor-alpha with insulin resistance. J Clin Endocrinol Metab 93:3165–3172
118. Kanda H, tateya S, Tamori Y et al (2006) MCP-1 contributes to macrophage infiltration into adipose tissue, insulin resistance, and hepatic steatosis in obesity. J Clin Invest 116:1494–1505
119. Kamei N, Tobek K, Suzuki R et al (2006) Overexpression of monocyte chemoattractant protein-1 in adipose tissues causes macrophage recruitment and insulin resistance. J Biol Chem 281:26602–26614
120. Furuichi K, Kaneko S, Wada T (2009) Chemokine/chemokine receptor-mediated inflammation regulates pathologic changes from acute kidney injury to chronic kidney disease. Clin Exp Nephrol 13:9–14
121. Tarabra E, Giunti S, Barutta F et al (2009) Effect of the monocyte chemoattractant protein-1/CC chemokine receptor 2 system on nephrin expression in streptozotocin-treated mice and human cultured podocytes. Diabetes 58:2109–2118
122. Ruiz Ortega M, Ruperez M, Lorenzo O et al (2002) Angiotensin II regulates the synthesis of proinflammatory cytokines and chemokines in the kidney. Kidney Int Suppl 82:S12–S22
123. Patel NS, Chatterjee PK, Di Paola R et al (2005) Endogenous interleukin-6 enhances the renal injury, dysfunction, and inflammation caused by ischemia/reperfusion. J Pharmacol Exp Ther 312:1170–1178
124. Hotamisligil GS (2003) Inflammatory pathways and insulin action. Int J Obes Relat Metab Disord 27(Suppl 3):S53–S55
125. Hotamisligil GS (2006) Inflammation and metabolic disorders. Nature 444:860–867
126. Makki K, Froguel P, Wolowczuk I (2013) Adipose tissue in obesity-related inflammation and insulin resistance: cells, cytokines, and chemokines. ISRN Inflamm 139239
127. Liao MT, Sung CC, Hung KC et al (2012) Insulin resistance in patients with chronic kidney disease. J Biomed Biotechnol 691369
128. Chen J, Muntner P, Hamm LL et al (2003) Insulin resistance and risk of chronic kidney disease in nondiabetic US adults. J Am Soc Nephrol 14:469–477
129. Parvanova AI, Trevisan R, Illiev IP et al (2006) Insulin resistance and microalbuminuria: a cross-sectional, case-control study of 158 patients with type 2 diabetes and different degrees of urinary albumin excretion. Diabetes 55(5):1456–1462
130. Adamczak M, Wiecek A (2013) The adipose tissue as an endocrine organ. Semin Nephrol 33:2–13
131. Cusumano AM, Bodkin NL, Hansen BC et al (2002) Glomerular hypertrophy is associated with hyperinsulinemia and precedes overt diabetes in aging rhesus monkeys. Am J Kidney Dis 40:1075–1085
132. Pete G, Walsh M, Hu Y et al (1996) Insulin-like growth factor-I decreases mean blood pressure and selectively increases regional blood flow in normal rats. Proc Soc Exp Biol Med 213:187–192

133. Hirschberg R, Adler S (1998) Insulin-like growth factor system and the kidney: physiology, pathophysiology, and therapeutic implications. Am J Kidney Dis 31:901–919
134. Abrass CK, Raugi GJ, Gabourel LS, Lovett DH (1988) Insulin and insulin-like growth factor I binding to cultured rat glomerular mesangial cells. Endocrinology 123:2432–2439
135. Kreisberg JI (1982) Insulin requirement for contraction of cultured rat glomerular mesangial cells in response to angiotensin II: Possible role for insulin in modulating glomerular hemodynamics. Proc Natl Acad Sci U S A 79:4190–4192
136. Abrass CK, Spicer D, Raugi GJ (1994) Insulin induces a change in extracellular matrix glycoproteins synthesized by rat mesangial cells in culture. Kidney Int 46:613–620
137. Abrass CK, Peterson CV, Raugi GJ (1988) Phenotypic expression of collagen types in mesangial matrix of diabetic and nondiabetic rats. Diabetes 37:1695–1702
138. Abrass CK, Spicer D, Raugi GJ (1995) Induction of nodular sclerosis by insulin in rat mesangial cells in vitro: studies of collagen. Kidney Int 47:25–37
139. Welsh GI, Hale LJ, Eremina V et al (2010) Insulin signaling to the glomerular podocyte is critical for normal kidney function. Cell Metab 12:329–340
140. Chavez JA, Knotts TA, Wang LP et al (2003) A role for ceramide, but not diacylglycerol, in the antagonism of insulin signal transduction by saturated fatty acids. J Biol Chem 278:10297–10303
141. Ahmed SB, Fisher ND, Stevanovic R, Hollenberg NK (2005) Body mass index and angiotensin-dependent control of the renal circulation in healthy humans. Hypertension 46:1316–1320
142. Kennefick TM, Anderson S (1997) Role of angiotensin II in diabetic nephropathy. Semin Nephrol 17:441–447
143. Zhuo JL, Li XC (2007) Novel roles of intracrine angiotensin II and signalling mechanisms in kidney cells. J Renin Angiotensin Aldosterone Syst 8:23–33
144. Griffin KA, Bidani AK (2006) Progression of renal disease: renoprotective specificity of renin-angiotensin system blockade. Clin J Am Soc Nephrol 1:1054–1065
145. Crowley SD, Gurley SB, Coffman TM (2007) AT(1) receptors and control of blood pressure: the kidney and more. Trends Cardiovasc Med 17:30–34
146. Ziyadeh FN (2004) Mediators of diabetic renal disease: the case for TGF-beta as the major mediator. J Am Soc Nephrol Suppl 1:S55–S57
147. Cooper R, McFarlane-Anderson N, Bennett FI et al (1997) ACE, angiotensinogen and obesity: a potential pathway leading to hypertension. J Hum Hypertens 11:107–111
148. Achard V, Boullu-Ciocca S, Desbriere R et al (2007) Renin receptor expression in human adipose tissue. Am J Physiol Regul Integr Comp Physiol 292:R274–R282
149. Bochud M, Nussberger J, Bovet P et al (2006) Plasma aldosterone is independently associated with the metabolic syndrome. Hypertension 48:239–245
150. Rossi GP, Belfiore A, Bernini G et al (2008) Body mass index predicts plasma aldosterone concentrations in overweight-obese primary hypertensive patients. J Clin Endocrinol Metab 93:2566–2571
151. Ehrhart-Bornstein M, Arakelyan K, Krug AW et al (2004) Fat cells may be the obesity–hypertension link: human adipogenic factors stimulate aldosterone secretion from adrenocortical cells. Endocr Res 30:865–870
152. Ohsawa M, Tamura K, Wakui H et al (2014) Deletion of the angiotensin II type 1 receptor-associated protein enhances renal sodium reabsorption and exacerbates angiotensin II-mediated hypertension. Kidney Int 86:570–581
153. Peti-Peterdi J, Warnock DG, Bell PD (2002) Angiotensin II directly stimulates ENaC activity in the cortical collecting duct via AT(1) receptors. J Am Soc Nephrol 13:1131–1135
154. Rozansky DJ (2006) The role of aldosterone in renal sodium transport. Semin Nephrol 26:173–181
155. Rahmoune H, Thompson PW, Ward JM et al (2005) Glucose transporters in human renal proximal tubular cells isolated from the urine of patients with non-insulin-dependent diabetes. Diabetes 54:3427–3434

156. Bautista R, Manning R, Martinez F et al (2004) Angiotensin Iidependent increased expression of Naþ-glucose cotransporter in hypertension. Am J Physiol Renal Physiol 286: F127–F133
157. Frederich RC Jr, Kahn BB, Peach MJ, Flier JS (1992) Tissue-specific nutritional regulation of angiotensinogen in adipose tissue. Hypertension 19:339–344
158. Schorr U, Blaschke K, Turan S et al (1998) Relationship between angiotensinogen, leptin and blood pressure levels in young normotensive men. J Hypertens 16:1475–1480
159. Xu ZG, Lanting L, Vaziri ND et al (2005) Upregulation of angiotensin II type 1 receptor, inflammatory mediators, and enzymes of arachidonate metabolism in obese Zucker rat kidney: reversal by angiotensin II type 1 receptor blockade. Circulation 111:1962–1969
160. Kim S, Soltani-Bejnood M, Quignard-Boulange A et al (2006) The adipose Renin-Angiotensin system modulates systemic markers of insulin sensitivity and activates the intrarenal renin-angiotensin system. J Biomed Biotechnol 27012
161. Yvan-Charvet L, Massiera F, Lamande N et al (2009) Deficiency of angiotensin type 2 receptor rescues obesity but not hypertension induced by overexpression of angiotensinogen in adipose tissue. Endocrinology 150:1421–1428
162. Bomback AS, Klemmer PJ (2008) Renal injury in extreme obesity: the important role of aldosterone. Kidney Int 74:1216
163. Nagase M, Fujita T (2008) Aldosterone and glomerular podocyte injury. Clin Exp Nephrol 12:233–242
164. de Paula RB, da Silva AA, Hall JE (2004) Aldosterone antagonism attenuates obesity-induced hypertension and glomerular hyperfiltration. Hypertension 43:41–47
165. Goodfriend TL, Ball DL, Egan BM et al (2004) Epoxy-keto derivative of linoleic acid stimulates aldosterone secretion. Hypertension 43:358–363
166. Jeon JH, Kim KY, Kim JH et al (2008) A novel adipokine CTRP1 stimulates aldosterone production. FASEB J 22:1502–1511
167. Eddy AA, Fogo AB (2006) Plasminogen activator inhibitor-1 in chronic kidney disease: evidence and mechanisms of action. J Am Soc Nephrol 17:2999–3012
168. Yang HC, Ma LJ, Ma J, Fogo AB (2006) Peroxisome proliferator-activated receptor-gamma agonist is protective in podocyte injury-associated sclerosis. Kidney Int 69:1756–1764
169. Ma L-J, Fogo AB (2009) PAI-1 and kidney fibrosis. Front Biosci 14:2028–2041
170. Amann K, Benz K (2013) Structural renal changes in obesity and diabetes. Semin Nephrol 33:23–33
171. Thethi T, Kamiyama M, Kobori H (2012) The link between the renin-angiotensin-aldosterone system and renal injury in obesity and the metabolic syndrome. Curr Hypertens Rep 14:160–169
172. Leehey DJ, Singh AK, Alavi N, Singh R (2000) Role of angiotensin II in diabetic nephropathy. Kidney Int Suppl 77:S93–S98
173. Lu H, Boustany-Kari CM, Daugherty A, Cassis LA (2007) Angiotensin II increases adipose angiotensinogen expression. Am J Physiol Endocrinol Metab 292:E1280–E1287
174. Ogihara T, Asano T, Ando K et al (2002) Angiotensin II-induced insulin resistance is associated with enhanced insulin signaling. Hypertension 40(6):872–879
175. Olivares-Reyes JA, Arellano-Plancarte A, Castillo-Hernandez JR (2009) Angiotensin II and the development of insulin resistance: implications for diabetes. Mol Cell Endocrinol 302:128–139
176. Kim JA, Jang HJ, Martinez-Lemus LA, Sowers JR et al (2012) Activation of mTOR/p70S6 kinase by ANG II inhibits insulin-stimulated endothelial nitric oxide synthase and vasodilation. Am J Physiol Endocrinol Metab 302:E201–E208
177. Blendea MC, Jacobs D, Stump CS et al (2005) Abrogation of oxidative stress improves insulin sensitivity in the Ren-2 rat model of tissue angiotensin II overexpression. Am J Physiol Endocrinol Metab 288:E353–E359
178. Sowers JR (2002) Hypertension, angiotensin II, and oxidative stress. N Engl J Med 346:1999–2001

179. Uysal KT, Wiesbrock SM, Marino MW, Hotamisligil GS et al (1997) Protection from obesity-induced insulin resistance in mice lacking TNF-alpha function. Nature 389:610–614
180. Taniyama Y, Hitomi H, Shah A et al (2005) Mechanisms of reactive oxygen species-dependent downregulation of insulin receptor substrate-1 by angiotensin II. Arterioscler Thromb Vasc Biol 25:1142–1147
181. Wahba IM, Mark RH (2007) Obesity and obesity initiated metabolic syndrome; mechanistic links to chronic kidney diseases. Clin J Am Soc Nephrol 2:550–562
182. Tanak H, Shiohira Y, Uezu Y et al (2006) Metabolic syndrome and chronic kidney disease in Okinawa, Japan. Kidney Int 69:369–374
183. Garland JS (2014) Elevated body mass index as a risk factor for chronic kidney disease: current perspectives. Diab Metab Syndr Obes Targets Therapy 7:347–355
184. Nashar K, Megan B (2014) Relationship between chronic kidney disease and metabolic syndrome: current perspectives. Diab Metab Syndr Obes Targets Therapy 7:421–435
185. Thomas G, Sehgal AR, Kashyap SR et al (2011) Metabolic syndrome and kidney disease: a systematic review and meta-analysis. Clin J Am Soc Nephrol 6:2364–2373
186. Alexander MP, Patel TV, Farag YM et al (2009) Kidney pathological changes in metabolic syndrome: a cross-sectional study. Am J Kidney Dis 53:751–759
187. Tsuboi N, Okabayashi Y, Shimizu A, Yokoo T (2017) The renal pathology of obesity. Kidney Int Rep 2:251–260
188. Lackland DT, Bendall HE, Osmond C et al (2000) Low birth weights contribute to high rates of early-onset chronic renal failure in the Southeastern United States. Arch Intern Med 160:1472–1476
189. Praga M, Hernandez E, Herrero JC et al (2000) Influence of obesity on the appearance of proteinuria and renal insufficiency after unilateral nephrectomy. Kidney Int 58:2111–2118
190. Bolignano D, Rastelli S, Agarwal R et al (2013) Pulmonary hypertension in chronic kidney disease. Am J Kidney Dis 61:612–622
191. Musso G, Gambino R, Tabibian JH et al (2014) Association of non-alcoholic fatty liver disease with chronic kidney disease: a systematic review and meta-analysis. PLoS Med 11: e1001680

Chapter 13
Consequences of Maternal Obesity on Neonatal Outcomes and Cardio-Metabolic Health in Infancy

Delphine Mitanchez and Pascale Chavatte-Palmer

Abstract The incidence of obesity (body mass index ≥ 30 kg/m^2) at the start of pregnancy has been rising worldwide. Although maternal diabetes and gestational weight gain constitute confounding factors, many studies report the independent effect of maternal obesity on fetal and neonatal complications, but also on the risk of childhood obesity and adverse cardio-metabolic profile. Although human cohort studies have shown an association between maternal obesity and altered cardio-metabolic health in childhood, it cannot establish causality, nor elucidate underlying mechanisms. Some molecular mechanisms of the programming effects of maternal obesity involving epigenetic traits have been investigated by human studies. But, the most important information has been provided by animal models that have been widely studied to decipher mechanisms linking maternal obesity and offspring health. Many studies in animal models demonstrate the importance of maternal obesity in the programming of offspring health and bring information on the underlying molecular mechanisms. Understanding the mechanisms is important in an attempt to improve preventive measures before and during pregnancy to limit the consequences in offspring and for disease prevention in the next generations.

Keywords Gestational diabetes · Gestational weight gain · Macrosomia · Congenital malformations · Metabolic syndrome · Animal models · Microbiote · Epigenetic

D. Mitanchez (✉)
INSERM UMR _S 938, Saint Antoine Research Center, Sorbonne University, Paris, France
e-mail: delphine.mitanchez@univ-tours.fr

Department of Neonatology, Bretonneau Hospital, CHRU, Tours, France

P. Chavatte-Palmer
UMR BDR, INRA, ENVA, Université Paris Saclay, 78350 Jouy en Josas, France

P. S. Tappia et al. (eds.), *Pathophysiology of Obesity-Induced Health Complications*, Advances in Biochemistry in Health and Disease 19,
https://doi.org/10.1007/978-3-030-35358-2_13

Introduction

Obesity is defined by the World Health Organization as abnormal or excessive fat accumulation threatening health. Obesity can be measured using the Body Mass Index (BMI, weight/(size in m)^2). Overweight is defined as a BMI range of 25–30 kg/m^2 and obesity is defined as BMI > 30 kg/m^2. Obesity is further subdivided into classes: 1—BMI range 30–34.9; 2—BMI range 35–39.9 and 3—BMI > 40. Although about one third of the obese population appears to be metabolically normal, obesity is often associated with diabetes in the context of the metabolic syndrome and it is in general difficult to discriminate the effects of obesity per se from those of abnormal metabolism during pregnancy [1].

Maternal Nutritional and Metabolic Situation

The deleterious effects of maternal obesity are exacerbated by the association with diabetes during pregnancy and by excessive gestational weight gain (GWG). As the diabetes epidemic is developing in parallel with the obesity epidemic, a growing number of women are diagnosed with type 2 diabetes before pregnancy, but there is also an increased number of women starting their pregnancy with undiagnosed type 2 diabetes or who develop gestational diabetes mellitus (GDM). Moreover, the increased risk of GDM in obese women was reported in many studies and analyzed by two systematic reviews. These found, respectively, unadjusted odds ratios (ORs) for developing GDM of 3.56 (95% CI 3.05–4.21) and 3.01 (95% CI 2.34–3.87) for women with moderate obesity and 8.56 (95% CI 5.07–16.04) and 5.55 (95% CI 4.27–7.21) for women with morbid obesity (BMI > 40) when compared with women with normal weight [2, 3]. So far, very few studies have examined the relative contribution of increased maternal BMI and diabetes to adverse pregnancy outcome. A report based on data from the Hyperglycemia and Adverse Pregnancy Outcome (HAPO) study showed that maternal GDM and obesity are independently associated with adverse pregnancy outcomes and that the combination of the two has a greater impact than either one alone [4].

The Institute of Medicine (IOM) recommends that GWG is limited to 5–9 kg for women with BMI $\geq$ 30 kg/m^2 because excessive GWG is associated with worse outcomes for the mother and the fetus. Despite this, 47–72% of obese women gain more weight than recommended [5]. In obese, pregnant women, fat accumulation predominantly occurs centrally in the visceral compartment [6]. Excess of visceral fat contributes to insulin resistance and is associated with higher metabolic risk factors during pregnancy [7].

Maternal Obesity and Perinatal Outcomes

Fetal Growth

Macrosomia or Large for Gestational Age (LGA)

The positive relationship between maternal obesity and macrosomia (defined as birth weight >90th percentile, or >+2DS, or >4 kg or >4.5 kg) was reported in many studies. Several meta-analyses have demonstrated that maternal obesity is associated with excess fetal growth, with an average increased risk of 2.5 to three-fold (Table 13.1) [8–10]. There is, however, substantial clinical heterogeneity between the studies included in the meta-analyses, notably concerning the method used for recording maternal weight and the inclusion of diabetic mothers. In a study on a smaller population, excluding maternal diabetes, the authors showed that there was no difference or a borderline increase in birth weight of neonates born to obese

Table 13.1 Summary for adverse fetal and neonatal outcomes in obese and very obese (BMI > 40 kg/m^2) women from meta-analyses. Figures are expressed as adjusted odds ratio 95% confidence intervals (95% CI); references are in the rectangular brackets

Outcomes	Obese versus normal weight	Very obese versus normal weight
Stillborn	1.81 (1.69–1.93) [8]	
Fetal compromise	1.62 (1.54–1.70) [21]	2.08 (1.92–2.25) [21]
Low Apgar score 5 min	1.57 (1.46–1.68) [21]	2.09 (1.87–2.35) [21]
	1.40 (1.27–1.54) [35]	1.71 (1.55–1.89) [35]
Presence of meconium at birth		1.57 (1.42–1.73) [8]
NICU admission	1.91 (1.60–2.29) [8]	
Post term birth >41–42 week	1.37 (1.33–1.41) [21]	1.56 (1.48–1.64) [21]
Birth weight >90th percentile	1.88 (1.67–2.11) [8]	
	2.08 (1.95–2.23) [10]	
	2.42 (2.16–2.72) [9]	
Birth weight <10th percentile	0.88 (0.78–0.99) [8]	
Low birth weight	<2 kg 1.24 (1.09–1.41) [8]	
	<2.5 kg 0.84 (0.75–0.95) [20]	
Preterm birth		
Spontaneous preterm birth <37 week	0.93 (0.85–1.01) [20]	
Induced preterm birth <37 week	1.56 (1.42–1.71) [20]	1.71 (1.50–1.94) [20]
All preterm birth <33 week	1.45 (1.23–1.71) [20]	1.82 (1.48–2.24) [20]
All preterm birth <32 week	1.48 (1.20–1.81) [32]	

non-diabetic women compared to neonates born to normal weight women. The fat mass of children born to these obese women was significantly increased, but not the lean mass [11]. Similar results were found in a prospective cohort comparing two groups of pregnant women, one with normal pregestational BMI ($18.5 \leq$ BMI < 25 kg/m^2) and one with pregestational BMI $\geq$ 30 kg/m^2, for whom screening and subsequent treatment for GDM were enhanced in order to minimize the potential effect of maternal hyperglycemia on neonatal anthropometrics. Birth weight was not different between the two groups but fat mass at birth and cord serum leptin were significantly higher in girls born to obese women, but not in boys [12].

The risk of macrosomia in maternal obesity is influenced by GWG. In a large meta-analysis, including 740,000 obese women, odds of LGA in all obese women increased by 23–87% across studies when GWG was above the IOM recommendations [5]. When weight gain was higher than recommended, the increased risk of macrosomia was more pronounced for fetuses in women with class 1 obesity compared to the others classes. On the other hand, weight loss or lower than recommended weight gain decreased the risk of macrosomia (respectively, OR 0.58 (95% CI 0.52–0.66) and 0.77 (95% CI 0.71–0.84)), the effect of weight loss being more pronounced for class 3 obesity (OR 0.53 (95% CI 0.41–0.67)) [13]. In one study including 5,930 women in Canada, GWG above the recommended limits contributed more to LGA births than being obese, with population attributable fractions of 15.9% and 8.9%, respectively [14].

Maternal metabolism is affected by obesity, and different metabolic factors may contribute to higher fat mass in neonates born to women with obesity. Obese pregnant women with normal glucose tolerance have a higher basal glycemia and HbA1c than normal weight pregnant women, thereby exposing the fetus to relative hyperglycemia [12, 15]. This means that regardless of GDM, deregulation of glucose metabolism is present in obese women and may contribute to fat mass in the neonates.

Furthermore, maternal insulin resistance, which is associated with high maternal circulating free fatty acid and triglyceride concentrations, contributes to neonatal adiposity. The metabolic environment of obese women is also characterized by low-grade inflammation that contributes to increased placental transfer of nutrients to the fetus [16].

The mechanisms underlying the sexual difference in the accumulation of fetal fat mass in offspring from obese women remain unclear and need further investigations. Being the active interface between the mother and the fetus, the placenta appears to be a relevant target to better understand the molecular link between maternal obesity, insulin-resistance and fetal growth. Recently, different studies in women with high BMI showed that placental adaptations to maternal environment, including placental biometry or histopathology, differ according to fetal sex, with significant changes so far only reported in girls [17, 18]. Nevertheless, Brass et al. showed that placental uptake of oleic acid was suppressed and that the expression of the placental transporter CD36 was lower in male but not in female newborn of obese women [19].

Small for Gestational Age (SGA) and Low Birth Weight

Data concerning the association between maternal obesity, SGA (birth weight <10th percentile) and low birth weight are controversial. Two meta-analyses reported a 15% risk reduction for low birth weight [20, 21] whereas another meta-analysis reported a 25% increase of the risk for birth weight less than 2 kg, although the risk for being SGA was decreased (Table 13.1) [8]. The risk of having an infant of very low birth weight (<1500 g) or extremely low birth weight (<1000 g) was increased for obese women. The heavier the woman, the higher the risk of having an extremely low birth weight infant, with relative risks in obese, and very obese women (BMI > 40) of 1.43 (95% CI 1.05–1.95), and 1.98 (95% CI 1.36–2.89), respectively.

In any case, the risk of SGA in overweight or obese women should also be evaluated regarding GWG. Indeed, in a meta-analysis, GWG below guidelines was associated with higher risk for SGA in overweight (OR 1.34 95% CI 1.24–1.44) and in obese (1.24 95% CI 1.06–1.45) women [13].

Perinatal Death

Many studies have reported an association between maternal obesity and increased risk of perinatal death. Indeed, one meta-analysis showed that the relative risk for each 5 unit increase in maternal BMI was 1.21 (95% CI 1.09–1.35) for fetal death, 1.24 (95% CI 1.18–1.30) for stillbirth, 1.16 (95% CI 1.00–1.35) for perinatal death, 1.15 (95% CI 1.07–1.23) for neonatal death and 1.18 (95% CI 1.09–1.28) for infant death [22]. Most studies did not consider whether GDM or other maternal or fetal complications were included in the analysis. In a large cohort study that excluded pregnancies with malformations and pre-gestational diabetes, maternal obesity was associated with increased risks of fetal and infant death with ORs of 2.32 (95% 1.64–3.28) and 1.97 (95% CI 1.13–3.45), respectively. For each one–point increase in maternal BMI over 23 kg/m^2, the risk increased by 7% for fetal death and by 6% for infant death [23]. The results were the same when women with hypertension were excluded.

Maternal overweight and obesity are the highest-ranking modifiable risk factors for stillbirth in high-income countries, with a population attributable fraction of 8–18%, contributing to around 8,000 stillbirths at 22 weeks of gestation per year. The population attributable fraction for pre-existing diabetes in high-income countries is much lower (3.3–4.7%), contributing to around 2,200 stillbirths per year [24]. The additional effect of GWG on neonatal mortality was reported in one study using a current national data set in US. Compared with normal weight women who gained 0.30–0.44 kg/week, for obese women who had weight gain of 0.30–0.44 kg/week and $\geq$0.45 kg/week, adjusted ORs for neonatal death were 1.82 (95% CI 1.1–3.0) and 3.5 (95% CI 2.34–5.25), respectively [25].

Malformations

The increased risk of fetal malformations in obese pregnant women has been mentioned for several years by various studies and was confirmed in a review of the literature with meta-analysis. It showed that newborn of obese women have a significantly increased risk of neural tube defects, spina bifida, cardiovascular malformations, cleft lip and/or palate, anorectal atresia, hydrocephalus and limb abnormalities (Table 13.2) [26]. With regards to abdominal wall abnormalities, the risk of laparoschisis was significantly decreased in obese women. This result is probably related to the influence of maternal age as the prevalence of obesity increases with age and it is among the younger women that the risk of laparoschisis is the highest. The risk of omphalocele was not studied in this meta-analysis, but different studies reported a significant increased risk of this malformation in relation to maternal obesity, with odds ranging from 1.6 to 3.3 [27–29]. Also, no significant association was observed with maternal obesity for the risk of diaphragmatic hernia in this meta-analysis, although this risk was significantly increased in two population studies, with odds ranging between 1.4 and 1.8 [27, 28].

Various hypotheses can be advanced to explain the association between maternal obesity and the risk of fetal malformations:

- The risk of non-detection of a malformation after a routine ultrasound was estimated to 1/250 in women of normal weight and 1/100 in women with obesity [30]. The decreased ability to perform antenatal detection could limit the number of pregnancy terminations and thus increase the prevalence of birth defects.
- A relative folic acid deficiency in obese women has been associated with neural tube defects. In a Canadian study, however, no beneficial effect of acid folic flour fortification was observed in obese women regarding the risk of these malformations [31].

Table 13.2 Adjusted odds ratio [95% CI] for the association between maternal obesity (BMI > 30 kg/m^2) and congenital malformations at birth from the meta-analysis by Stothard et al. [26]

Malformations	OR 95% CI
Hydrocephaly	1.68 [1.19–2.36]
Neural tube defects	1.87 [1.62–2.15]
Spina bifida	2.24 [1.86–2.69]
Hydrocephalus	1.68 [1.19–2.36]
Cleft palate	1.23 [1.03–1.47]
Cleft lip and palate	1.20 [1.03–1.40]
Cardiovascular anomalies	1.30 [1.12–1.51]
Ano-rectal atresia	1.48 [1.12–1.97]
Limb reduction defects	1.34 [1.03–1.73]
Gastroschisis	0.17 [0.10–0.30]
Diaphragmatic hernia	1.28 [0.95–1.71]

Most malformations associated with maternal obesity are also described in the case of pre-gestational diabetes, which raises the problem of an abnormal maternal glycaemic balance at the beginning of pregnancy, even of undiagnosed type 2 diabetes. Nevertheless, when cases with pre-gestational diabetes were excluded from the analysis, the risk of cardiac malformation or neural tube defect was little changed [26]. On the other hand, malformations such as omphalocele and diaphragmatic hernia are not usually reported in case of maternal diabetes. Thus, teratogenicity factors associated with maternal obesity, independently from diabetes, need to be further explored.

Preterm Birth

In meta-analyses, the risk of preterm birth before 37 weeks was not increased in obese women when only spontaneous preterm birth was considered. On the other hand, there was a 1.5-fold increase in the risk of induced preterm birth before 37 weeks and a 1.5 to 2-fold increase in the risk of preterm birth before 33 weeks (Table 13.1) [8, 20, 32]. These observations are related to maternal morbidity associated with obesity, particularly preeclampsia. In a study on more than 167,000 pregnancies, the risk of preterm birth before 32 weeks was increased only in obese nulliparous women but it was not further observed after excluding women with hypertension [33]. In a study including more than 24,000 nulliparous women, the risk of birth before 33 weeks was increased in obese women (BMI between 30 and 34.9 kg/m^2; OR 1.8 (95% CI: 1.4–2.3) and morbidly obese women (BMI > 35 kg/m^2, OR 2.8 (95% CI: 1.4–5.4)), but it was no longer significant in morbidly obese women after adjustment for hypertension and preeclampsia [34].

Other Neonatal Complications

Meta-analyses have shown that the neonatal risks associated with maternal obesity, excluding prematurity and malformations, are fetal compromise and perinatal asphyxia, post-term birth and increased risk of being admitted to neonatal intensive care unit (Table 13.1) [8, 21, 35]. Several large cohort studies have shown an increased risk for respiratory distress, resuscitation at birth, obstetric trauma, hypoglycaemia and cerebral palsy [36, 37]. These risks are particularly high when maternal BMI is above 40 kg/m^2.

The link between maternal obesity and the risk of neonatal hypoglycaemia is probably related to maternal diabetes frequently associated with obesity. In the HAPO study, the frequency of hypoglycaemia tended to increase with maternal BMI, but after adjusting for maternal blood glucose levels, the risk was not increased any more [38].

It appears that the number of failures to initiate and continue breastfeeding is higher in case of maternal obesity [39]. There are many reasons for this, including physical causes that make placement of the child difficult, psychological and social causes, as well as delayed lactogenesis, especially after caesarean section [39].

Maternal Obesity and Cardio Metabolic Health in Infancy

The concept of early origins of adult disease was first mentioned by David Barker in the late 1980s. He observed an association between low birth weight and cardiovascular risk at adulthood. Subsequently, a relationship between low birth weight and the risk of diabetes and metabolic syndrome (MS) was also demonstrated. This resulted in the Barker's hypothesis according to which undernutrition in utero alters tissue structures and functions, and hence metabolism, increasing the risk of cardiovascular and metabolic disorders in adulthood [40]. There is now much evidence that maternal obesity during pregnancy also influences the long-term health of the offspring [41]. Here, we will focus on the relative contribution of maternal obesity, GWG and diabetes on childhood obesity and cardio-metabolic outcome.

Maternal Obesity and GWG and the Risk of Childhood Overweight/Obesity

One meta-analysis showed that pre-pregnancy maternal overweight and obesity increased the risk of offspring overweight/obesity by 2 and 3-fold, respectively [10]. Another one showed that compared to normal weight women, maternal obesity was associated with higher risks of overweight/obesity throughout childhood, with stronger association for later ages (2–5 years: 2.43 (95% CI 2.24–2.64); 2–10 years: 3.12 (2.98–3.27); 10–18 years: 4.47 (3.99–5.23)). Furthermore, the risk increased further with higher classes of maternal obesity [42].

Two meta-analyses showed that excessive GWG according to IOM recommendations was associated with an increased risk of childhood obesity by 30 to 40% [43, 44]. In line with these two studies, another meta-analysis reported that excessive GWG according to IOM recommendations was related to a 39–72% increase in overweight/obesity throughout childhood, between 2 and 18 years. This study also concluded that in overweight or obese women, the additional effect of GWG is small. Overall, 21.7–41.7% of childhood overweight/obesity prevalence could be attributed to maternal overweight and obesity together, and 11.4–19.2% could be attributed to excessive GWG [42].

Several studies have shown that the timing of the GWG during pregnancy is important for the association with high BMI and adiposity in offspring childhood.

In one study, any weight gain in the first 14 weeks of gestation was incrementally associated with increased offspring adiposity, whereas between 14 and 36 weeks, only gestational weight gain >500 g/week was associated with increased offspring adiposity [45]. In the Generation R study, weight gain in early pregnancy was associated with childhood BMI and adiposity, independently from maternal pre-pregnancy BMI and later pregnancy weight gain [46]. The last study showed that each 200 g/week increase in first-trimester GWG was associated with higher risk of overweight/obesity from 2 years (1.25 95% CI 1.09–1.42) to 4 years of age (1.15 95% CI 1.05–1.25). Higher first trimester GWG was also associated with a 15% increased risk of high waist circumference and skinfold thickness at 4 years of age [47].

Maternal Obesity and GWG and Childhood Cardio-Metabolic Profile

Both maternal obesity and excessive GWG have also been associated with adverse cardio-metabolic profile in childhood, including increased blood pressure, impaired lipid profile, insulin resistance and elevated inflammatory markers [45, 46, 48–51]. This association, however, is not very strong and does not seem to be explained by birth weight but may be mainly mediated by offspring adiposity. Two studies reported an association between excessive first trimester GWG and adverse cardio-metabolic profile. In the Generation R study, compared to children from normal weight mothers, the clustering of cardio-metabolic risk factors was 1.7-fold higher in offspring from women who gained excess weight in early pregnancy [46]. In the "Rhea" study in Crete and Greece, each 200 g/week of weight gain in the first trimester of pregnancy was associated with higher diastolic blood pressure at 4 years of age [47]. On the other hand, in a UK prospective cohort, only GWG between 14 and 36 weeks was positively associated with adverse lipid and inflammatory profiles [45].

Contribution of Maternal Diabetes and Macrosomia to Offspring Health in the Context of Maternal Obesity

Intrauterine exposure to hyperglycemia has a negative impact on childhood health, but it is difficult to discriminate this effect from the effect of maternal obesity. In two meta-analyses, after adjusting on maternal prepregnancy BMI, the association between maternal diabetes and offspring BMI was no longer significant [52, 53]. A systematic review reported an association between exposure to maternal diabetes and systolic blood pressure in childhood. Nevertheless, there is some evidence that the association between maternal diabetes and childhood systolic blood pressure may be influenced by maternal prepregnancy BMI [54].

Macrosomia has been reported to increase the risk of obesity, type 2 diabetes and metabolic syndrome in offspring. Two meta-analyses reported a 2-fold increase in the risk of obesity in offspring when birth weight was more than 4 kg [55, 56]. There was a 36% increased risk for type 2 diabetes when birth weight was more than 4 kg compared to birth weight ranging from 2.5 to 4 kg [57]. The prevalence of metabolic syndrome in children between of ages 6–11 was 50% in those born LGA to diabetic mothers, 21% in children born to diabetic mothers with normal birth weight and 29% in LGA born to non-diabetic mothers. LGA status and maternal obesity each independently increased the risk of metabolic syndrome by approximately 2-fold, while maternal diabetes was not significant [58]. These results mean that maternal obesity on its own, exclusive of GDM diagnosis, may affect offspring outcome regardless of birth weight. Indeed, when analysing perinatal risk factors associated to childhood obesity, Catalano et al. showed that the strongest perinatal predictor for a child in the upper tertile for weight and for percentage body fat was maternal pregravid BMI > 30 (respectively 3.75 (95% CI 1.39–10.10) and 5.45 (95% CI: 1.62–18.41), independent of diabetes or birth weight [59]. Figure 13.1 summarizes the main results from the literature on fetal development and the outcome of the offspring of obese mothers.

Putative Mechanisms

Epidemiological and cohort studies that have shown an association between maternal obesity and altered cardio-metabolic health in childhood cannot establish causality, nor elucidate underlying mechanisms. To date, it is still difficult to determine the relative contribution of obesity, weight gain or diabetes in the programming of offspring health, based on human data. There is probably not one single mechanism responsible for the adverse long-term outcomes associated with maternal obesity as increased pre-pregnancy maternal insulin resistance and subsequent hyperinsulinemia, inflammation, and oxidative stress have been suggested to contribute to early placental and fetal dysfunction. One of the main factors is the excess nutrient supply to the fetus and subsequent excess fat at birth. Insulin resistance has been observed in neonates born to obese mothers and it has been proposed that the metabolic abnormalities of the mother could be transferred to the fetus via the intrauterine environment [60].

The mechanisms through which inflammatory signals in pregnancy may contribute to increased adiposity in the offspring are not known. Several studies have examined the association between inflammatory markers in maternal blood and long term outcomes of adiposity in the offspring but the results are controversial, depending, among others, on the markers assessed [61].

Questions also have been raised on the implication of maternal microbiota in the risk of offspring obesity. Gut microbiota is associated with body weight and weight gain in pregnant women [62] and the composition and development of infant gut

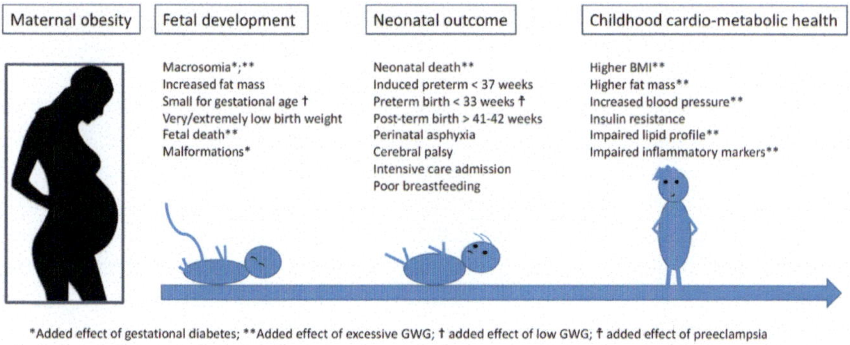

Fig. 13.1 Summary of major findings from meta-analyses and epidemiological studies on the effect of maternal obesity on fetal development, neonatal outcomes and childhood cardio-metabolic health. For each outcome, the additional effect of low or high gestational weight gain, gestational diabetes and preeclamsia are indicated by hallmark when they have been reported in the literature

microbiota are influenced by BMI and weight gain of mothers during pregnancy [63]. In recent years, increasing evidence indicate that microbes are present in the normal placenta and that placental microbiome is composed of non-pathogenic commensal microbiota. Also, it was shown that placental microbiota is altered among subject with GDM and that certain specific phylum were correlated with maternal GWG [64]. The effects of maternal dietary composition and nutritional status may be transferred to the fetus via the placenta and to the infant via lactation. There is growing evidence that the gut microbiota and its bacterial genome affect nutrient acquisition, energy regulation and fat storage. These findings raise the possibility that the gut microbiota plays a role in regulating host energy metabolism and may contribute towards the development of obesity and associated metabolic diseases [65].

Some molecular mechanisms of the programming effects of maternal obesity involving epigenetic traits have been investigated by human studies [66]. Global gene expression analysis of second trimester amniotic fluid suggested dysregulation of the expression of genes related to lipid metabolism and inflammatory signaling in fetuses of obese pregnant women compared to lean women [67]. Analyses of umbilical cord gene methylation have showed modifications in case of maternal obesity. Indeed, differential methylation of retinoid X receptor-a (RXRA) gene in umbilical cord tissue was shown to be associated with fat mass in children and with maternal carbohydrate intake early in pregnancy [68]. Maternal pre-pregnancy BMI was positively correlated with methylation in the umbilical cord of the peroxisome proliferator-activated receptor gamma (PPARγ) promoter, a gene involved in the lipids anabolic pathways [69]. Gene methylation in the placenta is also modified by maternal obesity, affecting pathways linked to immunity, cell-adhesion and metabolism [70]. It seems anyway that maternal early pregnancy obesity is only modestly associated with modulation of DNA methylation in the offspring at birth [71], although it was showed that maternal obesity programs protein expression of important metabolic regulatory genes [72]. Also it was reported that GWG has effects independent of maternal obesity on offspring DNA methylation patterns at birth [71].

A specific placental miRNA profile was observed in maternal obesity and these miRNAs were related to cell proliferation and insulin signaling pathways. Also, one of them, miR-296, was present in maternal second-trimester plasma samples and was associated with placental expression and prenatal and postnatal growth parameters, suggesting that placental miRNAs deregulated in maternal obesity may be involved in mediation of growth-promoting effects of maternal obesity on offspring [73].

Can We Prevent Adverse Outcome in Offspring Born to Obese Mother?

Dietary or Physical Activity Intervention

Improving maternal health and lifestyle before pregnancy and the control of weight gain during pregnancy according to IOM recommendations are common sense measures to limit the short- and long-term complications. Multiple lifestyle intervention trials consisting of weight management by various diets and/or increased physical activity have been employed to prevent excessive GWG and improve perinatal outcomes. In a meta-analysis, these trials were shown to have achieved modest success in decreasing excessive GWG, the risk of GDM, preeclampsia or excessive fetal growth [74].

Bariatric Surgery

Bariatric surgery is thought to be an effective intervention to sustain weight loss and is increasingly being used as an effective treatment for obesity. Bariatric surgery procedures are generally categorized into three groups. Restrictive procedures (laparoscopic adjustable gastric banding and sleeve gastrectomy) lead to weight loss by reducing gastric capacity which in turn restricts energy intake. Malabsorptive procedure (biliopancreatic diversion) leads to weight loss by restricting absorption of nutrients. Malabsorptive and restrictive procedure (Roux-en-Y gastric bypass) reduces stomach capacity, thereby causing malabsorption and a certain degree of restriction of food intake. Bariatric surgery reduces weight before pregnancy and GWG. In a meta-analysis, bariatric surgery improved pregnancy outcomes with a lower risk of preeclampsia, GDM and macrosomia but a higher risk of small neonates and preterm birth [75]. Currently, there are very few data on the long-term effects of bariatric surgery in offspring. In the only long-term follow-up study available to date, the authors compared the prevalence of obesity in children aged 2 to 18 years and born to obese mothers (BMI: 31 ± 9 kg/m^2) with substantial weight loss after biliopancreatic bypass surgery with same-age siblings who were born before maternal surgery (mothers' BMI: 48 ± 8 kg/m^2) and with current population standards. They showed that after maternal surgery, the prevalence of obesity in the offspring decreased by 52% and that of severe obesity by 45.1%. Moreover, the prevalence of overweight and obesity became similar to that in the general population, with no increase in the prevalence of underweight [76]. Further investigations in the same children aged 2.5–26 years demonstrated improvements in cardio-metabolic markers after maternal surgery sustained into adolescence with greater insulin sensitivity, improved lipid profile, lower C-reactive protein and leptin compared to those born before maternal surgery [77]. The investigators also found 5,698 differentially methylated genes in children born before or after maternal surgery. Functional analysis showed that these genes were mainly glucoregulatory, inflammatory and vascular disease genes [78].

Anyway, further clinical data are needed to consider the benefit for offspring health of such weight management interventions, particularly because of the two-fold increased risk of small babies [75] that is by itself a risk factor for adverse long-term outcome.

Neonatal Nutrition

The onset of obesity in childhood and adulthood is probably multifactorial in origin, with a combination of genetic factors, and the pre- and postnatal environment. Nutrition during the first few months after birth is one of these factors. Beyond the neonatal period, weight gain up to one year of age was reported to be associated with an increased risk of obesity [79].

Breast-feeding compared to bottle feeding has long-term beneficial effects on glucose tolerance, hypertension and dyslipidemia. The most convincing evidence was obtained for the effect on the risk of obesity in the long-term [80]. There is a "dose" effect in that a longer duration of breastfeeding is associated with a lower risk of obesity [81]. Several studies have shown the benefit of breastfeeding on the risk of obesity and diabetes in children exposed in utero to maternal diabetes [82, 83].

Beyond infancy, family food habits and lifestyle also play a predominant role. Feeding practice of obese mothers, in conjunction with psychosocial environment, may shape the development of appetite, energy intakes and food preferences during early infancy and the period of adoption of family diet [84].

Insight from Animal Models

As developed above, cohort prospective studies highlight associations between maternal obesity and gestational weight gain and the risk of obesity or adverse metabolic profile in childhood but they do not support conclusions about causality. Indeed, other factors involved in offspring health such as maternal smoking or vitamin D deficiency have been associated with childhood obesity. These factors are often observed during obese pregnancy but are mostly overlooked [85]. Moreover, the importance of pre- or perinatal exposition [86] as well as many postnatal expositions, in particular breastfeeding duration and age at solid food introduction, influence the risk of childhood higher adiposity [87].

Due to the experimental control of developmental conditions and the short life cycle of many animals, animal models have been widely studied to try and decipher mechanisms linking maternal and offspring obesity. Nevertheless, extreme diets, such as diets with 60% fat in mice, are usually needed to induce maternal obesity in rodents, as model animals will often adjust their feed intake to maintain body condition. Fatty acid composition of the diet is also essential, as fatty acids are involved in inflammatory processes known to be associated with obesity and cardiovascular diseases [88]. Also, amino-acids are known to have diverse effects on metabolism [89]. The precise amino-acid composition of proteins composing the animals' diets is often not known and may even vary during the experimental procedure, which may explain that some experimental results are not necessarily confirmed in subsequent studies.

Moreover, it is critical to understand the physiological similarities and differences between humans and animal models, including non-rodent animals, in order to infer significant information from studies generated in animals [90]. Since considerable data are available in animal models, the aim of the following section is to shed light on key observations and treatment opportunities suggested by studies using animal models without attempting to exhaustively review the literature.

General Observations on the Effects of Maternal Obesity and Diabetes During Gestation in Animal Models

In contrast to what is commonly expected, excess maternal nutrition or maternal obesity in animals usually leads to fetal growth restriction rather than excess fetal growth, as a result of impaired placental function [91], although macrosomia and/or increased neonatal BMI was observed in offspring born to obese mothers in sheep [92] and non-human primates [93]. Nevertheless, feeding an obesogenic diet to females before and during pregnancy generally increases the incidence of obesity and insulin resistance in offspring [94, 95]. Important information is that GDM does not occur spontaneously in most animals, although increased insulin resistance is a common feature during pregnancy. Thus, GDM may be induced through pharmaceutical intervention with use of streptozotocin or alloxan to destroy maternal pancreatic beta-cells before pregnancy, genetic modification or maternal infusion with glucose at specific stages of pregnancy. As observed for maternal obesity, macrosomia is not a consistent finding in offspring born to diabetic dams, but impaired glucose homeostasis and beta cell dysfunction are reported in offspring, (for review [90]).

Key Observations

Sexual Dimorphism

Among the key observations in animal models are the sexual dimorphism in feto-placental response and long-term offspring outcome in response to maternal obesity [96]. The placenta, of fetal origin, is the active maternal-fetal interface and a key actor of programming through its nutrient sensing, nutrient transfer and endocrine function [97]. The sexually dimorphic gene expression responses demonstrated in mouse placenta in response to maternal diet [98] have now been confirmed in other species, together with physiological observations [96]. It is interesting to observe that not so many of the above cited studies in humans take into consideration fetal or child sex and omission of this factor may wipe out important observations. Indeed, we showed in a prospective human cohort study that neonatal fat mass and cord serum leptin concentrations were increased in girls born to obese women compared to normal weight women, but not in boys [12].

Importance of the Peri-conceptional Period

Many studies in animals have elegantly demonstrated the importance of the pre- and peri-conceptional periods in the programming of offspring health [99, 100]. In particular, maternal obesity is associated with distinct gene expression profiles in the uterus of obese female rats, affecting inflammatory and lipid metabolism pathways and associated with lipid accumulation [101]. Gene expression in early and late pre-implantation embryo development is also affected by maternal obesity and a high fat diet both in rats and rabbits [102, 103]. Embryo transfer experiments in mice have elegantly demonstrated that pre-gestational and peri-conceptional maternal obesity independently affect offspring development [104], thus highlighting the importance of these periods when considering potential corrective interventions.

Independent of obesity, in a rat model, transient hyperglycemia in the early stages of pregnancy was shown to increase fetal and placental weight at term and to affect System-A amino-acid transporter activity, whereas transient hyperglycemia in late pregnancy did not affect fetal nor placental growth [105]. In mice, oocyte quality is adversely affected by maternal diabetes [106]. In rabbits, alloxan-induced type-1 diabetes on week prior to conception increases maternal plasma triglyceride and glucose concentrations, delaying blastocyst development and impairing intracellular accumulation of lipids [107, 108]. When these blastocysts are transferred to control dams, they are hyperglycemic, dyslipidemic and growth-retarded in late gestation, with perturbed placental function [109], further demonstrating the importance of peri-conceptional metabolism. So far, to our knowledge, the peri-conceptional effects of obesity associated with hyperglycemia has not been evaluated in animals.

Treatment or Prophylactic Opportunities

Although pre-conceptional obesity directly affects offspring development, weight loss before pregnancy was shown to restore normal feto-placental phenotype, despite differential placental expression of several epigenetic machinery enzymes compared to the lean control group [110]. Similarly, in Japanese macaques, when females were fed a Western style diet and subsequently fed a control diet before and during a subsequent pregnancy, most fetal consequences of maternal obesity were reversed [111].

Modifications in dietary macro- and micro-nutriments to modify placental function and possibly modulate fetal growth and metabolism were also explored in animals. Since high-fat and obesogenic diets were shown to alter epigenetic marks in the fetus and placenta [112], maternal intake of methyl donors such as folic acid and methionine could modify the methylation of fetal metastable epialleles, as demonstrated both in mice and humans [113, 114].

Maternal obesity affects placental vascularization and function. Thus, maternal supplementation with amino-acids such as citrulline or arginine, that improve placental function through multiple mechanisms, have been explored [115]. In rats, supplementation with citrulline was shown to improve placental function and enhance fetal growth [116], whereas arginine supplementation improves insulin sensitivity dams fed a high-fat diet [117]. In horses, maternal supplementation with arginine in the last third of pregnancy modified maternal metabolism and increased foal birth weight [118]. Preliminary clinical data already indicates that arginine supplementation may also be beneficial for obese pregnant women [119].

Maternal obesity is also clearly accompanied with systemic inflammation. Increasing the n-3/n-6 polyunsaturated fatty acid ratio in the maternal plasma in obese mice was shown to reduce maternal inflammation and prevent adverse offspring metabolic outcomes [120]. In a rat model of maternal diabetes, maternal poly-unsaturated fatty acid supplementation was also shown to normalize placental nutrient sensor signaling in their female offspring that developed gestational diabetes, suggesting that maternal treatment may also protect second generation fetal offspring development [121]. Resveratrol supplementation has been used with some success as a maternal treatment to prevent oxidative stress in the dam and offspring [111, 122] and its potential use needs to be explored [123].

Finally, very encouraging results have been obtained using maternal exercise before and/or during pregnancy in obese female rats, improving offspring outcomes until old age in male offspring [124]. Implementation of exercising strategies in obese women has been attempted, although compliance may be an issue.

Conclusion

In conclusion, effects of maternal obesity and associated gestational diabetes and GWG on fetal, neonatal and childhood health are confirmed but there still are considerable gaps of knowledge regarding the role of specific nutriments, the importance of pre- and perinatal nutrition and metabolism and potential therapeutic treatments [125]. Moreover, the emergence of the paternal phenotype [126] and of the gut microbiota [127] as important actors of offspring programming complexify the picture and claim for more research to tackle this important problem and for disease prevention in the next generations.

Conflict of interest The authors have no conflicts of interest to declare.

References

1. Engin A (2017) The definition and prevalence of obesity and metabolic syndrome. In: Engin AB, Engin A (eds) Obesity and lipotoxicity advances in experimental medicine and biology, vol 960. Springer, Cham, pp 1–17
2. Chu SY, Callaghan WM, Kim SY et al (2007) Maternal obesity and risk of gestational diabetes mellitus. Diabetes Care 30:2070–2076
3. Torloni MR, Betran AP, Horta BL et al (2009) Prepregnancy BMI and the risk of gestational diabetes: a systematic review of the literature with meta-analysis. Obes Rev 10:194–203
4. Catalano PM, McIntyre HD, Cruickshank JK et al (2012) The hyperglycemia and adverse pregnancy outcome study: associations of GDM and obesity with pregnancy outcomes. Diabetes Care 35:780–786
5. Faucher MA, Barger MK (2015) Gestational weight gain in obese women by class of obesity and select maternal/newborn outcomes: A systematic review. Women Birth 28:e70–e79
6. Ehrenberg HM, Huston-Presley L, Catalano PM (2003) The influence of obesity and gestational diabetes mellitus on accretion and the distribution of adipose tissue in pregnancy. Am J Obstet Gynecol 189:944–948
7. Bartha JL, Marin-Segura P, Gonzalez-Gonzalez NL et al (2007) Ultrasound evaluation of visceral fat and metabolic risk factors during early pregnancy. Obesity (Silver Spring) 15:2233–2239
8. Liu P, Xu L, Wang Y et al (2016) Association between perinatal outcomes and maternal pre-pregnancy body mass index. Obes Rev 17:1091–1102
9. Gaudet L, Ferraro ZM, Wen SW, Walker M (2014) Maternal obesity and occurrence of fetal macrosomia: a systematic review and meta-analysis. Biomed Res Int 2014:640291
10. Yu Z, Han S, Zhu J et al (2013) Pre-pregnancy body mass index in relation to infant birth weight and offspring overweight/obesity: a systematic review and meta-analysis. PLoS ONE 8:e61627

11. Hull HR, Dinger MK, Knehans AW et al (2008) Impact of maternal body mass index on neonate birthweight and body composition. Am J Obstet Gynecol 198(416):e1–e6
12. Mitanchez D, Jacqueminet S, Nizard J et al (2017) Effect of maternal obesity on birthweight and neonatal fat mass: a prospective clinical trial. PLoS ONE 12:e0181307
13. Goldstein RF, Abell SK, Ranasinha S et al (2017) Association of gestational weight gain with maternal and infant outcomes: a systematic review and meta-analysis. JAMA 317:2207–2225
14. Dzakpasu S, Fahey J, Kirby RS et al (2015) Contribution of prepregnancy body mass index and gestational weight gain to adverse neonatal outcomes: population attributable fractions for Canada. BMC Pregnancy Childbirth 15:21
15. Harmon KA, Gerard L, Jensen DR et al (2011) Continuous glucose profiles in obese and normal-weight pregnant women on a controlled diet: metabolic determinants of fetal growth. Diabetes Care 34:2198–2204
16. Catalano PM, Hauguel-De Mouzon S (2011) Is it time to revisit the Pedersen hypothesis in the face of the obesity epidemic? Am J Obstet Gynecol 204:479–487
17. Mando C, Calabrese S, Mazzocco MI et al (2016) Sex specific adaptations in placental biometry of overweight and obese women. Placenta 38:1–7
18. Leon-Garcia SM, Roeder HA, Nelson KK et al (2016) Maternal obesity and sex-specific differences in placental pathology. Placenta 38:33–40
19. Brass E, Hanson E, O'Tierney-Ginn PF (2013) Placental oleic acid uptake is lower in male offspring of obese women. Placenta 34:503–509
20. McDonald SD, Han Z, Mulla S, Beyene J (2010) Overweight and obesity in mothers and risk of preterm birth and low birth weight infants: systematic review and meta-analyses. BMJ 341:c3428
21. Heslehurst N, Simpson H, Ells LJ et al (2008) The impact of maternal BMI status on pregnancy outcomes with immediate short-term obstetric resource implications: a meta-analysis. Obes Rev 9:635–683
22. Aune D, Saugstad OD, Henriksen T, Tonstad S (2014) Maternal body mass index and the risk of fetal death, stillbirth, and infant death: a systematic review and meta-analysis. JAMA 311:1536–1546
23. Tennant PW, Rankin J, Bell R (2011) Maternal body mass index and the risk of fetal and infant death: a cohort study from the North of England. Hum Reprod 26:1501–1511
24. Flenady V, Koopmans L, Middleton P et al (2011) Major risk factors for stillbirth in high-income countries: a systematic review and meta-analysis. Lancet 377:1331–1340
25. Chen A, Feresu SA, Fernandez C, Rogan WJ (2009) Maternal obesity and the risk of infant death in the United States. Epidemiology 20:74–81
26. Stothard KJ, Tennant PW, Bell R, Rankin J (2009) Maternal overweight and obesity and the risk of congenital anomalies: a systematic review and meta-analysis. JAMA 301:636–650
27. Blomberg MI, Kallen B (2010) Maternal obesity and morbid obesity: the risk for birth defects in the offspring. Birth Defects Res A Clin Mol Teratol 88:35–40
28. Waller DK, Shaw GM, Rasmussen SA et al (2007) Prepregnancy obesity as a risk factor for structural birth defects. Arch Pediatr Adolesc Med 161:745–750
29. Watkins ML, Rasmussen SA, Honein MA et al (2003) Maternal obesity and risk for birth defects. Pediatrics 111:1152–1158
30. Dashe JS, McIntire DD, Twickler DM (2009) Effect of maternal obesity on the ultrasound detection of anomalous fetuses. Obstet Gynecol 113:1001–1007
31. Ray JG, Wyatt PR, Vermeulen MJ et al (2005) Greater maternal weight and the ongoing risk of neural tube defects after folic acid flour fortification. Obstet Gynecol 105:261–265
32. Torloni MR, Betran AP, Daher S et al (2009) Maternal BMI and preterm birth: a systematic review of the literature with meta-analysis. J Matern Fetal Neonatal Med 22:957–970
33. Cnattingius S, Bergstrom R, Lipworth L, Kramer MS (1998) Prepregnancy weight and the risk of adverse pregnancy outcomes. N Engl J Med 338:147–152

34. Bhattacharya S, Campbell DM, Liston WA, Bhattacharya S (2007) Effect of Body Mass Index on pregnancy outcomes in nulliparous women delivering singleton babies. BMC Public Health 7:168
35. Zhu T, Tang J, Zhao F et al (2015) Association between maternal obesity and offspring Apgar score or cord pH: a systematic review and meta-analysis. Sci Rep 5:18386
36. Blomberg M (2013) Maternal obesity, mode of delivery, and neonatal outcome. Obstet Gynecol 122:50–55
37. Villamor E, Tedroff K, Peterson M et al (2017) Association between maternal body mass index in early pregnancy and incidence of cerebral palsy. JAMA 317:925–936
38. HAPO Study Cooperative Research Group (2010) Hyperglycaemia and Adverse Pregnancy Outcome (HAPO) Study: associations with maternal body mass index. BJOG 117:575–584
39. Amir LH, Donath S (2007) A systematic review of maternal obesity and breastfeeding intention, initiation and duration. BMC Pregnancy Childbirth 7:9
40. Barker DJ (1990) The fetal and infant origins of adult disease. BMJ 301:1111
41. Godfrey KM, Reynolds RM, Prescott SL et al (2017) Influence of maternal obesity on the long-term health of offspring. Lancet Diabetes Endocrinol 5:53–64
42. Voerman E, Santos S, Patro Golab B et al (2019) Maternal body mass index, gestational weight gain, and the risk of overweight and obesity across childhood: an individual participant data meta-analysis. PLoS Med 16:e1002744
43. Tie HT, Xia YY, Zeng YS et al (2014) Risk of childhood overweight or obesity associated with excessive weight gain during pregnancy: a meta-analysis. Arch Gynecol Obstet 289:247–257
44. Mamun AA, Mannan M, Doi SA (2014) Gestational weight gain in relation to offspring obesity over the life course: a systematic review and bias-adjusted meta-analysis. Obes Rev 15:338–347
45. Fraser A, Tilling K, Macdonald-Wallis C et al (2010) Association of maternal weight gain in pregnancy with offspring obesity and metabolic and vascular traits in childhood. Circulation 121:2557–2564
46. Gaillard R, Steegers EA, Franco OH et al (2015) Maternal weight gain in different periods of pregnancy and childhood cardio-metabolic outcomes. The Generation R Study. Int J Obes (Lond) 39:677–685
47. Karachaliou M, Georgiou V, Roumeliotaki T et al (2015) Association of trimester-specific gestational weight gain with fetal growth, offspring obesity, and cardiometabolic traits in early childhood. Am J Obstet Gynecol 212(502):e1–e14
48. Daraki V, Georgiou V, Papavasiliou S et al (2015) Metabolic profile in early pregnancy is associated with offspring adiposity at 4 years of age: the Rhea pregnancy cohort Crete, Greece. PLoS One 10:e0126327
49. Kaar JL, Crume T, Brinton JT et al (2014) Maternal obesity, gestational weight gain, and offspring adiposity: the exploring perinatal outcomes among children study. J Pediatr 165:509–515
50. Oostvogels AJ, Stronks K, Roseboom TJ et al (2014) Maternal prepregnancy BMI, offspring's early postnatal growth, and metabolic profile at age 5–6 years: the ABCD Study. J Clin Endocrinol Metab 99:3845–3854
51. Gaillard R, Steegers EA, Duijts L et al (2014) Childhood cardiometabolic outcomes of maternal obesity during pregnancy: the Generation R Study. Hypertension 63:683–691
52. Kim SY, England JL, Sharma JA, Njoroge T (2011) Gestational diabetes mellitus and risk of childhood overweight and obesity in offspring: a systematic review. Exp Diabetes Res 2011:541308
53. Philipps LH, Santhakumaran S, Gale C et al (2011) The diabetic pregnancy and offspring BMI in childhood: a systematic review and meta-analysis. Diabetologia 54:1957–1966
54. Aceti A, Santhakumaran S, Logan KM et al (2012) The diabetic pregnancy and offspring blood pressure in childhood: a systematic review and meta-analysis. Diabetologia 55:3114–3127

55. Yu ZB, Han SP, Zhu GZ et al (2011) Birth weight and subsequent risk of obesity: a systematic review and meta-analysis. Obes Rev 12:525–542
56. Schellong K, Schulz S, Harder T, Plagemann A (2012) Birth weight and long-term overweight risk: systematic review and a meta-analysis including 643,902 persons from 66 studies and 26 countries globally. PLoS ONE 7:e47776
57. Harder T, Rodekamp E, Schellong K et al (2007) Birth weight and subsequent risk of type 2 diabetes: a meta-analysis. Am J Epidemiol 165:849–857
58. Boney CM, Verma A, Tucker R, Vohr BR (2005) Metabolic syndrome in childhood: association with birth weight, maternal obesity, and gestational diabetes mellitus. Pediatrics 115:e290–e296
59. Catalano PM, Farrell K, Thomas A et al (2009) Perinatal risk factors for childhood obesity and metabolic dysregulation. Am J Clin Nutr 90:1303–1313
60. Catalano PM, Presley L, Minium J, Hauguel-de Mouzon S (2009) Fetuses of obese mothers develop insulin resistance in utero. Diabetes Care 32:1076–1080
61. Catalano PM, Shankar K (2017) Obesity and pregnancy: mechanisms of short term and long term adverse consequences for mother and child. BMJ 356:j1
62. Santacruz A, Collado MC, Garcia-Valdes L et al (2010) Gut microbiota composition is associated with body weight, weight gain and biochemical parameters in pregnant women. Br J Nutr 104:83–92
63. Collado MC, Isolauri E, Laitinen K, Salminen S (2010) Effect of mother's weight on infant's microbiota acquisition, composition, and activity during early infancy: a prospective follow-up study initiated in early pregnancy. Am J Clin Nutr 92:1023–1030
64. Zheng J, Xiao XH, Zhang Q et al (2017) Correlation of placental microbiota with fetal macrosomia and clinical characteristics in mothers and newborns. Oncotarget 8:82314–82325
65. Gerard P (2016) Gut microbiota and obesity. Cell Mol Life Sci 73:147–162
66. Neri C, Edlow AG (2015) Effects of maternal obesity on fetal programming: molecular approaches. Cold Spring Harb Perspect Med 6:a026591
67. Edlow AG, Vora NL, Hui L et al (2014) Maternal obesity affects fetal neurodevelopmental and metabolic gene expression: a pilot study. PLoS ONE 9:e88661
68. Godfrey KM, Sheppard A, Gluckman PD et al (2011) Epigenetic gene promoter methylation at birth is associated with child's later adiposity. Diabetes 60:1528–1534
69. Gemma C, Sookoian S, Alvarinas J et al (2009) Maternal pregestational BMI is associated with methylation of the PPARGC1A promoter in newborns. Obesity (Silver Spring) 17:1032–1039
70. Mitsuya K, Parker AN, Liu L et al (2017) Alterations in the placental methylome with maternal obesity and evidence for metabolic regulation. PLoS ONE 12:e0186115
71. Thakali KM, Faske JB, Ishwar A et al (2017) Maternal obesity and gestational weight gain are modestly associated with umbilical cord DNA methylation. Placenta 57:194–203
72. Thakali KM, Saben J, Faske JB et al (2014) Maternal pregravid obesity changes gene expression profiles toward greater inflammation and reduced insulin sensitivity in umbilical cord. Pediatr Res 76:202–210
73. Carreras-Badosa G, Bonmati A, Ortega FJ et al (2017) Dysregulation of placental miRNA in maternal obesity is associated with pre- and postnatal growth. J Clin Endocrinol Metab 102:2584–2594
74. Thangaratinam S, Rogozinska E, Jolly K et al (2012) Interventions to reduce or prevent obesity in pregnant women: a systematic review. Health Technol Assess 16:iii–iv, 1–191
75. Yi XY, Li QF, Zhang J, Wang ZH (2015) A meta-analysis of maternal and fetal outcomes of pregnancy after bariatric surgery. Int J Gynaecol Obstet 130:3–9
76. Kral JG, Biron S, Simard S et al (2006) Large maternal weight loss from obesity surgery prevents transmission of obesity to children who were followed for 2 to 18 years. Pediatrics 118:e1644–e1649
77. Smith J, Cianflone K, Biron S et al (2009) Effects of maternal surgical weight loss in mothers on intergenerational transmission of obesity. J Clin Endocrinol Metab 94:4275–4283

78. Guenard F, Deshaies Y, Cianflone K et al (2013) Differential methylation in glucoregulatory genes of offspring born before vs. after maternal gastrointestinal bypass surgery. Proc Natl Acad Sci U S A 110:11439–11444
79. Druet C, Stettler N, Sharp S et al (2012) Prediction of childhood obesity by infancy weight gain: an individual-level meta-analysis. Paediatr Perinat Epidemiol 26:19–26
80. Lanigan J, Singhal A (2009) Early nutrition and long-term health: a practical approach. Proc Nutr Soc 68:422–429
81. Harder T, Bergmann R, Kallischnigg G, Plagemann A (2005) Duration of breastfeeding and risk of overweight: a meta-analysis. Am J Epidemiol 162:397–403
82. Mayer-Davis EJ, Rifas-Shiman SL, Zhou L et al (2006) Breast-feeding and risk for childhood obesity: does maternal diabetes or obesity status matter? Diabetes Care 29:2231–2237
83. Crume TL, Ogden L, Maligie M et al (2011) Long-term impact of neonatal breastfeeding on childhood adiposity and fat distribution among children exposed to diabetes in utero. Diabetes Care 34:641–645
84. Thompson AL (2013) Intergenerational impact of maternal obesity and postnatal feeding practices on pediatric obesity. Nutr Rev 71(Suppl 1):S55–S61
85. Robinson SM, Crozier SR, Harvey NC et al (2015) Modifiable early-life risk factors for childhood adiposity and overweight: an analysis of their combined impact and potential for prevention. Am J Clin Nutr 101:368–375
86. Fleming TP, Watkins AJ, Velazquez MA et al (2018) Origins of lifetime health around the time of conception: causes and consequences. Lancet 391:1842–1852
87. Woo Baidal JA, Locks LM, Cheng ER et al (2016) Risk factors for childhood obesity in the first 1,000 days: a systematic review. Am J Prev Med 50:761–779
88. Mennitti LV, Oliveira JL, Morais CA et al (2015) Type of fatty acids in maternal diets during pregnancy and/or lactation and metabolic consequences of the offspring. J Nutr Biochem 26:99–111
89. Lin G, Wang X, Wu G et al (2014) Improving amino acid nutrition to prevent intrauterine growth restriction in mammals. Amino Acids 46:1605–1623
90. Chavatte-Palmer P, Tarrade A, Rousseau-Ralliard D (2016) Diet before and during pregnancy and offspring health: the importance of animal models and what can be learned from them. Int J Environ Res Public Health 13:586
91. Myatt L, Maloyan A (2016) Obesity and placental function. Semin Reprod Med. 34:42–49
92. Zhang L, Long NM, Hein SM et al (2011) Maternal obesity in ewes results in reduced fetal pancreatic beta-cell numbers in late gestation and decreased circulating insulin concentration at term. Domest Anim Endocrinol 40:30–39
93. Farley D, Tejero ME, Comuzzie AG et al (2009) Feto-placental adaptations to maternal obesity in the baboon. Placenta 30:752–760
94. Zambrano E, Ibanez C, Martinez-Samayoa PM et al (2016) Maternal obesity: lifelong metabolic outcomes for offspring from poor developmental trajectories during the perinatal period. Arch Med Res 47:1–12
95. Wankhade UD, Thakali KM, Shankar K (2016) Persistent influence of maternal obesity on offspring health: Mechanisms from animal models and clinical studies. Mol Cell Endocrinol 435:7–19
96. Tarrade A, Panchenko P, Junien C, Gabory A (2015) Placental contribution to nutritional programming of health and diseases: epigenetics and sexual dimorphism. J Exp Biol 218:50–58
97. Sferruzzi-Perri AN, Camm EJ (2016) The programming power of the placenta. Front Physiol 7:33
98. Mao J, Zhang X, Sieli PT et al (2010) Contrasting effects of different maternal diets on sexually dimorphic gene expression in the murine placenta. Proc Natl Acad Sci U S A 107:5557–5562
99. Watkins AJ, Lucas ES, Fleming TP (2010) Impact of the periconceptional environment on the programming of adult disease. J Dev Orig Health Dis 1:87–95

100. Dahlhoff M, Pfister S, Blutke A et al (2014) Peri-conceptional obesogenic exposure induces sex-specific programming of disease susceptibilities in adult mouse offspring. Biochim Biophys Acta 1842:304–317
101. Shankar K, Zhong Y, Kang P et al (2011) Maternal obesity promotes a proinflammatory signature in rat uterus and blastocyst. Endocrinology 152:4158–4170
102. Picone O, Laigre P, Fortun-Lamothe L et al (2011) Hyperlipidic hypercholesterolemic diet in prepubertal rabbits affects gene expression in the embryo, restricts fetal growth and increases offspring susceptibility to obesity. Theriogenology 75:287–299
103. Tarrade A, Rousseau-Ralliard D, Aubriere MC et al (2013) Sexual dimorphism of the feto-placental phenotype in response to a high fat and control maternal diets in a rabbit model. PLoS ONE 8:e83458
104. Sasson IE, Vitins AP, Mainigi MA et al (2015) Pre-gestational vs gestational exposure to maternal obesity differentially programs the offspring in mice. Diabetologia 58:615–624
105. Ericsson A, Saljo K, Sjostrand E et al (2007) Brief hyperglycaemia in the early pregnant rat increases fetal weight at term by stimulating placental growth and affecting placental nutrient transport. J Physiol 581:1323–1332
106. Wang Q, Ratchford AM, Chi MM et al (2009) Maternal diabetes causes mitochondrial dysfunction and meiotic defects in murine oocytes. Mol Endocrinol 23:1603–1612
107. Ramin N, Thieme R, Fischer S et al (2010) Maternal diabetes impairs gastrulation and insulin and IGF-I receptor expression in rabbit blastocysts. Endocrinology 151:4158–4167
108. Schindler M, Pendzialek M, Navarrete Santos A et al (2014) Maternal diabetes leads to unphysiological high lipid accumulation in rabbit preimplantation embryos. Endocrinology 155:1498–1509
109. Rousseau-Ralliard D, Couturier-Tarrade A, Thieme R et al (2019) A short periconceptional exposure to maternal type-1 diabetes is sufficient to disrupt the feto-placental phenotype in a rabbit model. Mol Cell Endocrinol 480:42–53
110. Panchenko PE, Voisin S, Jouin M et al (2016) Expression of epigenetic machinery genes is sensitive to maternal obesity and weight loss in relation to fetal growth in mice. Clin Epigenetics 8:22
111. Pound LD, Comstock SM, Grove KL (2014) Consumption of a Western-style diet during pregnancy impairs offspring islet vascularization in a Japanese macaque model. Am J Physiol Endocrinol Metab 307:E115–E123
112. Aagaard-Tillery KM, Grove K, Bishop J et al (2008) Developmental origins of disease and determinants of chromatin structure: maternal diet modifies the primate fetal epigenome. J Mol Endocrinol 41:91–102
113. Waterland RA, Kellermayer R, Laritsky E et al (2010) Season of conception in rural gambia affects DNA methylation at putative human metastable epialleles. PLoS Genet 6:e1001252
114. Waterland RA, Travisano M, Tahiliani KG et al (2008) Methyl donor supplementation prevents transgenerational amplification of obesity. Int J Obes (Lond) 32:1373–1379
115. Herring CM, Bazer FW, Johnson GA, Wu G (2018) Impacts of maternal dietary protein intake on fetal survival, growth, and development. Exp Biol Med (Maywood) 243:525–533
116. Bourdon A, Parnet P, Nowak C et al (2016) L-Citrulline supplementation enhances fetal growth and protein synthesis in rats with intrauterine growth restriction. J Nutr 146:532–541
117. Miczke A, Suliburska J, Pupek-Musialik D et al (2015) Effect of L-arginine supplementation on insulin resistance and serum adiponectin concentration in rats with fat diet. Int J Clin Exp Med. 8:10358–10366
118. Chavatte-Palmer P, Derisoud E, Robles M et al (2018) Effects of dietary arginine supplementation in pregnant mares on maternal metabolism and foal birthweight. J Equine Vet Sci 66:225
119. Suliburska J, Bogdanski P, Szulinska M et al (2014) Changes in mineral status are associated with improvements in insulin sensitivity in obese patients following L-arginine supplementation. Eur J Nutr 53:387–393

120. Heerwagen MJ, Stewart MS, de la Houssaye BA et al (2013) Transgenic increase in N-3/n-6 fatty acid ratio reduces maternal obesity-associated inflammation and limits adverse developmental programming in mice. PLoS ONE 8:e67791
121. Capobianco E, Fornes D, Roberti SL et al (2018) Supplementation with polyunsaturated fatty acids in pregnant rats with mild diabetes normalizes placental PPARgamma and mTOR signaling in female offspring developing gestational diabetes. J Nutr Biochem 53:39–47
122. Vega CC, Reyes-Castro LA, Rodriguez-Gonzalez GL et al (2016) Resveratrol partially prevents oxidative stress and metabolic dysfunction in pregnant rats fed a low protein diet and their offspring. J Physiol 594:1483–1499
123. Tain YL, Hsu CN (2018) Developmental programming of the metabolic syndrome: can we reprogram with resveratrol? Int J Mol Sci 19:2584
124. Nathanielsz PW, Ford SP, Long NM et al (2013) Interventions to prevent adverse fetal programming due to maternal obesity during pregnancy. Nutr Rev 71(Suppl 1):S78–S87
125. Friedman JE (2018) Developmental programming of obesity and diabetes in mouse, monkey, and man in 2018: where are we headed? Diabetes 67:2137–2151
126. Soubry A (2018) POHaD: why we should study future fathers. Environ Epigenet 4:dvy007
127. Devaux CA, Raoult D (2018) The microbiological memory, an epigenetic regulator governing the balance between good health and metabolic disorders. Front Microbiol 9:1379

Chapter 14
The Developmental Mechanisms of Obesity by Maternal Obesity

Long T. Nguyen, Carol A. Pollock and Sonia Saad

Abstract Obesity is a major global concern due to its alarming prevalence and associated risks for multiple diseases. The rate of obesity has nearly tripled in the last four decades and amounting evidence is implying a critical role of developmental factors before, during and after pregnancy in promoting this global pandemic. Maternal obesity in particular has been associated with large-for-gestational age babies and increased risk of obesity in adulthood, thus generating a vicious cycle. Studies in animal models demonstrated that such effects of maternal obesity can be detected in the offspring across up to three generations, suggesting a profound transgenerational impact. This chapter will discuss critical windows for developmental programming of obesity and possible mechanisms involved such as oxidative stress, mitochondrial dysfunction, placental insults, intrauterine overnutrition, appetite dysregulation and microbiome. A special focus will be put on epigenetic regulation and the role of sirtuins, which have been suggested to play a central role in the metabolic programming process. Finally, the prospective of intervention therapies for maternal obesity-induced developmental programming will be briefly discussed.

Keywords Obesity · Metabolic disorders · Pregnancy · Developmental programming · Epigenetic · Sirtuin

Introduction

Overweight and obesity, which are generally characterised by greater energy intake than energy expenditure state, with excess energy being stored as fats in adipose tissues, are affecting 25% of the global population, leading to diabetes, hypertension and other chronic diseases. Despite the fundamental role of genetic background in

L. T. Nguyen (✉) · C. A. Pollock · S. Saad
Renal Medicine, Kolling Institute Level 9, Royal North Shore Hospital,
The University of Sydney, Pacific Hwy, St. Leonard, NSW 2065, Australia
e-mail: long.t.nguyen@sydney.edu.au

P. S. Tappia et al. (eds.), *Pathophysiology of Obesity-Induced Health Complications*, Advances in Biochemistry in Health and Disease 19,
https://doi.org/10.1007/978-3-030-35358-2_14

the development of overweight/obesity and related disorders, less than 5% of differences in body weight can be explained purely by genetic variation. This suggests that environmental factors including diets and lifestyles play a dominant role in shaping individual metabolic phenotypes. As such, it is tempting to presume that to a great extent are our own metabolic fates controlled by our behaviours.

Interestingly, amounting evidence suggests that what we are exposed to in our lifetime not only defines our own obesity risk, but also has significant impacts on the development of obesity and related metabolic disorders in our descendants. The phenomenon, termed developmental programming, was first described in 1986, by Barker, based on an observation that increased mortality due to ischemic heart disease coincided with infant mortality and low birth weight (LBW) in England and Wales populations during the same period [1, 2]. Subsequent studies demonstrated the association of LBW with higher risk of obesity, insulin resistance intolerance, dyslipidaemia, and hypertension in adulthood. Although the mechanism for such effect remains not completely understood, it is believed that maternal malnutrition-induced intrauterine growth restriction (IUGR) followed by postnatal 'catch-up' growth is the main driving mechanism. Developmental programming is therefore referred to as a permanent or long-term change in the structure or function of an organism resulting from a stimulus or insult acting at a critical period of early life [3].

Although maternal undernutrition remains a popular topic in the field of developmental origin of obesity, the focus has gradually shifted to maternal overnutrition mainly due to the dramatic rise of obesity incidence around the world. According to a recent systemic review, overweight and obesity account for 40–60% of women of reproductive ages in developed countries, and the figure in developing countries is 30–40% [4]. Importantly, maternal obesity is one of strongest risk factor of childhood obesity and related metabolic complications including hyperglycaemia, dyslipidaemia, renal and cardiovascular diseases, leading to a vicious cycle of exacerbation of the global obesity pandemic. The mechanism for such effect is not clearly understood.

Apart from maternal malnutrition and obesity, other maternal factors such as stress, lifestyles (e.g. lack of physical activities), environmental exposure (e.g. to endocrine-disrupting chemicals) and diseases (e.g. polycystic ovary syndrome, gestational diabetes mellitus) have also been associated with a higher risk of obesity in the offspring, as seen in Fig. 14.1 [5]. Paternal obesity also contributes to developmental programming of obesity but to a lesser extent [6, 7]. Due to limitation in length, only the effects of maternal obesity will be extensively discussed in this chapter.

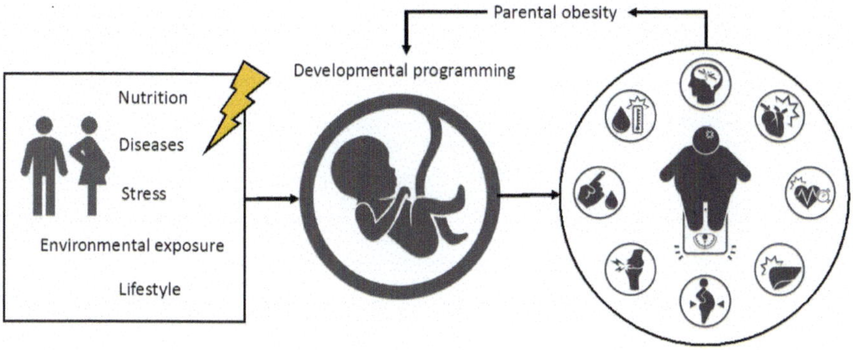

Fig. 14.1 Developmental programming of obesity by maternal factors

Critical Developmental Windows and Programming Factors by Maternal Obesity

Periconception—From Germ Cells to Embryonic Defects

Conception is the fertilisation process of an ovum by a sperm cell, leading to the formation, implantation and subsequent development of the embryos. Ultimately, the genetic and non-genetic characteristics of the gametes participating in such process, which are dependent of the preconceptional phenotypes of the parents, define the nature of the foetus/offspring. It is well-known that maternal obesity and hyperglycaemia induce oxidative stress and inflammation in different tissues including reproductive organs [8], which has been implicated in impaired oogenesis and maturation [9–12], infertility and embryonic retardation [13]. In animals, obese dams generally produce smaller litter with significantly higher postnatal mortality [14]. In human, significantly higher risks for in vitro fertilization (IVF) cycle cancellation and obstetric complications in obese patients have been reported [15]. These reproductive defects has been associated with impaired glucose metabolism and increased endogenous triglyceride levels in the resulting blastocysts [16], suggesting a strong association between oocyte quality and foetal metabolic programming [14, 17–19].

Maternal obesity is associated with increased oxidative stress and mitochondrial dysfunction in the oocytes [18, 19]. Oxidative stress is induced by the relative abundance of reactive oxygen species (ROS) compared to antioxidant capacity, and because the majority of ROS are produced by mitochondria, this organelle is highly prone to oxidative damage (Table 14.1). Importantly, foetal mitochondria are maternally inherited, which means that damaged mitochondria in the oocyte as the result of preconceptional obesity will be transmitted to the foetus upon fertilisation, subsequently contribute to ROS production and metabolic disorders in the offspring. Indeed, mitochondrial abnormalities and impaired glucose homeostasis were

Table 14.1 Critical windows of developmental programming of obesity and involved mechanisms in different periods

<table>
<tr><th>Developmental window</th><th>Before conception</th><th>Gestation</th><th>Lactation</th></tr>
<tr><td rowspan="8">Programming factors</td><td colspan="3">Oocyte defects (oxidative damage, mitochondrial dysfunction)</td></tr>
<tr><td></td><td>Placental defects</td><td></td></tr>
<tr><td></td><td>Intrauterine overnutrition</td><td></td></tr>
<tr><td></td><td colspan="2">Endocrine and inflammatory factors</td></tr>
<tr><td colspan="2"></td><td>Milk composition</td></tr>
<tr><td colspan="2"></td><td>Circadian rhythm disruption</td></tr>
<tr><td colspan="2"></td><td>Microbiome colonisation</td></tr>
<tr><td colspan="3">Epigenetic modification</td></tr>
</table>

shown in offspring of rats fed a fat-rich diet in pregnancy [17]. In another study, similar effects on mitochondrial disorders can be found across three generations of mice [20].

Gestation—The Roles of Placental Stress and Intrauterine Overnutrition

In gestation, the maternal body undergoes a number of physiological adaptations with the ultimate goal is to meet the high energy demands of foetal development. In early gestation, insulin sensitivity in maternal tissues is increased to encourage energy storage in the form of lipogenesis and adiposity, which leads to hypoglycaemia. In late gestation, insulin resistance is established so that glucose can be maintained at high levels in the maternal circulation to ensure energy supply for foetal exponential growth, hence the likelihood to develop gestational diabetes in the third trimester. Maternal obesity is associated with hyperglycaemia and hyperlipidaemia even in the first trimester, which unnecessarily advances foetal development and induces metabolic pressure on the immature foeto-placental unit. Indeed, maternal obesity has been suggested to induce placental lipotoxicity, oxidative stress, inflammation and vasculopathy [21–24], which interfere with the selective permeability of the blood-placental barrier for excess nutrients, metabolic hormones and proinflammatory cytokines [23, 25].

Placentas exposed to maternal HFD consumption showed increased expression of glucose transporter (GLUT)1, fatty acid transporter (FATP) and lipoprotein lipase (LPL) on the syncytiotrophoblasts [26, 27], Together with the increased abundance of glucose and lipid in the obese mothers, foetuses are exposed to intrauterine overnutrition and tend to overgrow. The inflammatory environment in the uterus also render the foetus more vulnerable to tissue developmental remodelling in early ages and dysfunction later in life [28]. Although exercise during pregnancy has shown beneficial effects on offspring metabolism in animals [29], the same approach in clinical trials have achieved limited improvements in pregnancy and neonatal outcomes [30], suggesting persisted programming effects. Another possibility is that the overwhelming physiological responses during pregnancy have suppressed the effects of external interventions. Arguably, gestation is the least modifiable window in foetal programming despite its utmost importance.

Lactation—The Roles of Milk Composition, Appetite Dysregulation, Circadian Rhythm and Microbiome

A number of cross-fostering studies in high-fat diet (HFD)-fed animals have been conducted to understand the relative contribution of maternal obesity in prenatal and postnatal periods to developmental programming process of offspring obesity. Consistent data suggest that although both control offspring cross-fostered by HFD-fed dams and HFD offspring suckled by chow-fed dams presented metabolic complications such as increased adiposity, hyperglycaemia, insulin resistance and non-alcoholic fatty liver disease (NAFLD), the former appeared to be more susceptible [31–34], suggesting that postnatal nutrition is a stronger driving factor of maternal obesity-induced foetal programming.

Although maternal obesity has been associated with reduced milk production in both human and animals [35–37], the offspring of obese dams have higher milk intake [38, 39]. Moreover, breast milk from obese individuals contains higher fat and glucose content [40–42] and altered metabolome [43], which contribute to higher energy intake and metabolic disorders in the offspring. Epidemiologic studies have demonstrated that energy intake and dietary composition in offspring in early ages significantly correlate with their mothers' [44], suggesting a strong programming effect of maternal obesity on offspring appetite preferences and feeding behaviours.

Hyperphagia in the offspring of obese mothers can be partially explained by leptin resistance and dysregulation of neuropeptides in the hypothalamus [45], the central region of appetite regulation and energy homeostasis. Studies in rat models of maternal obesity showed that offspring exposed to maternal and postnatal high-fat diet (HFD) exhibit increased density of orexigenic peptide-expressing neurons in the hypothalamus [46], which are hyperactive in response to fasting

[47–49], thus stimulating food intake. Breast milk from obese mothers has increased levels of leptin [50], which has been associated with higher weight gain [42].

It is well-known that feeding, sleep-wake cycle, metabolic responses and all other physiological aspects of a developed organism are influenced by its own circadian rhythm or biological clock. A disruption of which has been associated with sleep/eating disorders and metabolic diseases. In human, circadian rhythm starts developing after three months of age, which is during lactation. Growing evidence in animals suggest that maternal obesity during gestation and lactation can lead to alteration of the key circadian regulators such as CLOCK and BMAL1 in the offspring liver in association with metabolic and epigenetic programming [51–53].

During and after birth, infants are rapidly colonised by maternal microbes and surrounding environment. Different maternal diets and levels of energy intake result in different maternal gut microbiomes, the colonies of which are likely to be transmitted to neonates, leading to immunometabolic disorders [54]. Offspring of non-human primate dams fed a HFD during gestation and lactation showed a significant reduction of bacterial diversity, which was not entirely corrected when weaning onto a control diet. Germ-free mice colonised with stool from infants born to obese mothers were more prone to obesity and NAFLD [55].

Epigenetic Modification

Epigenetic modification has been suggested to play a central role in the mechanism of developmental programming. It consists of DNA methylation, histone acetylation and micro RNAs, which interfere with gene expression without altering the genetic imprint. Most epigenetic marks are developed as the result of normal cellular growth and differentiation, while the others are highly dynamic and cumulative as per environmental changes. Maternal high diet consumption and maternal obesity have been associated with epigenomic alterations in the offspring. In animals, in utero exposure to HFD has been shown to global DNA hypermethylation [56] as well as specific changes in methylation at 3360 loci, a great number of which are associated with transcriptional changes and/or have been implicated in metabolic regulation [57]. Mice born to obese dams also demonstrated persistent histone modifications up to 5 weeks of age independently of postnatal feeding [53]. Consistently, human studies using cord blood samples found that global DNA methylation are higher in premature and extreme birthweight newborns compared with controls and associated with increased adiposity in later life [58]. Besides, a few studies have demonstrated altered expression of several miRNAs in the offspring tissues due to maternal obesity [59–61]. Among which, the increased level of mir126 in adipose tissue is believed to induce suppression of Insulin receptor substrate 1 (Irs1) [61], leading to reduced insulin sensitivity in offspring adipocytes.

Particularly, maternal body mass index (BMI) has been correlated with the methylation level of Peroxisome proliferator-activated receptor gamma co-activator

1-α (PGC-1α) gene in cord blood [62], which plays an essential role in mitochondrial biogenesis, fatty acid oxidation and metabolic homeostasis. A deficiency of which has been associated with obesity, diabetes and many other metabolic comorbidities. Blood samples from obese mothers and their offspring at birth showed higher levels of leptin, which has been correlated with DNA hypomethylation of Leptin gene promoter [63]. Leptin is produced by adipocytes and hyperleptinaemia is often an indicator of increased adiposity and/or hypothalamic leptin resistance. In another study, genome-wide interrogation of cord blood samples from more than a thousand mother–child pairs identified multiple CpG sites in the offspring of obese mothers, a number of which persisted until adulthood [64]. Among the affected genes, SUCLG2 is a mitochondrial matrix enzyme, of which deficiency has been shown to result in decreased mitochondrial DNA and oxidative phosphorylation-dependent growth [65]. This is consistent with our data showing reduced mitochondrial copy number in offspring mice born to obese dams. Apart from the intrauterine environment, breastfeeding also represents a critical period of epigenetic programming. In a recent study examining whole blood samples of 120 mother–child pairs, a negative correlation between breastfeeding duration and leptin methylation was found at 17 months post-partum [66].

As epigenetic marks are to be erased after fertilization, it is believed that epigenetic modifications in the offspring are largely attributed to a harmful intrauterine environment and suboptimal lactation. In the case of maternal obesity where overnutrition is coupled with oxidative stress and inflammation, alterations to the foetal epigenome are required to titrate down mitochondrial oxidative phosphorylation because it is the main source of ATP and ROS. From this perspective, special consideration needs to be taken to whether epigenetic changes due to maternal obesity precede the development of offspring obesity and whether these changes reflect an actual causal relationship. It has been argued that true epigenetic programming is transgenerational inheritance of permanent epigenetic changes that are exempt from erasure and persist across more than one generation [67]. Intrauterine overnutrition may not only program foetal somatic tissues but also germ cells, leading to abnormalities in F2 and further generations, which can be seen in Fig. 14.2. Indeed, emerging evidence suggested some epigenetic changes are resistant to this process and can be transmitted to the subsequent generations [67]. In animals, residual effects of maternal obesity including increased birthweight, adiposity, leptin/adiponectin expression, tissue inflammation and mitochondrial impairment, can be detected across up to three generations even though F2 offspring were not directly exposed to intrauterine overnutrition environment [20, 68, 69], suggesting a transgenerational effect. Some of the metabolic changes were associated with altered patterns of histone acetylation and methylation at the genes encoding leptin and adiponectin and their expression in white adipose tissue [70]. While supporting evidence is mounting in animal models, evidence in humans is scarce, in part because epidemiologic studies spanning multiple generations are challenging and exposures are often cumulative and difficult to isolate in time [55].

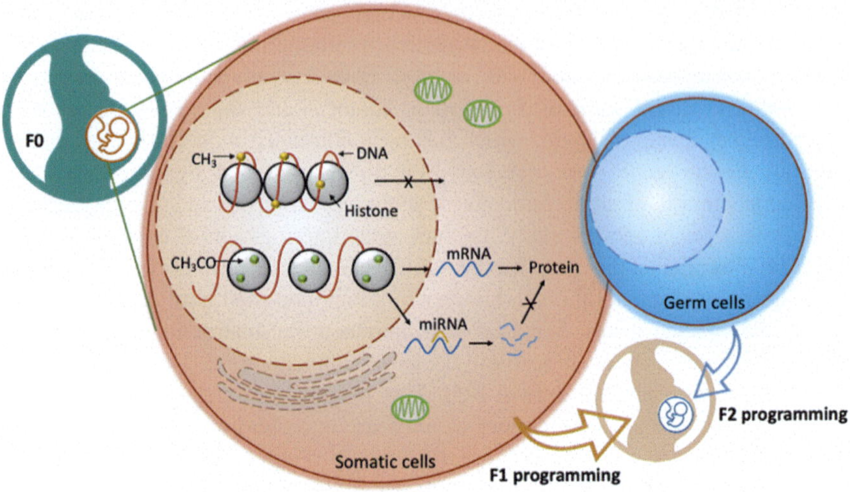

Fig. 14.2 Epigenetic regulation by maternal obesity in foetal somatic and germ cells leads to developmental programming in the first and second generation respectively

Sirtuins—Key Regulators of Developmental Programming

Sirtuin (SIRT)s are a group of deacetylases and mono-ADP-ribosyl transferases whose functions are strictly dependent on the intracellular level of nicotinamide adenine dinucleotide (NAD^+), an essential coenzyme in the electron transport chain for ATP synthesis. Seven isoforms of Sirtuins (SIRT1-7) have been described in mammals with a variety of roles in metabolic regulation, stress responses and senescence [71]. Among these isoforms, SIRT1 and SIRT3 are two of the most studied with regard to foetal programming, as seen in Fig. 14.3. In obesity, the expression of SIRT1/3 has been shown to be downregulated in various tissues [72–76], including oocytes [18]. Particularly, SIRT3 is mainly expressed in the mitochondria, and because mitochondria are maternally inherited, SIRT3 deficiency in the oocyte can directly result in SIRT3 deficiency in the offspring at early developmental stages. Indeed, SIRT3 deficiency has been associated with increased oxidative stress [18], and p53-mediated developmental arrest of the embryos [77], as well as reduced fatty acid oxidation in offspring liver at weaning due to maternal obesity [78]. SIRT3 overexpression has been shown to reverse these germline defects [18, 79].

Compared to SIRT3, SIRT1 has a broader role in the mechanism of foetal programming. Apart from the direct implication in foetal metabolism and stress responses, SIRT1 also suppresses PPARγ and lead to downregulation of nutrient receptors and transporters such as lipoprotein lipase (LPL) and fatty acid transporter (FATP) in the placenta [23, 27], leading to increased nutrient supply to the foetus. Our unpublished data also suggest that maternal SIRT1 overexpression can partially

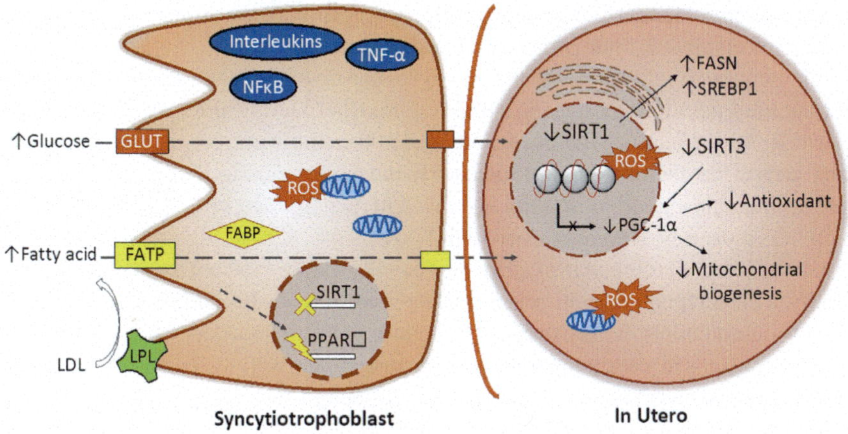

Fig. 14.3 The role of placental and embryonic sirtuins in regulating intrauterine nutrition, mitochondrial function, oxidative stress and inflammation

rescue maternal obesity-induced metabolic disorders in mice offspring. SIRT1 also acts as a regulator of epigenetic traits during foetal programming. It has been demonstrated that the hepatic histone H3 acetylation level, a target of SIRT1, was significantly increased in the offspring born to obese mothers [51]. In another study, the methylation levels of histone H3 (H3K4me3 and H3K27me3) were also significantly altered in association with a reduction of hepatic SIRT1 in the offspring [44]. In addition, we also demonstrated that hypothalamic expression of SIRT1 and leptin receptor are concomitantly suppressed in the offspring born to obese dams [39], suggesting a role of SIRT1 in appetite regulation. Postnatal administration of SIRT1 activators such as SRT1720 and NMN, has been shown to attenuate maternal obesity-induced metabolic programming [80, 81] and kidney disorders [82, 83].

Conclusion

Evidence in both human and animals is sufficient to conclude that maternal factors such as obesity during early developmental windows including periconception, gestation and lactation can program the development of obesity and related diseases later in life. A number of programming mechanisms have been suggested, providing a multifaceted perspective regarding this complicated phenomenon. Together with recent advances in next gen sequencing and system biology, it is now possible to discover specific epigenetic and microbiota variations in the offspring born to obese mothers at a much faster rate, which is critical for deep mechanistic understanding of the process. Despite the difficulty for longitudinal studies in human that have led to the scarcity of intergenerational data, ongoing studies on non-human primates are expected to shed clinical-relevant insights to the transmission of epigenetic traits across multiple generations.

Intervention is the main challenge regarding developmental origin of obesity because traditional weight management approaches during pregnancy have been shown to be ineffective. Gestation, especially the first trimester, is undoubted the most critical developmental window for foetal programming. However, intervention during this period is limited by the fact that intrauterine environment is a relatively closed system and maternal body is physiologically programmed to prioritise foetal development during pregnancy. Safety concern is another aspect preventing therapeutic approaches during pregnancy. In comparison, dietary/ physical interventions before pregnancy appear to be the most appealing, supported by epidemiological data showing that interpregnancy weight loss has been associated with reduced risk of large-for-gestational-age babies [84, 85], as well as experimental data from animal models [86]. Bariatric surgery induced weight loss, on the other hand, was not effective to reduce BMI in early childhood [87, 88]. Postnatal therapies are rare in human but recent intervention studies in animals have shown positive results, for example by administration of SIRT1 activators [80, 81]. Finally, despite the identification of an increasing number of epigenetic markers associated with maternal obesity, the lack of a specific epigenetic editing tool is restraining the development of novel therapies.

References

1. Barker DJ, Osmond C (1986) Infant mortality, childhood nutrition, and ischaemic heart disease in England and Wales. Lancet 1(8489):1077–1081
2. Barker DJ et al (1989) Weight in infancy and death from ischaemic heart disease. Lancet 2 (8663):577–580
3. Lucas A (1991) Programming by early nutrition in man. Child Environ Adult Dis 1991:38–55
4. Black RE et al (2013) Maternal and child undernutrition and overweight in low-income and middle-income countries. Lancet 382(9890):9427–9451
5. Padmanabhan V, Cardoso RC, Puttabyatappa M (2016) Developmental programming, a pathway to disease. Endocrinology 157(4):1328–1340
6. Fullston T et al (2013) Paternal obesity initiates metabolic disturbances in two generations of mice with incomplete penetrance to the F2 generation and alters the transcriptional profile of testis and sperm microRNA content. FASEB J 27(10):4226–4243
7. Linabery AM et al (2013) Stronger influence of maternal than paternal obesity on infant and early childhood body mass index: the F els L ongitudinal S tudy. Pediatric obesity 8(3):159–169
8. Shankar K et al (2011) Maternal obesity promotes a proinflammatory signature in rat uterus and blastocyst. Endocrinology 152(11):4158–4170
9. Jungheim ES et al (2010) Diet-induced obesity model: abnormal oocytes and persistent growth abnormalities in the offspring. Endocrinology 151(8):4039–4046
10. Shah DK et al (2010) Oocyte and embryo quality in obese patients undergoing in vitro fertilization (IVF). Fertil Steril 94(4):S51
11. Zhang L et al (2015) Sirt3 prevents maternal obesity-associated oxidative stress and meiotic defects in mouse oocytes. Cell Cycle 14(18):2959–2968

12. Wang H et al (2018) Loss of TIGAR induces oxidative stress and meiotic defects in oocytes from obese mice. Mol Cell Proteomics 17(7):1354–1364
13. Han L et al (2018) Embryonic defects induced by maternal obesity in mice derive from Stella insufficiency in oocytes. Nat Genet 50(3):432–442
14. Igosheva N et al (2010) Maternal diet-induced obesity alters mitochondrial activity and redox status in mouse oocytes and zygotes. PLoS ONE 5(4):e10074
15. Dokras A et al (2006) Obstetric outcomes after in vitro fertilization in obese and morbidly obese women. Obstet Gynecol 108(1):61–69
16. Leary C, Leese HJ, Sturmey RG (2014) Human embryos from overweight and obese women display phenotypic and metabolic abnormalities. Hum Reprod 30(1):122–132
17. Taylor PD et al (2005) Impaired glucose homeostasis and mitochondrial abnormalities in offspring of rats fed a fat-rich diet in pregnancy. Am J Physiol-Regul Integr Comp Physiol 288(1):R134–R139
18. Zhang L et al (2015) Sirt3 prevents maternal obesity-associated oxidative stress and meiotic defects in mouse oocytes. Cell Cycle 14:2959–2968
19. Wu LL et al (2015) Mitochondrial dysfunction in oocytes of obese mothers: transmission to offspring and reversal by pharmacological endoplasmic reticulum stress inhibitors. Development 142(4):681–691
20. Saben JL et al (2016) Maternal metabolic syndrome programs mitochondrial dysfunction via germline changes across three generations. Cell Reports 16(1):1–8
21. Liang C, DeCourcy K, Prater MR (2010) High–saturated-fat diet induces gestational diabetes and placental vasculopathy in C57BL/6 mice. Metabolism 59(7):943–950
22. Li H-P, Chen X, Li M-Q (2013) Gestational diabetes induces chronic hypoxia stress and excessive inflammatory response in murine placenta. Int J Clin Exp Pathol 6(4):650
23. Zhu MJ et al (2010) Maternal obesity markedly increases placental fatty acid transporter expression and fetal blood triglycerides at midgestation in the ewe. Am J Physiol-Regul Integr Comp Physiol 299(5):R1224–R1231
24. Saben J et al (2014) Maternal obesity is associated with a lipotoxic placental environment. Placenta 35(3):171–177
25. Jones HN et al (2009) High-fat diet before and during pregnancy causes marked up-regulation of placental nutrient transport and fetal overgrowth in C57/BL6 mice. FASEB J 23(1):271–278
26. Magnusson-Olsson A et al (2006) Gestational and hormonal regulation of human placental lipoprotein lipase. J Lipid Res 47(11):2551–2561
27. Qiao L et al (2015) Maternal high fat feeding increases placenta lipoprotein lipase activity by reducing Sirt1 expression in mice. Diabetes 64(9):3111–3120
28. Heerwagen MJ et al (2010) Maternal obesity and fetal metabolic programming: a fertile epigenetic soil. Am J Physiol-Regul Integr Comp Physiol 299(3):R711–R722
29. Vega CC et al (2015) Exercise in obese female rats has beneficial effects on maternal and male and female offspring metabolism. Int J Obes (Lond) 39(4):712–719
30. Catalano P (2015) Maternal obesity and metabolic risk to the offspring: why lifestyle interventions may have not achieved the desired outcomes. Int J Obes 39(4):642
31. Shankar K et al (2008) Maternal obesity at conception programs obesity in the offspring. Am J Physiol-Regul Integr Comp Physiol 294(2):R528–R538
32. Oben JA et al (2010) Maternal obesity during pregnancy and lactation programs the development of offspring non-alcoholic fatty liver disease in mice. J Hepatol 52(6):913–920
33. Sun B et al (2012) Maternal high-fat diet during gestation or suckling differentially affects offspring leptin sensitivity and obesity. Diabetes 61(11):2833–2841
34. Desai M et al (2014) Maternal obesity and high-fat diet program offspring metabolic syndrome. Am J Obstet Gynecol 211(3):237. e1–237. e13
35. Rasmussen KM (2007) Association of maternal obesity before conception with poor lactation performance. Annu Rev Nutr 27:103–121

36. Leonard SA et al (2011) Associations between high prepregnancy body mass index, breast-milk expression, and breast-milk production and feeding–. Am J Clin Nutr 93(3):556–563
37. Saben JL et al (2014) Maternal obesity reduces milk lipid production in lactating mice by inhibiting acetyl-CoA carboxylase and impairing fatty acid synthesis. PLoS ONE 9(5):e98066
38. Chen H, Morris MJ (2009) Differential responses of orexigenic neuropeptides to fasting in offspring of obese mothers. Obesity 17(7):1356–1362
39. Nguyen LT et al (2019) SIRT1 overexpression attenuates offspring metabolic and liver disorders as a result of maternal high-fat feeding. J Physiol 597(2):467–480
40. Purcell RH et al (2011) Maternal stress and high-fat diet effect on maternal behavior, milk composition, and pup ingestive behavior. Physiol Behav 104(3):474–479
41. Bautista CJ et al (2016) Changes in milk composition in obese rats consuming a high-fat diet. Br J Nutr 115(3):538–546
42. Fields DA, Demerath EW (2012) Relationship of insulin, glucose, leptin, IL-6 and TNF-α in human breast milk with infant growth and body composition. Pediatric Obesity 7(4):304–312
43. Isganaitis E et al (2019) Maternal obesity and the human milk metabolome: associations with infant body composition and postnatal weight gain. Am J Clin Nutr
44. Brion MJ et al (2010) Maternal macronutrient and energy intakes in pregnancy and offspring intake at 10 y: exploring parental comparisons and prenatal effects. Am J Clin Nutr 91 (3):748–756
45. Morris MJ, Chen H (2009) Established maternal obesity in the rat reprograms hypothalamic appetite regulators and leptin signaling at birth. Int J Obes 33(1):115
46. Chang G-Q et al (2008) Maternal high-fat diet and fetal programming: increased proliferation of hypothalamic peptide-producing neurons that increase risk for overeating and obesity. J Neurosci 28(46):12107–12119
47. Férézou-Viala J et al (2007) Long-term consequences of maternal high-fat feeding on hypothalamic leptin sensitivity and diet-induced obesity in the offspring. Am J Physiol-Regul Integr Comp Physiol 293(3):R1056–R1062
48. Page KC et al (2009) Maternal and postweaning diet interaction alters hypothalamic gene expression and modulates response to a high-fat diet in male offspring. Am J Physiol-Regul Integr Comp Physiol 297(4):R1049–R1057
49. Chen H et al (2008) Maternal and postnatal overnutrition differentially impact appetite regulators and fuel metabolism. Endocrinology 149(11):5348–5356
50. Schuster S et al (2011) Leptin in maternal serum and breast milk: association with infants' body weight gain in a longitudinal Study over 6 months of lactation. Pediatr Res 70:633
51. Suter MA et al (2012) A maternal high-fat diet modulates fetal SIRT1 histone and protein deacetylase activity in nonhuman primates. FASEB J 26(12):5106–5114
52. Borengasser SJ et al (2014) High fat diet and in utero exposure to maternal obesity disrupts circadian rhythm and leads to metabolic programming of liver in rat offspring. PLoS ONE 9 (1):e84209
53. Suter MA et al (2014) In utero exposure to a maternal high-fat diet alters the epigenetic histone code in a murine model. Am J Obstet Gynecol 210(5):463. e1–463. e11
54. Osorio JS et al (2013) Effect of the level of maternal energy intake prepartum on immunometabolic markers, polymorphonuclear leukocyte function, and neutrophil gene network expression in neonatal Holstein heifer calves1. J Dairy Sci 96(6):3573–3587
55. Friedman JE (2018) Developmental programming of obesity and diabetes in mouse, monkey, and man in 2018: where are we headed? Diabetes 67(11):2137–2151
56. Yu H-L et al (2015) Global DNA methylation was changed by a maternal high-lipid, high-energy diet during gestation and lactation in male adult mice liver. Br J Nutr 113 (7):1032–1039
57. Seki Y et al (2017) In utero exposure to a high-fat diet programs hepatic hypermethylation and gene dysregulation and development of metabolic syndrome in male mice. Endocrinology 158(9):2860–2872

58. Michels KB, Harris HR, Barault L (2011) Birthweight, maternal weight trajectories and global DNA methylation of LINE-1 repetitive elements. PLoS ONE 6(9):e25254
59. Benatti R et al (2014) Maternal high-fat diet consumption modulates hepatic lipid metabolism and microRNA-122 (miR-122) and microRNA-370 (miR-370) expression in offspring. Br J Nutr 111(12):2112–2122
60. Yan X et al (2013) Maternal obesity downregulates microRNA let-7g expression, a possible mechanism for enhanced adipogenesis during ovine fetal skeletal muscle development. Int J Obes 37(4):568
61. Fernandez-Twinn DS et al (2014) Downregulation of IRS-1 in adipose tissue of offspring of obese mice is programmed cell-autonomously through post-transcriptional mechanisms. Mol Metab 3(3):325–333
62. Laker RC et al (2014) Exercise prevents maternal high-fat diet–induced hypermethylation of the Pgc-1α gene and age-dependent metabolic dysfunction in the offspring. Diabetes 63 (5):1605–1611
63. Lesseur C et al (2013) Tissue-specific Leptin promoter DNA methylation is associated with maternal and infant perinatal factors. Mol Cell Endocrinol 381(1–2):160–167
64. Sharp GC et al (2015) Maternal pre-pregnancy BMI and gestational weight gain, offspring DNA methylation and later offspring adiposity: findings from the Avon Longitudinal Study of Parents and Children. Int J Epidemiol 44(4):1288–1304
65. Miller C et al (2011) The interplay between SUCLA2, SUCLG2, and mitochondrial DNA depletion. Biochimica et Biophysica Acta (BBA)-Mol Basis Dis 1812(5):625–629
66. Obermann-Borst SA et al (2013) Duration of breastfeeding and gender are associated with methylation of the LEPTIN gene in very young children. Pediatr Res 74(3):344
67. Skinner MK (2008) What is an epigenetic transgenerational phenotype?: F3 or F2. Reprod Toxicol 25(1):2–6
68. Tsoulis MW et al (2016) Maternal high-fat diet-induced loss of fetal oocytes is associated with compromised follicle growth in adult rat offspring. Biol Reprod 94(4):94, 1–11
69. Cheong Y et al (2014) Diet-induced maternal obesity alters ovarian morphology and gene expression in the adult mouse offspring. Fertil Steril 102(3):899–907
70. Masuyama H et al (2015) The effects of high-fat diet exposure in utero on the obesogenic and diabetogenic traits through epigenetic changes in adiponectin and leptin gene expression for multiple generations in female mice. Endocrinology 156(7):2482–2491
71. Chang H-C, Guarente L (2014) SIRT1 and other sirtuins in metabolism. Trends Endocrinol Metab 25(3):138–145
72. Gao Z et al (2011) Sirtuin 1 (SIRT1) protein degradation in response to persistent c-Jun N-terminal kinase 1 (JNK1) activation contributes to hepatic steatosis in obesity. J Biol Chem 286(25):22227–22234
73. dos Santos Costa C et al (2010) SIRT1 transcription is decreased in visceral adipose tissue of morbidly obese patients with severe hepatic steatosis. Obes Surg 20(5):633–639
74. Çakir I et al (2009) Hypothalamic Sirt1 regulates food intake in a rodent model system. PLoS ONE 4(12):e8322
75. Hasegawa K et al (2013) Renal tubular Sirt1 attenuates diabetic albuminuria by epigenetically suppressing Claudin-1 overexpression in podocytes. Nat Med 19(11):1496–1504
76. Lappas M et al (2011) SIRT1 is a novel regulator of key pathways of human labor. Biol Reprod 84(1):167–178
77. Kawamura Y et al (2010) Sirt3 protects in vitro–fertilized mouse preimplantation embryos against oxidative stress–induced p53-mediated developmental arrest. J Clin Investig 120 (8):2817
78. Borengasser SJ et al (2011) Maternal obesity during gestation impairs fatty acid oxidation and mitochondrial SIRT3 expression in rat offspring at weaning. PLoS ONE 6(8):e24068
79. Di Emidio G et al (2014) SIRT1 signalling protects mouse oocytes against oxidative stress and is deregulated during aging. Hum Reprod 29(9):2006–2017

80. Nguyen LT et al (2018) SRT1720 attenuates obesity and insulin resistance but not liver damage in the offspring due to maternal and postnatal high-fat diet consumption. Am J Physiol-Endocrinol Metab 315(2):E196–E203
81. Uddin GM et al (2017) Nicotinamide mononucleotide (NMN) supplementation ameliorates the impact of maternal obesity in mice: comparison with exercise. Sci Rep 7(1):15063
82. Nguyen LT et al (2017) SIRT1 reduction is associated with sex-specific dysregulation of renal lipid metabolism and stress responses in offspring by maternal high-fat diet. Sci Rep 7 (1):8982
83. Nguyen LT et al (2019) SIRT1 attenuates kidney disorders in male offspring due to maternal high-fat diet. Nutrients 11(1):146
84. Jain AP et al (2013) The impact of interpregnancy weight change on birthweight in obese women. Am J Obstet Gynecol 208(3):205. e1–205. e7
85. Oteng-Ntim E et al (2018) Interpregnancy weight change and adverse pregnancy outcomes: a systematic review and meta-analysis. BMJ Open 8(6):e018778
86. Zambrano E et al (2010) RAPID REPORT: dietary intervention prior to pregnancy reverses metabolic programming in male offspring of obese rats. J Physiol 588(10):1791–1799
87. Willmer M et al (2013) Surgically induced interpregnancy weight loss and prevalence of overweight and obesity in offspring. PLoS ONE 8(12):e82247
88. Berglind D et al (2014) Differences in gestational weight gain between pregnancies before and after maternal bariatric surgery correlate with differences in birth weight but not with scores on the body mass index in early childhood. Pediatr Obes 9(6):427–434

Chapter 15
Diet Induced Maternal Hypercholesterolemia and *In Utero* Fetal Programming

V. S. Jayalekshmi and Surya Ramachandran

Abstract In recent years, obesity has increased prominently around the world—contributing to both short-term and long-term health effects for women of child bearing age as well as for their offspring. On the basis of data from several epidemiological as well as clinical studies, obesity has been established as a predisposing risk factor for cardiovascular diseases, hypertension, stroke, type 2 diabetes and certain types of cancers. Moreover, obesity during pregnancy is associated with maternal hypercholesterolemia, placental vascular dysfunction, alterations in placental transporters activity, placental inflammation leading to adverse cardio-metabolic profile in childhood. In maternal hypercholesterolemia, the increased transport of low-density lipoprotein (LDL) from mother to fetus via receptors and transporters of syncytiotrophoblasts and placental endothelial cells results in fetal programming that ultimately predispose offspring to atherosclerosis in their adulthood.

Keywords Maternal hypercholesterolemia · Cholesterol transporters · Fetal programming · Atherosclerosis

Obesity and Chronic Diseases in Women

Incidences of obesity has increased dramatically in recent years around the world. By the end of 2016, 1.9 billion adults above 18 years of age were reported overweight. Among this population, 650 million are obese [1]. Along with rise in obesity, the problem of obesity-associated diseases have stridently increased [2]. Obesity is a well-established risk factor for not only coronary heart disease, ventricular dysfunction, congestive heart failure, cardiac arrhythmias, hypertension, stroke and diabetes mellitus, but also certain types of cancer, including colorectal,

V. S. Jayalekshmi · S. Ramachandran (✉)
Cardiovascular Diseases and Diabetes Biology, Rajiv Gandhi Centre for Biotechnology, Thiruvananthapuram, India
e-mail: suryaramachandran@rgcb.res.in

P. S. Tappia et al. (eds.), *Pathophysiology of Obesity-Induced Health Complications*, Advances in Biochemistry in Health and Disease 19,
https://doi.org/10.1007/978-3-030-35358-2_15

liver, pancreatic, endometrial, renal cell carcinoma, esophageal adenocarcinoma and postmenopausal breast cancer [3].

A causal link between obesity and cardiometabolic diseases was provided by a recent Mendelian randomization analysis within the UK Biobank [4]. On the basis of data from 119,859 individuals, significant positive associations were seen between genetically linked higher body mass index (BMI) and risk of hypertension, coronary heart disease (CHD), and type 2 diabetes [3]. The pathophysiological basis for these associations is not yet fully understood. In recent times, the indicators of body fat distribution such as waist circumference, waist-hip ratio, waist-height ratio has emerged as important indicators of cardiometabolic risk than body mass index which represents obesity of the individual [5].

Women are particularly prone to obesity and associated cardiometabolic disorders including insulin resistance and atherosclerosis [6]. Obesity is associated with detrimental health effects on women of childbearing age and their offspring [7]. Chronic obesity is also associated with the development of chronic diseases such as diabetes and cardiovascular disease later in life [8]. Several factors predispose women of childbearing age to obesity. These include genetics, junk foods, medications, availability of food and excessive consumption of foods rich in simple sugars [9]. Age, gender, psychological framework, lifestyle, pregnancy, environmental factors, emotional disorders such as depression can also contribute to development of obesity in would be mothers [10]. Certain less common causes of obesity are presence of medical conditions such as Cushing syndrome, depression, polycystic ovarian syndrome (PCOS) and eating disorders, dependency on medications such as steroids, birth control pills or antidepressants [10]. Obesity and overweight women are predisposed to several risks in pregnancy such as caesarean section, chest, genital and urinary tracts infections, cholecystitis, depression, diabetes, difficult surgical access, caesarean sections, failed induction of labour and preeclampsia [11]. Other complications in the mother include hemorrhage, maternal mortality, sleep apnoea, preterm birth, reduced breastfeeding, and thromboembolic disease [11]. There occurs fetal/neonatal consequences on the newborn such as neural tube defects and congenital heart diseases [12].

Maternal Obesity and Cardiovascular Diseases

An optimal range of weight gain during pregnancy based on the mother's body mass index before pregnancy has been recognized as seen in Table 15.1 [13]. Both obesity in mothers before pregnancy and unwarranted weight gain during gestation have long-lasting effects on childhood outcomes [14] and are important risk factors for several adverse fetal outcomes. Maternal prepregnancy obesity correlates with increased placental weight, vascular dysfunction, inflammation and alterations in placental transporters activity as well as mitochondrial activity [15–21].

Table 15.1 Recommended weight gain during pregnancy based on body mass index

	Normal weight	Overweight	Obese
Body mass index (kg/m^2)	18.5–24.9	25–29.9	>30
Recommended weight gain (kg)	11.2–15.9	6.8–11.2	6.8

Furthermore, the offspring are associated with an increased risk of obesity during adolescence [22]. A meta-analysis demonstrated a threefold higher risk of childhood obesity in obese mothers. Weight gain in the first 14 weeks of pregnancy is associated with more adipose tissue in the offspring by 9 years of age [23]. It is also responsible for an unfavorable cardio-metabolic profile during childhood [24]. Maternal first trimester weight gain was also associated with higher childhood diastolic blood pressure [25]. A maternal weight gain of 7 kg in the first 20 weeks of gestation has been associated with the risk of offspring becoming overweight by 16 years of age [26]. Thus, weight gain during early pregnancy, when fat accumulation is the major component of gestational weight gain, may be the cause for an adverse childhood cardiovascular risk profile. The adiposity levels of such offspring in adulthood are also reported to be high [27–31] resulting in cardiovascular disease and mortality in the adult offspring.

Impact of Obesity on Maternal Vascular Adaptations to Pregnancy

The maternal cardiovascular system undergoes significant changes to meet the increased oxygen and nutritional demands of the growing fetus [32]. This includes a 40% increase in blood volume which is required to supply the uterus and developing fetus [33, 34]. There is marked vasodilation which is dependent on endothelial cell and smooth muscle cell leading to a reduction in systemic vascular resistance [35]. In obese mothers, the basement membrane thickness and vessel stiffness increases [36]. These changes are due to elevation in blood pressure [37]. As obesity progresses, atrophy of the microvascular walls occur, leading to narrowing of the microvessels [38, 39] and tissue ischemia [40]. Obese women are at risk of endothelial dysfunction [40], which initiates atherosclerosis [41]. Atherosclerosis is a progressive disease. Fatty streak initiates at the age of 11–12 and plaque formation begins between 15 and 30 years of age. Fatty streaks are composed of inflammatory cells, smooth muscle cells, fibrous connective tissue and lipids.

Effect of Maternal Obesity on the Placental Phenotype

The placenta attaches to the wall of uterus usually to the top, side, front or back of the uterus and rarely in the lower uterine region. Human placenta is chorioallantoic placenta that forms from the chorion and allantois weighing approximately 500 g. It has mainly two compartments; fetal placenta (Chorion frondosum), which develops from the same blastocyst that forms the fetus, and the maternal placenta (Decidua basalis), which develops from the maternal uterine tissue. The umbilical cord which contains two umbilical arteries and one umbilical vein inserts into the chorionic plate in the fetal side and vessels branch out over the surface of the placenta to form chorionic villi on the maternal side.

The placental development starts when blastocyst attaches to the endometrium. The placenta develops throughout pregnancy. The outer layer of the blastocyst becomes the outer layer of the placenta called as trophoblast, comprising of the underlying cytotrophoblast layer and the overlying syncytiotrophoblast layer. The multinucleated syncytiotrophoblast that covers the surface of the placenta is formed from cytotrophoblast cells through the process that continues throughout placental development. The syncytiotrophoblast is responsible for the nutrient transport and barrier function of the placenta.

The placenta is now known as a dynamic organ which undergoes changes in molecular structure and function throughout pregnancy. Placenta mediates interactions of the fetus with the maternal immune system. It is also exposed to the compounds in the mother's blood. It serves as a neuro-endocrine organ producing hormones to encourage development if the fetus. In essence, placenta is the master regulator of the fetal development. The fully functional placenta is unique in that maternal blood enters the intervillous spaces of placenta, directly flooding the syncytialized trophoblasts 3–4 times in each minute. The early hypothesis that maternal cholesterol could not cross the placental barrier was disputed when high fetal cholesterol levels correlated with maternal levels in mid-pregnancy, suggesting that maternal-fetal cholesterol transport occurs at during pregnancy. These mechanisms can be receptor-dependent/independent processes, using the help of scavenger receptors and transporters such as ABCA1 and ABCG2. The exchange between the maternal and fetal circulation takes place either via placental permeability or via active transport mechanisms. Placenta takes up lipoproteins through receptors like LDL receptor, LDL Receptor related protein (LRP), scavenger receptor class B type I (SR-B1). Cholesterol transporters such as ABCA1 and ABCG1 are up regulated by Liver X Receptor (LXR) activation resulting in increased cholesterol transport in vivo. It is possible that they are regulated differently in different stages of pregnancy either by fetal demand or by high maternal cholesterol levels [42–45]. This suggests that extensive maternal hypercholesterolemia possibly results in an increased transfer of cholesterol into the fetal circulation. In addition to affecting placental permeability and transport mechanism, maternal factors may also end up altering mRNA and protein expression or activities of transcription products in the syncytiotrophoblast or endothelial cells of the fetal circulation which can influence mother or fetus [46].

The placenta mediates maternal and fetal interactions and therefore plays a central role in fetal organ growth and development. The impact of maternal obesity on function of the placenta particularly during development of fetal cardiovascular system has been studied. The mechanisms however, that are altered in the placenta in response to an adverse in utero environment are incompletely understood. Obese mothers have 50% increased placental lipid droplet accumulation than lean women, decreased angiogenesis regulators such as angiopoietin, vascular endothelial growth factor A, and hypoxia inducible factor 1. Their levels of interleukins and chemokine receptors are higher [47], demonstrating a pro inflammatory placental phenotype due to maternal obesity. In non-human primate (NHP) models fed with high fat diet, increased expression of inflammatory cytokines and infarction [48, 49] has been reported. Importantly, in vivo studies have proved that [50, 51] the vascular adaptations cause structural and physiological changes which possibly influence cardiovascular development.

Fetal Programming in the Placenta

According to the Barker hypothesis, organ structures undergo programming during fetal lifetime which may determine the metabolic responses that they carry forward into adulthood. Therefore, any prolonged stimulus during the developmental stage can result in variations that induce permanent metabolic changes in the individual. These changes may predispose the individual to cardiovascular disease in adult life [52].

The hypothesis also called thrifty phenotype hypothesis was proposed by the British epidemiologist David Barker in 1990. He stated that intrauterine growth retardation, low birth weight, and premature birth have a causal relationship to the origins of hypertension, coronary heart disease, and non-insulin-dependent diabetes, in middle age. The evidence is presented in Barker's book *Fetal and Infant Origins of Adult Disease* (1992).

The developmental origins of health and disease (DOHaD) are also based on the concept of fetal programming. It evolved from epidemiological studies of infant and adult mortality. Several research studies are focused on the molecular mechanisms underlying fetal programming to provide targeted therapy to the population at risk. In addition to altered maternal nutrition and reduced utero–placental blood flow, fetal programming can be caused by a number of different trepidations in the maternal placenta, such as maternal hypercholesterolemia and hyperglycemia; however, the underlying mechanisms remain to be fully established. These stimuli in the maternal environment must be transferred across the placenta in order to affect the fetus.

Although the human placenta is a transient organ that persists only nine months, the effects of this organ on the offspring remains for a lifetime. The placental adaptations help the fetus in development and growth and drive the proportions of metabolic substrates to be supplied to the fetus. The flow of such substrates and nutrients program the physiological systems from the gene to system levels. Inadequacies and over supply of certain metabolites and nutrients can cause permanent functional and physiological changes that lead to metabolic diseases particularly with increasing age and unhealthy lifestyle in adulthood. Altered expression of imprinted genes, altered enzymatic activity, or altered efficiencies in nutrient transport can imprint epigenetic markings in such individuals.

Although, Barker's hypothesis was primarily controversial, scores of studies across diverse populations and experimental animals exposed to various insults during gestation period has proven the onset of chronic and metabolic diseases to be predetermined by the environment in the womb [53–55]. Furthermore, there has been an effort to identify the factors which control intrauterine growth and are involved in early-life programming of health in adulthood.

Maternal Hypercholesterolemia and Fetal Programming

Apart from oxygen, amino acids, glucose, and fatty acids are essential nutrients for fetal growth. The placenta transports these nutrients to the fetus by passive diffusion, transporter-mediated processes and active transport mechanisms [56]. Oxygen, urea and carbon dioxide cross the placenta by passive diffusion. Lactate and glucose are transported by transporter proteins in the plasma membrane. Amino acids are actively transported using membrane proteins. Transfer of fatty acids to the fetus is less understood. It possibly involves the release of fatty acids from maternal triglycerides and cytoplasmic and membrane carrier proteins for diffusion [57, 58]. Changes in any of these transport mechanisms in the placenta may affect fetal growth and result in health and disease in adulthood [59]. The transport capacity of placental fatty acid is enhanced by excess maternal dietary fat, and is associated with increased accumulation of fat within the fetus [60].

Maternal lipid profile is one of the metabolic components known to influence fetal developmental processes, such as fetal growth [61]. Lipid metabolism changes during pregnancy. There is an accumulation of lipids in maternal tissue in the first two trimesters and increased lipolysis in the last trimester resulting in maternal hyperlipidemia [61]. Thus, gestational hyperlipidemia was thought to be a common physiological phenomenon during pregnancy characterized by a significant increase in circulating lipid levels [62]. In addition to that, it is entirely different from inherited lipid disorders like familial hypercholesterolemia. Familial hypercholesterolemia is a genetic disorder caused by a defect on chromosome 19 and the condition begins at birth. This is an autosomal dominant disorder in which abnormal gene from only one parent (heterozygous condition) is enough to inherit the disease; but individuals with abnormal gene from both parents (homozygous condition) can also have familial hypercholesterolemia where they rarely live into their 20 s. This abnormal chromosome has the mutation in the gene for the LDL cholesterol receptor, which is involved in passing LDL from the body. As a consequence, LDL levels in the blood remain very high, above 190 milligrams per deciliter (mg/dL) of blood. Mutations in other genes involved in the cholesterol mechanism, PCSK9 and Apolipoprotein B can also cause familial hypercholesterolemia.

As discussed earlier, maternal hypercholesterolemia during pregnancy has not been given importance clinically for a considerably long time. The first indication that it may be pathogenic was by the observation of increased fatty streak formation in the aortas of 6-month old fetuses of mothers with chronic hypercholesterolemia [63]. The Fate of Early Lesions in Children (FELIC) study has established that maternal hypercholesterolemia is associated with increased aortic atherosclerosis in normocholesterolemic children [64]. This study confirmed that changes induced by or associated with maternal hypercholesterolemia during pregnancy or formation of fatty streaks that do not fully regress predetermine the rate of progression of atherosclerosis throughout childhood. This finding implies an understanding of early pathophysiological events in atherogenesis and its prevention. The development of atherosclerosis, a progressive disease characterized by the accumulation of lipids and fibrous elements in the arteries possibly begins in early life, in some cases before birth [65]. Adverse intrauterine environment and impaired fetal growth can contribute to early development to atherosclerosis with along latency period. Exposure of the fetus to maternal adiposity, dyslipidemia, and hyperglycemia are important determinants of long-term cardiometabolic outcome among adult offspring.

Fetal Cholesterol Metabolism and Maternal Hypercholesterolemia

Cholesterol is an essential lipid for human cells and is a major sterol synthesized by all animals. Liver regulates the cholesterol synthesis and levels in the body as the excess of cholesterol can be toxic and can contribute to atherosclerotic vascular disease. Cholesterol biosynthesis occurs through 37 steps process regulated by the key enzyme called HMG CoA reductase, which is subjected to a negative feedback regulation. Cholesterol has to reach its site of action such as plasma membrane and other compartments from endoplasmic reticulum, cytosolic lipid droplets and endocytic compartments; where it occurs via receptor mediated endocytosis.

Brown and Goldstein in the 1970s discovered the mechanism of normal cells to extract the cholesterol of LDL with a cell surface receptor (LDL receptor) activated by sterol regulatory element binding protein (SREBP) transcription factors termed as receptor mediated endocytosis for which they were awarded with Nobel prize in Physiology or Medicine in 1985. They explained the receptor-mediated internalization of LDL and its transport to the lysosome. The clathrin coated pits contains LDL receptors to which the free LDL in blood binds. The coated pits with receptor bound LDL molecule pinch off from the surface of the cell to form endosomes and the low pH in the endosomes dissociate the receptors from their ligands. These endosomes fuse with lyzosomes for the enzymatic digestion of the bound LDL molecules. Later the unbound LDL receptors are recycled back to the surface of the cells for the next cargo which happen in every 10 min. LDL receptors are less expressed when cell cholesterol increases establishing a regulatory mechanism. This will leads to the reduction in HMG CoA reductase and endogenous synthesis of cholesterol, decreasing the free cholesterol load in the blood. This mechanism is impaired in the familial hypercholesterolemia patients where they have a defect in the gene encoding LDL receptor that makes the enzyme resistant to the feedback regulation by LDL cholesterol.

Plasma cholesterol concentrations are comparatively low in newborn. Like adults, fetuses also have two sources of cholesterol; endogenous and exogenous cholesterol. The endogenous cholesterol is from de novo synthesis where fetal liver synthesize it. The cholesterol synthesis rate is high in fetus compared to adults due to the high requirement by cells and the SREBP transcription factor is expressing high leading to a fully active sterol biosynthetic pathway [66]. Conversely to the adult exogenous cholesterol from diet, fetal exogenous cholesterol comes from the mother. It is believed that maternal cholesterol is transported across the placenta and fetal cholesterol level varies throughout the gestation; despite the much debate on the maternal transport of cholesterol to the fetus. It's found that there is no correlation between maternal and newborn cholesterol levels in term or late preterm infants but do in early gestation [63] suggesting the cholesterol transport is in early gestation but not in late gestation. During the gestation, serum lipids, including total cholesterol (TC), low-density lipoprotein cholesterol (LDL-C), high-density lipoprotein cholesterol (HDL-C) levels, triglyceride (TG), and were significantly elevated from the second trimester and reached a maximum in the third trimester [67] with plasma cholesterol levels increased by 25–50%. This is necessary to improve the availability of essential metabolites for fetal growth [61]. In early gestation, the secondary yolk sac is considered to be responsible for the transport of maternal cholesterol to the fetus. This is because placenta becomes fully functional only by the end of the first trimester and cholesterol transport occurs through the receptor mediated as well as receptor independent processes in the yolk sac. The secondary yolk sac regresses when the placenta becomes functional.

Obesity is the over nourished state and the placenta is sensitive to this condition during pregnancy. It can elevate placental inflammatory cytokine expression and oxidative stress creating a pro-inflammatory state on placental function. This can influence the nutrient transport to the fetus and the modified nutrient transport mechanism may drive excess cholesterol to the fetus which ultimately results in atherosclerosis.

Epigenetics of Maternal Hypercholesterolemia

In maternal hypercholesterolemia the oxidation of low density lipoprotein (LDL) and fatty streak formation occurs during fetal development [45]. As epigenome of placenta is varying throughout the gestation and can change in response to environmental stimuli such as maternal diet and lipid levels, the differential gene expression patterns in the womb as a reflection of developmental plasticity can lead to early onset of diseases in adulthood. The researches on animals on this aspect have enough data to prove the importance of placenta in regulating the development of cardiovascular system. However, the epidemiological evidences supporting this concept should also accumulate to firmly establish its role in the etiology of maternal hypercholesterolemia. In light of the scientific evidence accumulated in recent years, epigenetic mechanisms have been recognized as important mediators of the atherosclerotic process and the subsequent cardiovascular diseases during the gestational age [68]. Global hypermethylation, one of the epigenetic programming has been associated with inflammation.

Maternal hypercholesterolemia may impact offspring DNA methylation changes at birth, however, whether these changes impact on development of later adverse outcomes in the offspring remains unclear. Several DNA methylation patterns are tissue- and cell- specific [69], so the relevance of findings from DNA extracted from cord or peripheral blood leukocytes remains unclear [70]. Ideally epigenetic profiling at the placental tissue level are required for obtaining a clear understanding of epigenetic programming during gestation. Classic epigenetic changes such as DNA methylation and histone acetylation leading to irreversible changes in the transcriptional machinery [71] can result in changes in function and structure of genes involved in LDL metabolism and cholesterol transport between mother and fetus. This can show inter-individual variability at birth that may be stable throughout the life course and can be likely drivers of early life programming of adult-onset diseases such as atherosclerosis.

Conclusion

Cholesterol transport mechanisms across the placenta in hypercholesterolemic conditions of the mother are relatively unknown. Placenta takes up lipoproteins through receptors like LDL receptor, LDL Receptor related protein (LRP), scavenger receptor class B type I (SRB1). Cholesterol transporters such as ABCA1 and ABCG1 are up regulated by Liver X Receptor (LXR) activation resulting in increased cholesterol transport in vivo. This suggests that extensive maternal hypercholesterolemia possibly results in an increased transfer of cholesterol into the fetal circulation as noted in Fig. 15.1. Establishing a causal role of maternal/fetal hypercholesterolemia in accelerated fetal or postnatal lesion formation would not only add to our understanding of the pathogenesis of atherosclerosis in young

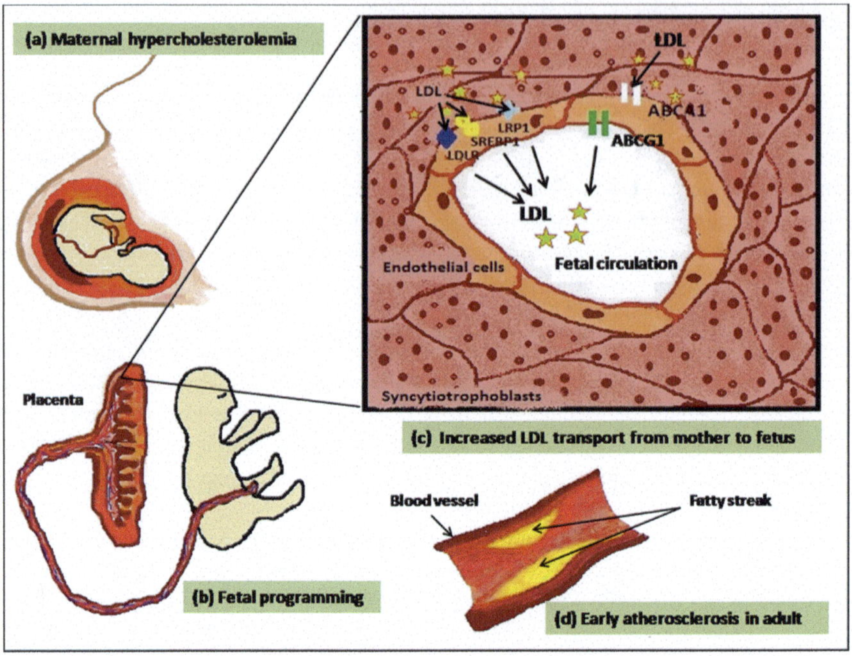

Fig. 15.1 Maternal hypercholesterolemia and atherosclerosis in adulthood. **a** Maternal blood with high levels of cholesterol enters placenta for exchange of nutrients, gases and other requirements for growth of the fetus. **b** High levels of low density lipoprotein (LDL) cholesterol in maternal blood leads to increased LDL transport to the fetus inducing in utero programming. **c** Inset showing cross section of chorionic villi in placenta. Receptors such as LDL receptor (LDLR), Sterol regulatory element binding protein 1 (SREBP1), LDL receptor related protein 1 (LRP1) on syncytiotrophoblasts and ATP-binding cassette transporters such as ABCA1, ABCG1 on endothelial cells of placenta transport excess LDL to the fetal circulation. **d** The increased LDL transport between hypercholesterolemic mothers to fetus predisposes the offspring to early atherosclerosis in their adulthood

adults, but can also have important clinical implications. Identifying the LDL transport mechanisms and epigenetic changes underlying this programming can help in developing population-based lifestyle strategies for women in their child bearing years to identify those at risk and direct lipid specific nutritional and life style interventions or therapy. As pregnancy is the time in which offspring develops inside the woman's womb, treatment for maternal obesity is not quite easy and may end up in severe complications. Moreover, sufficient nutrition is important for pregnant women; so as to dieting and interventions to treat obesity cannot be implemented during gestational phase. Hence, obese women are recommended to lose weight before becoming pregnant and are advised for getting enough folic acid, quitting smoking, and avoiding alcohol to avoid the possible risks. Women are also recommended to have appropriate calorie intake and exercise adequately.

Acknowledgements Jayalekshmi V. S. is supported by DST-INSPIRE Ph.D. FELLOWSHIP (IF170711), Ministry of Science and Technology. The authors acknowledge Indian Council of Medical Research, Ministry of Health, Government of India for financial support.

References

1. Organization W HO (2016)
2. Collaborators GO (2017) Health effects of overweight and obesity in 195 countries over 25 years. N Engl J Med 377:13–27
3. Pischon T, Nimptsch K (2016) Obesity and cancer. Springer, pp 1–15
4. Lyall DM et al (2017) Association of body mass index with cardiometabolic disease in the UK Biobank: a Mendelian randomization study. JAMA Cardiol 2:882–889
5. Eckel RH, Alberti KG, Grundy SM, Zimmet PZ (2010) The metabolic syndrome. The Lancet 375:181–183
6. Bergman RN et al (2011) A better index of body adiposity. Obesity 19:1083–1089
7. Zaadstra BM et al (1993) Fat and female fecundity: prospective study of effect of body fat distribution on conception rates. BMJ 306:484–487
8. Health UD, Services H (2001) The Surgeon General s call to action to prevent and decrease overweight and obesity [Internet]. Pubic Health Service, Office of the Surgeon General, US Department of Health and Human Services, United States Government Printing Office, Washington (DC)
9. Ojiegbe I (2016) Impacts of obesity on the health of women of childbearing age: a call for action. Int J Med Biomed Res 5:19–27
10. Galletta GM (2012) Obesity. eMedicine health, expert of everyday emergencies. Available at: http://www.emedicinehealth.com/obesity/article_em.htm 12 January 2012
11. Gunatilake RP, Perlow JH (2011) Obesity and pregnancy: clinical management of the obese gravida. Am J Obstet Gynecol 204:106 119
12. Stothard KJ, Tennant PW, Bell R, Rankin J (2009) Maternal overweight and obesity and the risk of congenital anomalies: a systematic review and meta-analysis. JAMA 301:636–650
13. Mathiowetz N, Wunderlich G (2000). National Academies Press (US), Washington (DC)
14. Gaillard R, Felix JF, Duijts L, Jaddoe VW (2014) Childhood consequences of maternal obesity and excessive weight gain during pregnancy. Acta Obstet Gynecol Scand 93:1085–1089
15. Ouyang F et al (2013) Placental weight mediates the effects of prenatal factors on fetal growth: the extent differs by preterm status. Obesity 21:609–620
16. Ditchfield A et al (2015) Maternal obesity is associated with a reduction in placental taurine transporter activity. Int J Obes 39:557
17. Mele J, Muralimanoharan S, Maloyan A, Myatt L (2014) Impaired mitochondrial function in human placenta with increased maternal adiposity. Am J Physiol-Endocrinol Metab 307: E419–E425
18. Huang L et al (2014) Maternal prepregnancy obesity is associated with higher risk of placental pathological lesions. Placenta 35:563–569
19. Aye IL et al (2014) Increasing maternal body mass index is associated with systemic inflammation in the mother and the activation of distinct placental inflammatory pathways. Biol Reprod 90(129):121–129
20. Malti N et al (2014) Oxidative stress and maternal obesity: feto-placental unit interaction. Placenta 35:411–416
21. Zera CA et al (2014) The association of body mass index with serum angiogenic markers in normal and abnormal pregnancies. Am J Obstet Gynecol 211:247.e241–247.e247
22. Yu Z et al (2013) Pre-pregnancy body mass index in relation to infant birth weight and offspring overweight/obesity: a systematic review and meta-analysis. PLoS ONE 8:e61627

23. Fraser A et al (2010) Association of maternal weight gain in pregnancy with offspring obesity and metabolic and vascular traits in childhood. Circulation 121:2557
24. Gaillard R, Steegers E, Franco O, Hofman A, Jaddoe V (2015) Maternal weight gain in different periods of pregnancy and childhood cardio-metabolic outcomes. The Generation R Study. Int J Obes 39:677
25. Karachaliou M et al (2015) Association of trimester-specific gestational weight gain with fetal growth, offspring obesity, and cardiometabolic traits in early childhood. Am J Obstet Gynecol 212:502.e501–502.e514
26. Laitinen J et al (2012) Maternal weight gain during the first half of pregnancy and offspring obesity at 16 years: a prospective cohort study. BJOG Int J Obstet Gynaecol 119:716–723
27. Hochner H et al (2012) Associations of maternal prepregnancy body mass index and gestational weight gain with adult offspring cardiometabolic risk factors: the Jerusalem Perinatal Family Follow-up Study. Circulation 125:1381–1389
28. Schack-Nielsen L, Michaelsen KF, Gamborg M, Mortensen EL, Sørensen TI (2010) Gestational weight gain in relation to offspring body mass index and obesity from infancy through adulthood. Int J Obes 34:67
29. Hrolfsdottir L et al (2015) Gestational weight gain in normal weight women and offspring cardio-metabolic risk factors at 20 years of age. Int J Obes 39:671
30. Rooney BL, Mathiason MA, Schauberger CW (2011) Predictors of obesity in childhood, adolescence, and adulthood in a birth cohort. Matern Child Health J 15:1166–1175
31. Mamun AA et al (2010) Associations of excess weight gain during pregnancy with long-term maternal overweight and obesity: evidence from 21 y postpartum follow-up. Am J Clin Nutr 91:1336–1341
32. Reynolds LP, Redmer DA (1995) Utero-placental vascular development and placental function. J Anim Sci 73:1839–1851
33. Hart MV, Morton MJ, Hosenpud JD, Metcalfe J (1986) Aortic function during normal human pregnancy. Am J Obstet Gynecol 154:887–891
34. Longo L (1983) Maternal blood volume and cardiac output during pregnancy: a hypothesis of endocrinologic control. Am J Physiol-Regul Integr Comp Physiol 245:R720–R729
35. van Drongelen J, Hooijmans CR, Lotgering FK, Smits P, Spaanderman ME (2012) Adaptive changes of mesenteric arteries in pregnancy: a meta-analysis. Am J Physiol Heart Circ Physiol 303:H639–H657
36. Zebekakis PE et al (2005) Obesity is associated with increased arterial stiffness from adolescence until old age. J Hypertens 23:1839–1846
37. van den Berkmortel F et al (2001) Progressive arterial wall stiffening in patients with increasing diastolic blood pressure. J Hum Hypertens 15:685
38. Seifalian AM, Filippatos TD, Joshi J, Mikhailidis DP (2010) Obesity and arterial compliance alterations. Curr Vasc Pharmacol 8:155–168
39. Frisbee JC (2005) Hypertension-independent microvascular rarefaction in the obese Zucker rat model of the metabolic syndrome. Microcirculation 12:383–392
40. Denison FC, Roberts KA, Barr SM, Norman JE (2010) Obesity, pregnancy, inflammation, and vascular function. Reproduction 140:373–385
41. Kelishadi R, Poursafa P (2014) A review on the genetic, environmental, and lifestyle aspects of the early-life origins of cardiovascular disease. Curr Prob Pediatr Adolesc Health Care 44:54–72
42. Wyne KL, Woollett LA (1998) Transport of maternal LDL and HDL to the fetal membranes and placenta of the Golden Syrian hamster is mediated by receptor-dependent and receptor-independent processes. J Lipid Res 39:518–530
43. Burke KT et al (2009) Transport of maternal cholesterol to the fetus is affected by maternal plasma cholesterol concentrations in the golden Syrian hamster. J Lipid Res 50:1146–1155
44. Stefulj J et al (2009) Human endothelial cells of the placental barrier efficiently deliver cholesterol to the fetal circulation via ABCA1 and ABCG1. Circ Res 104:600–608
45. Palinski W (2009). Am Heart Assoc

46. Thornburg K, O'tierney P, Louey S (2010) The placenta is a programming agent for cardiovascular disease. Placenta 31:S54–S59
47. Saben J et al (2014) Maternal obesity is associated with a lipotoxic placental environment. Placenta 35:171–177
48. Frias AE et al (2011) Maternal high-fat diet disturbs uteroplacental hemodynamics and increases the frequency of stillbirth in a nonhuman primate model of excess nutrition. Endocrinology 152:2456–2464
49. Roberts VH et al (2014) Beneficial and cautionary outcomes of resveratrol supplementation in pregnant nonhuman primates. FASEB J 28:2466–2477
50. Roberts VH et al (2012) Restriction of placental vasculature in a non-human primate: a unique model to study placental plasticity. Placenta 33:73–76
51. Wladimiroff J, Degani S, Noordam M, Tonge H (1987) Cerebral and umbilical arterial blood flow velocity waveforms in normal and growth-retarded pregnancies. Obstet Gynecol 69: 705–709
52. Kwon EJ, Kim YJ (2017) What is fetal programming?: a lifetime health is under the control of in utero health. Obstet Gynecol Sci 60:506–519
53. McMillen IC, Robinson JS (2005) Developmental origins of the metabolic syndrome: prediction, plasticity, and programming. Physiol Rev 85:571–633
54. Hanson MA, Gluckman PD (2008) Developmental origins of health and disease: new insights. Basic Clin Pharmacol Toxicol 102:90–93
55. Susser E, St Clair D (1982)Prenatal famine and adult mental illness: interpreting concordant and discordant results from the Dutch and Chinese Famines. Soc Sci Med 97:325
56. Duttaroy AK (2009) Transport of fatty acids across the human placenta: a review. Prog Lipid Res 48:52–61
57. Hay JC et al (1995) ATP-dependent inositide phosphorylation required for Ca2+-activated secretion. Nature 374:173
58. Jansson T, Powell TL (2006) Human placental transport in altered fetal growth: does the placenta function as a nutrient sensor?–a review. Placenta 27:91–97
59. Fowden A, Forhead A, Coan P, Burton G (2008) The placenta and intrauterine programming. J Neuroendocrinol 20:439–450
60. Rebholz CM et al (2011) Mortality from suicide and other external cause injuries in China: a prospective cohort study. BMC Pub Health 11:56
61. Herrera E (2002) Lipid metabolism in pregnancy and its consequences in the fetus and newborn. Endocrine 19:43–55
62. Emet T et al (2013) Plasma lipids and lipoproteins during pregnancy and related pregnancy outcomes. Arch Gynecol Obstet 288:49–55
63. Napoli C et al (1997) Fatty streak formation occurs in human fetal aortas and is greatly enhanced by maternal hypercholesterolemia. Intimal accumulation of low density lipoprotein and its oxidation precede monocyte recruitment into early atherosclerotic lesions. J Clin Investig 100:2680–2690
64. Napoli C et al (1999) Influence of maternal hypercholesterolaemia during pregnancy on progression of early atherosclerotic lesions in childhood: Fate of Early Lesions in Children (FELIC) study. The Lancet 354:1234–1241
65. Skilton MR, Evans N, Griffiths KA, Harmer JA, Celermajer DS (2005) Aortic wall thickness in newborns with intrauterine growth restriction. Lancet 365:1484–1486
66. Woollett L, Heubi JE (2016) Endotext [Internet], MDText. com, Inc
67. Qureshi I-A et al (1999) Hyperlipidemia of normal pregnancy in Karachi-Pakistan. Kaohsiung J Med Sci 15:529–535

68. Lorenzen JM, Martino F, Thum T (2012) Epigenetic modifications in cardiovascular disease. Basic Res Cardiol 107:245
69. De Bustos C et al (2009) Tissue-specific variation in DNA methylation levels along human chromosome 1. Epigenetics Chromatin 2:7
70. Soubry A et al (2015) Newborns of obese parents have altered DNA methylation patterns at imprinted genes. Int J Obes 39:650
71. Bernstein BE et al (2010) The NIH roadmap epigenomics mapping consortium. Nat Biotechnol 28:1045

Part IV
Strategies for the Prevention of Obesity-Induced Complications

Chapter 16
Modified Denouement in Bariatric Surgery Due to Genetic Polymorphism

Bhoomika M. Patel, Shuchi H. Dave and Ramesh K. Goyal

Abstract Obesity is a major issue and has a prevalence of about 13% around the world. Obese patients are prone to have certain diseases like cardiovascular, metabolic disorders and cancers such as breast, endometrial and colon. There are certain treatments available for obesity like exercise, lifestyle modification, pharmacological treatment and surgery. Most people now opt for the surgical intervention for eradicating obesity. Bariatric surgery is carried out on the stomach and/ or intestine to reduce obesity in patients having BMI more than 35 kg/m^2. Reports suggest that people going for bariatric surgery have greater variations in weight loss after the surgery. Variable weight loss is experienced post surgery primarily due to polymorphism of genes. Genes such as PGC-1α and GHRL, as well as mitochondrial uncoupling protein are related to polymorphism in bariatric surgery. The current review will familiarize readers with types of bariatric surgeries and their association with obesity and genetic polymorphism.

Keywords Obesity · Bariatric surgery · Weight loss · Genetic polymorphism, single nucleotide polymorphism

Introduction

Obesity is growing worldwide and has become a considerable healthcare burden worldwide. It's incidence has doubled since 1980. The number of people suffered from being overweight in 2014 was 1.9 billion, and amongst them, 600 million people were obese. Out of these 600 million people 13% who are obese were adults.

B. M. Patel · S. H. Dave
Institute of Pharmacy, Nirma University, Ahmedabad, India

R. K. Goyal (✉)
Delhi Pharmaceutical Sciences & Research University, M.B. Road, Pushp Vihar, Sector 3, Delhi 110017, India
e-mail: goyalrk@gmail.com

P. S. Tappia et al. (eds.), *Pathophysiology of Obesity-Induced Health Complications*, Advances in Biochemistry in Health and Disease 19,
https://doi.org/10.1007/978-3-030-35358-2_16

Table 16.1 Classification of obesity

Body mass index (BMI) kg/m^2	Classification
18.5–24.9	Normal
25.0–29.9	Overweight
30.0–34.9	Obesity (type I)
35.0–39.9	Obesity (type II)
≥ 40	Morbid Obesity
≥ 50	Super Obesity

The World Health Organization defines a person as obese when his/her Body Mass Index is greater than or equal to 30, and overweight when the BMI is more than or equal to 25. If a person is suffering from obesity, he/she will be prone to many diseases like cardiovascular, metabolic, musculoskeletal and some cancers like breast, endometrial and colon [1]. Table 16.1 describes the classification of obesity. There are several factors that causes obesity including energy imbalance with increased intake of energy, increase in workload with physical inactivity, change in the modes of transport, sedentary nature of work and urbanization [1]. Obesity can be prevented by adapting to certain life style modifications, exercise, pharmacological treatment and surgical options. Lifestyle modification includes engaging in a healthy diet and exercise. Dietary interventions demand low calorie diet and liquid meal according to Melanson et al. [2]. Physical activity has a role in decreasing weight. A program of walking 45–60 min can make a difference but without calorie restriction, weight loss cannot be achieved [3]. Pharmacological treatment involves use of drugs such as: Orlistat, Rimonnabant and Sibutramine. Orlistat is approved by FDA (Food and Drug administration) and inhibits pancreatic lipase which in turn reduces fat absorption by 30% [4]. Side effects reported include changes in bowel function because of the unabsorbed fat, fatty/oily stool, persistent nausea, vomiting, severe stomach-ache, dark urine. [5]. Rimonabant is a diet pill and works by blocking the binding of endogenous cannabinoid to CB1 neuronal receptors. This drug causes GI, skin, musculoskeletal and CNS problems. Moreover, it has not been effective in patients for weight loss and was not approved in US because of its side effects [6]. Sibutramine, a serotonin norepinephrine reuptake inhibitor, is associated with the feeling of fullness i.e., satiation. Side effects caused by Sibutramine include dizziness, dry mouth, inflammation of nose, nervousness, and abnormal liver functions. [7]. Despite these approaches, obesity does not seem to be controlled in many individuals. Hence, an approach which has been in use within the last few years is surgery. Surgery is suitable for patients who have a BMI (body mass index) score greater than 35 or 40 and with or without co-morbidities like cardiovascular or metabolic disease. This surgery is often preferred by patients. Many bariatric surgeries are performed including Roux-en-Y Gastric bypass, Gastric banding, Sleeve gasterectomy, vertical banded Gastroplasty, and biliopancreatic. [8, 9]. Weight loss is major goal of bariatric surgery. There are some problems associated with the selection criteria of the bariatric patients because the result is not homogenous and no specific markers are found which predicts best outcomes [10]. The identification of these markers may help doctors to select their patients for the surgery.

Polymorphism is the change in the sequencing of DNA in one person or in a large amount of people [11]. Sources of genetic polymorphism include single nucleotide polymorphism, replication in the sequence, deletion and recombination. The genetic polymorphism is the result of the chance processes or may have been induced by external agents such as viruses or radiation [12]. Weight loss is observed in all the patients, but variation occurs from surgery to surgery. Weight loss occurs at variable speeds dependent on the patient. This can be explained by polymorphism. Much research has been carried out on the identification of genes in which polymorphism has been observed. Several SNPs are reported which are responsible for variation in weight loss after bariatric surgery. In this review, we shall briefly describe the various kinds of bariatric surgeries. The second part of review shall familiarise readers with the types of polymorphism associated with alternate patterns of weight loss in patients undergoing bariatric surgery.

Bariatric Surgery

"Surgery which is performed on the gut or intestine for the remission of obesity on extremely obese patients" is defined as bariatric surgery [13]. There are certain types of bariatric surgeries performed on patients having BMI more than 35 [13]. This surgery started in 1954, though it was not successful. In 1970, it had its first success in the treatment for obesity. There are three broad categories of bariatric surgery: the first is based on the pure gastric limitation (restriction), the second is gastric restriction with malabsorption and the third is a combination of both [14]. Table 16.2 shows various types of Bariatric Surgeries.

Gastric Bypass

The first surgery is gastric bypass, also called Roux-en-Y- Gastric Bypass (RYGB), one of the best procedures worldwide. In the procedure, a small pouch of the stomach of (about 30 ml) is created by segmentation of the stomach with staples. Then proximal part of small intestine is separated, and the bottom end of the segmented small intestine is joined to the newly created pouch. This will bypass most of the stomach and also bypass the gastric acid and enzymes present.

Table 16.2 Comparison of different bariatric surgery

Parameters	RYGB	Sleeve gasterectomy	Biliopancreatic with Duodenal switch	Adjustable gastric banding
Type of Surgery	Restrictive	Malabsorptive	Malabsorptive	Restrictive
% weight loss[26]	50–70%	65–75%	65–75%	Extremely variable average is about 50% and ranges from 25 to 80%
Length of hospitalization[27]	1–5 days	1–2 days	2–7 days	1–3 days
Blood loss	0.6%[28]	1–6%[29]	Not reported	Blood clots are observed but very less amount of blood loss
Complication rate[26]	Up to 15%	Up to 10%	Up to 24%	Up to 33%
Chances of re operation	Less[30]	Less than 5% of the patients[31]	Not common[32]	Greater risk of re-operation[33], 30–50%
Mortality rate %[34]	2–5 in 1000	2 in 1000	2 in 1000	1 in 1000
Hormonal changes	Ghrelin: variable GLP-1 and PYY: meal induced in both[35,36,37]	GLP and PYY: Elevated[38,39]	GLP-1 and PYY: elevated[40,41,42,43]	GLP-1, PYY: no change[44,45]
Other comorbid conditions which improves after bariatric surgeries	Hypertension and diabetes, Dyslipidaemia	Metabolic disorders	Hypertension, diabetes and dyslipidaemia	Hypertension, diabetes and dyslipidaemia

Advantages:

- Significant long-term weight reduction
- Restrict food consumption
- Increase in energy depletion
- Produces favourable changes in the gut and will reduce stomach hormones which reduces hunger hormones
- >50% excess weight loss [15].

Disadvantages:

- Complex operation relative to other surgeries
- It can lead to certain deficiencies after some time like vitamin B_{12}, iron, calcium and folate
- Hospitalization time increases
- Follow up compliance is needed ('Bariatric Surgery Procedures—ASMBS', n.d.).

Gasterectomy

In this procedure approximately 80% of the stomach is removed. The new pouch holds a considerably smaller volume of food. As a result, the patient can consume less food and therefore will lose weight. This surgery has an effect on the gut hormones, which also facilitate weight reduction. This surgery is done laparoscopically.

Advantages:

- It will restrict the amount of the food
- Induces rapid weight reduction similar to RYGB
- No gastric bypass
- Relatively short stay in hospital
- Suppresses hormones and induces changes in gut hormones.

Disadvantages:

- Non-reversible procedure
- It may cause long-term vitamin deficiencies
- It has more complications than Adjustable Gastric Bypass.

Adjustable Gastric Bypass

Adjustable gastric bypass results in a decrease in food consumption. The reduction in food absorption is observed because of the placement of a small bracelet-like band around the upper part of the stomach. This will result in the reduction of the stomach size and then it can be adjusted by the surgeon by placing a balloon inside the band [16].

Advantages:

- Reduce the extent of food that stomach can hold
- Around 40–50% weight loss is observed after the surgery

- Stomach part is not removed after the surgery; the intestine is also not disturbed
- Short hospital stay
- Reversible and adjustable
- Lowest postoperative complications
- Least deficiencies observed.

Disadvantages:

- Slow weight reduction
- Patients fail to reduce more than 50% of their weight
- It requires a foreign body to remain after surgery
- It can have mechanical problems with band and tubes in certain patients
- Requires strict diet after the surgery
- Highest rate of re-operation [15].

Biliopancreatic Diversion with Duodenal Switch

It is popularly known as "duodenal switch". Firstly, removal of a large portion of the stomach takes place, which makes the patient feel full and will allow him to eat less. Secondarily re-routing is conducted from the stomach to the small intestines, resulting in less absorption of the food. Digestive and bile salts may change the body's digestion process and ability of the food and calorie absorption [16].

Advantages:

- It will cause more weight loss than other surgeries (70%)
- It will allow patients to eat normal meals post-surgery
- Reduce the absorption by 70%
- It will cause the changes in gut hormones
- It is most effective compared to other surgeries for diabetic patient.

Disadvantages:

- More complications after the surgery
- Longer hospital stay
- May cause protein deficiency and long-term vitamin and mineral deficiencies
- Follow-up is necessary and that may require patient compliance [16].

Genetic Polymorphism and Weight Loss in Bariatric Surgery

Genetic polymorphism is associated with different weight loss outcomes in patients undergoing bariatric surgery. One study was carried out to examine markers associated with eleven obesity related genes in which maximum weight loss and maximum weight gain was observed after the bariatric surgery. In this study 11 genes were taken i.e. ADIPOQ, BDNF, FTO, G1NB3, LEP, LEPR, MC4R, NR3C1, PPARG, PPARGC1A and TNF. After the study, it was found that the minor allele at the FTO rs16945088 locus is associated with maximum weight reduction after banding surgery. An allele of FTO single nucleotide polymorphism rs8050136 was correlated with higher body mass index and high chances of obesity. None of the other genotypes were related to weight gain or loss after bariatric surgery [17]. Table 16.3 describes about the summery of gene variations in bariatric

Table 16.3 Different genes and their effect on weight loss

Name of peptide	Associated gene	Polymorphism	Implication
PGC-1α	Ppargc1a	Gly482Ser	Waist ratio is more reduced in Gly/ser + Ser/Ser group compared to Gly/Gly group
Ghrelin	GHSR gene	Single nucleotide polymorphism (SNPs) rs9819506 and rs490683	Patinets with homozygous for rs490683 showed highest weight loss after the surgery in 30 months
Amide hydrolase gene	Fatty acid amide hydrolase gene	C358A	Allele of A358C of fatty acid amide hydrolase had better initial weight loss after 9–12 months
Leptin	LEPR gene	Lys656Asn	Weight loss is higher in mutant group i.e. Lys656Asn and Asn656Asn compared to wild group i.e. Lys656Lys
Glucagon like peptide-1	GLP-1R	Rs6923761 GLP-1R	BMI, weight and waist circumference were lower in non A allele carrier than A allele. Initial weight loss was higher in patients with GG genotype than A allele carrier
Interleukin -6	174 G>C	174 G>C polymorphism of IL-6	The C(−) carrier had more weight loss than the C(+) carrier
Mitochondrial uncoupling protein	UCP-2	rs660339 (Ala55Val)	Patients with either TT or CT genotype on rs660339 experienced more weight loss than patients with CC genotype

surgery. Details of several genetic polymorphism associated with different types of bariatric surgeries are provided in the following section.

Genetic Polymorphism in Gastric Bypass

PGC-1

Peroxisome proliferator activated receptor gamma co-activator (PGC)-1α is a co-activator in the body responsible for the transcription involved in the regulation of metabolic and non-metabolic disorders. The hypothesis was made that Gly482Ser polymorphism of the ppargc1a (peroxisome proliferator activated receptor gamma co-activator 1 alpha) gene would predict differently after bariatric surgery than other genes. The association was found between the ppargc1a (gene coding for PGC1α) polymorphism and bariatric surgery. The two groups were selected i.e. Gly/Gly and Gly/Ser + Ser/Ser. The inflammatory parameters, anthropometric, C-IMT and metabolic parameters were performed in the intervals of 1, 6 and 12 months. The polymerase chain reaction restriction fragment length polymorphism assay was performed for the primers Gly482Ser sense and Gly482Ser antisense. It was found that Gly482Ser might be an important marker for the complications and amelioration of the disease. The waist ratio was the anthropometric parameter that was most significantly reduced. The Gly/Ser + Ser/Ser group had more reduction in hip/waist ratio than the Gly/Gly group. It was concluded that Gly482Ser polymorphism may improve the metabolic and inflammatory outcomes that lead to reduction in the structural marker of atherosclerosis in obese patients undergoing gastric bypass surgery. Bariatric surgery showed improvement in various outcomes such as metabolic, inflammatory and vascular in the patients having polymorphism in the PGC1α gene Gly482Ser [18].

Ghrelin

Ghrelin is a hormone which has a major influence on energy balance. It is a natural ligand of the growth hormone secretogogue receptor, thus stimulating growth hormone secretion and playing a role of appetite inducer. The genetic correlation has been found between the ghrelin gene and bariatric surgery. Single nucleotide polymorphism in the promoter region of its receptor gene has some correlation with weight loss. Association between the genotypes for single nucleotide polymorphisms (SNPs) rs9819506 and rs490683 in the promoter region of the Growth hormone secretogogue receptor (GHSR) gene and weight loss outcome was performed for 30 months. Three models were taken: additive, recessive and dominant. Patients who are homozygous for rs490683 (GHSR194) C/C genotype showed the most weight loss in the additive model [19]. The study suggested that patients with C/C genotype lost 5% more weight than those who do not have this genotype.

Genetic Polymorphism in Biliopancreatic Surgery

Amide Hydrolaze

Fatty acid amide hydrolase is a member of the serine hydrolase family of enzymes. This enzyme is responsible for the hydrolysis of anandamide, an endocannabinoid. In biliopancriatic surgery the polymorphism was checked for C358A of the fatty acid amide hydrolase gene on the clinical outcomes for one year. It was noted that the allele A358C of the fatty acid amide hydrolase was associated with a better initial percentage of excess weight loss 9 and 12 months after biliopencriatic diversion [20].

Leptin

Leptin is a hormone that plays major role in the energy balance. It is a mediator for the long-term regulation of energy and food suppression thus facilitating weight loss. Leptin is strongly associated with bariatric surgery because BMI levels are indicators of the decrease or increase of leptin levels. Two groups were taken for the study i.e. mutant group and wild group. Patients with Lyc656/Asn656 and Asn6566/Asn656 genotype were included in the mutant group and patients with Lys656/Lys656 genotype were included in the wild group. After the bariatric surgery, the influence was checked for Lys656Asn polymorphism of leptin receptor gene in biliopancreatic diversion. The study concluded that weight loss was higher in mutant group i.e. Lys656Asn and Asn656Asn higher than the wild type Lys656Lys group after the surgery [21].

GLP-1 Gene

Glucagon, like peptide-1, is a 30-amino acid peptide hormone. GLP- 1 stimulates the insulin secretion (i.e. to act like incretin hormone) and to inhibit the glucagon secretion. GLP-1 also has a role in the regulation of appetite and food intake. Studies of Glucagon like peptide-1 (GLP-1) receptor (GLP-1R) have been done to identify the relation between the polymorphism of the gene GLP-1R and pathogenesis of the obesity. Investigations were made on the polymorphism of rs6923761 GLP-1R gene and its outcome after the biliopancreatic diversion. 137 patients were admitted. Parameters including weight, blood pressure, basal glucose and lipid profiles were measured at the initial visit and each following visit (3, 9, 12 and 18 months). The BMI, weight and waist circumference were lower in the non-A allele carrier than A carriers after 18 months. The initial weight loss was higher in patients with GG genotype than allele carriers [22].

Genetic Polymorphism in Adjustable Gastric Banding

174 G>C IL-6

In laparoscopic adjustable gastric banding the polymorphism was checked in the 174 G>C IL-6 for the bioelectrical parameters in the obese subjects. It was thought that 174 G>C IL-6 polymorphism may play a role in variable therapeutic responses to the LAGB surgery. The C allele frequency was 35% and the GG, GC, and CC genotype frequency were 45%, 40%, and 15% respectively for IL-6. The two carriers were selected i.e. C(−) and C(+) carriers. Weight loss and BMI were significantly reduced in the C(−) carrier than C(+ carrier) [23].

UCP2 Gene

Mitochondrial uncoupling proteins (UCP2) are the members of a larger family of mitochondrial anion carrier proteins. They separate the oxidative phosphorylation from ATP synthesis with energy dissipated as heat. This is highly expressed in the skeletal muscles. This gene plays a role in thermogenesis, obesity and diabetes [24]. One study selected four SNPs i.e. rs660339, -866G/A, rs17132534 and rs643064. It is associated with diabetes and the Ala55Val polymorphism on the UCP2 gene may predict greater weight loss in morbidly obese patients who undergo the Gastric banding. All the subjects were genotyped for four SNPs on the gene. The SNPs associated with obesity are the markers of the weight change. rs660339 on exon 4 was associated with the morbid obesity. Patients with TT or CT genotypes experienced more weight loss compared to CC after the laparoscopic adjustable Gastric Banding (LABG) [25].

Prediction of Outcomes Using Genetic Risk Score

The collective impact of various SNPs, which are likely to underwrite the risk, can be determined by a genetic risk score (GRS). The GRS creates one continuous variable which indicates the chance of developing any specific parameter, outcome or disease, and in case of bariatric surgery, the outcome being weight loss. The importance of the same outcome is also depicted in a study by Bandstein et al. [26]. Waist-to-hip ratio associated SNP and BMI associated SNP was used to prepare weighted GRS with weight loss. Patients with least GRS scores had maximum weight loss suggesting that GRS can be used to predict weight loss in different polymorphism.

Conclusions

Obesity has become one of the most worrisome conditions leading to increased risk for several disorders like cardiovascular diseases and metabolic abnormalities. Bariatric surgery is useful not only for weight loss but also for the co-morbid conditions like diabetes, cardiovascular disorders and other metabolic diseases. In current scenario, there are large number of non-obese patients who are undergoing bariatric surgery for remission of conditions like diabetes, polycystic ovarian syndrome etc. However, despite this, many patients do not lose enough weight after bariatric surgery. Additionally, some patients do not exhibit remission of co-morbid conditions. Genetic polymorphism is one of the major factors for such inconsistent and inadequate response to bariatric surgery. With recent advancements in the field of molecular biology and biotechnology, several polymorphisms are identified to be associated with weight loss. With such basic knowledge about polymorphism, future studies should be directed towards development of functional genomic diagnostic kits which will facilitate the identification of patients who will respond to surgery before the actual surgery itself. Prima facie, such kits may appear to be a costly diagnostic procedure. However, given that bariatric surgery is also a costly affair, this increases the overall economic burden on the patient and on society as well.

Obesity also contributes significantly to psychological burden with most obese patients facing mild to moderate depressive-like conditions. Bariatric surgery offers them hope. However, if they do not lose weight after bariatric surgery, the psychological impact remains, and patients become more depressive. Hence, with the help of some diagnostic procedures, patients can be made quite aware that bariatric surgery would not be beneficial for them. This will prevent development of severe psychological impact.

Disclosure of potential conflicts of interest The authors declare no conflict of interests.

Ethical Standards Not applicable as this is a review article.

References

1. WHO | Obesity and overweight. (n.d.). (Cited 29 Sept 2015). Available from http://www.who.int/mediacentre/factsheets/fs311/en/
2. Melanson K, Dwyer J (2002) Popular diets for treatment of overweight and obesity. In: Wadden TA, Stunkard AJ et al (eds) Handbookof obesity treatment. Guilford Press, New York, pp 249–275
3. Pronk NP, Wing RR (1994) Physical activity and long-term maintenance of weight loss. Obes Res 2(6):587–599
4. Borgström B (1988) Mode of action of tetrahydrolipstatin: a derivative of the naturally occurring lipase inhibitor lipstatin. Biochim Biophys Acta 962(3):308–316

5. Orlistat oral: Uses, Side Effects, Interactions, Pictures, Warnings & Dosing. [Internet]. [Cited 2015 Dec 26]. Available from http://www.webmd.com/drugs/2/drug-17220/orlistat-oral/details
6. Leanbodylook.com. Dangerous Side Effects of Rimonabant (Acomplia). [internet].[Cited 2015 Dec 12]. Available from http://www.leanbodylook.com/dangerous-side-effects-of-rimonabant-acomplia/
7. Webmd.com. Common and Rare Side Effects for sibutramine oral. [Internet], [Cited 2015 Dec 26]. Available from http://www.webmd.com/drugs/2/drug-5405/sibutramine-oral/details/list-sideeffects
8. niddk.nih.gov. Bariatric Surgery for Severe Obesity. [Internet] U.S. Department of Health and Human Service, [Cited 2016 Feb 9]. Available from http://www.niddk.nih.gov/health-information/health-topics/weight-control/bariatric-surgery-severe-obesity/Pages/bariatric-surgery-for-severe-obesity.aspx
9. Latifi R, Kellum JM, De Maria EJ, Sugerman HJ (2002) Surgical treatment of obesity. In: Wadden TA, Stunkard AJ et al (eds) Handbookof obesity treatment. Guilford Press, New York, pp 339–356
10. Blackburn GL, Hutter MM, Harvey AM, Apovian CM, Boulton HR, Cummings S, Fallon JA, Greenberg I, Jiser ME, Jones DB, Jones SB, Kaplan LM, Kelly JJ, Kruger RS Jr, Lautz DB, Lenders CM, Lonigro R, Luce H, McNamara A, Mulligan AT, Paasche-Orlow MK, Perna FM, Pratt JS, Riley SM Jr, Robinson MK, Romanelli JR, Saltzman E, Schumann R, Shikora SA, Snow RL, Sogg S, Sullivan MA, Tarnoff M, Thompson CC, Wee CC, Ridley N, Auerbach J, Hu FB, Kirle L, Buckley RB, Annas CL (2009) Expert panel on weight loss surgery: executive report update. Obesity (Silver Spring, Md.) 17(5):842–862
11. Maggert KA (2012) Genetics: polymorphisms, epigenetics, and something in between. Genet Res Int 867951
12. Genetic polymorphism and SNP, Genotyping, Haplotype Assembly Problem, Haplotype Map, Functional Genomics and Proteomics. Available from http://www.cs.mcgill.ca/~kaleigh/compbio/snp/snp_summary.html
13. medicinenet.com. Medical Definition of Bariatric surgery. [Internet]. medterms medical dictionary a-z list [Cited 2015 Oct 8]. Available from http://www.medicinenet.com/script/main/art.asp?articlekey=23436
14. Li JF, Lai DD, Lin ZH, Jiang TY, Zhang AM, Dai JF (2014) Comparison of the long-term results of Roux-en-Y gastric bypass and sleeve gastrectomy for morbid obesity: a systematic review and meta-analysis of randomized and nonrandomized trials. Surg Laparosc Endosc Percutan Tech 24(1):1–11
15. asmbs.org Bariatric Surgery Procedures - ASMBS. [Internet]. American Society for Metabolic and Bariatric Surgery [Cited 2015 Oct 12]. Available from https://asmbs.org/patients/bariatric-surgery-procedures
16. Bariatric Surgery for Severe Obesity. [internet]., U.S. Department of Health and Human Service [Cited 2015 Dec 12], Available from http://www.niddk.nih.gov/health-information/health-topics/weight-control/bariatric-surgery-severe-obesity/Pages/bariatric-surgery-for-severe-obesity.aspx
17. Sarzynski MA, Jacobson P, Rankinen T, Carlsson B, Sjöström L, Bouchard C, Carlsson LMS (2011) Associations of markers in 11 obesity candidate genes with maximal weight loss and weight regain in the SOS bariatric surgery cases. Int J Obes (Lond) 35(5):676–683
18. Geloneze SR, Geloneze B, Morari J, Matos-Souza JR, Lima MM, Chaim EA, Velloso LA (2012) PGC1α gene Gly482Ser polymorphism predicts improved metabolic, inflammatory and vascular outcomes following bariatric surgery. Int J Obes (Lond) 36(3):363–368
19. Matzko ME, Argyropoulos G, Wood GC, Chu X, McCarter RJM, CD, Gerhard GC (2012) Association of Ghrelin receptor promoter polymorphisms with weight loss following Roux-en-Y gastric bypass surgery. Obes Surg 22(5):783–790
20. de Luis DA, Sagrado MG, Pacheco D, Terroba MC, Martin T, Cuellar L, Ventosa M (2010) Effects of C358A missense polymorphism of the endocannabinoid degrading enzyme fatty acid amide hydrolase on weight loss and cardiovascular risk factors 1 year after biliopancreatic diversion surgery. Surg Obes Relat Dis 6(5):516–520

21. de Luis DA, Aller R, Sagrado MG, Izaola O, Terroba MC, Cuellar L, Martin T (2010) Influence of lys656asn polymorphism of leptin receptor gene on surgical results of biliopancreatic diversion. J Gastrointest Surg 14(5):899–903
22. de Luis DA, Pacheco D, Aller R, Izaola O (2014) Role of the rs6923761 gene variant in glucagon-like peptide 1 receptor gene on cardiovascular risk factors and weight loss after biliopancreatic diversion surgery. Ann Nutr Metab 65(4):259–263
23. Di Renzo L, Carbonelli MG, Bianchi A, Domino E, Migliore MR, Rillo G, De Lorenzo A (2012) Impact of the -174 G > C IL-6 polymorphism on bioelectrical parameters in obese subjects after laparoscopic adjustable gastric banding. J Obes 2012(208953):1–7
24. UCP2 uncoupling protein 2 (mitochondrial, proton carrier) humans. [Internet] Gene ID: 7351, updated on 6th December 2015, NCBI, [Cited 2016 Jan 2]. Available from http://www.ncbi.nlm.nih.gov/gene/7351
25. Chen HH, Lee WJ, Wang W, Huang MT, Lee YC, Pan WH (2007) Ala55Val polymorphism on UCP2 gene predicts greater weight loss in morbidly obese patients undergoing gastric banding. Obes Surg 17(7):926–933
26. Bandstein M, Voisin S, Nilsson EK et al (2016) A genetic risk score is associated with weight loss following Roux-en-Y gastric bypass surgery. Obes Surg 29:2183–2189

Chapter 17
Anti-inflammatory Components from Functional Foods for Obesity

Sunil K. Panchal and Lindsay Brown

Abstract Obesity, defined as excessive fat accumulation that may impair health, has been described throughout human history, but it has now reached epidemic proportions with the WHO estimating that 39% of the world's adults over 18 years of age were overweight or obese in 2016. Obesity is a chronic low-grade inflammatory state leading to organ damage with an increased risk of common diseases including cardiovascular and metabolic disease, non-alcoholic fatty liver disease, osteo-arthritis and some cancers. This inflammatory state may be influenced by adipose tissue hypoxia and changes in the gut microbiota. There has been an increasing focus on functional foods and nutraceuticals as treatment options for obesity as drug treatments are limited in efficacy. This chapter summarises the importance of anthocyanin-containing fruits and vegetables, coffee and its components, tropical fruit and food waste as sources of phytochemicals for obesity treatment. We emphasise that preclinical studies can form the basis for clinical trials to determine the effectiveness of these treatments in humans.

Keywords Obesity · Inflammation · Functional foods · Anthocyanins · Coffee · Tropical fruits · Food waste

Obesity—The Extent of the Problem

Obesity is often referred to as a global health challenge. The literature on the epidemiology, causes, co-morbidities and potential treatments of obesity is enormous. As an example, a PubMed search from 2013 to 2019 for "obesity" lists about

S. K. Panchal · L. Brown
Functional Foods Research Group, University of Southern Queensland, Toowoomba, QLD 4350, Australia

L. Brown (✉)
School of Health and Wellbeing, University of Southern Queensland, Toowoomba, QLD 4350, Australia
e-mail: Lindsay.Brown@usq.edu.au

P. S. Tappia et al. (eds.), *Pathophysiology of Obesity-Induced Health Complications*, Advances in Biochemistry in Health and Disease 19,
https://doi.org/10.1007/978-3-030-35358-2_17

125,000 references. This chapter will provide a background on the disease risk in obesity and the role of inflammation before examining some examples of functional foods that may reduce obesity by their anti-inflammatory actions.

The World Health Organisation defines overweight and obesity as abnormal or excessive fat accumulation that may impair health [1] and provides the key facts on overweight and obesity that are listed in Table 17.1.

The increase in the number of people with overweight and obesity has been described as an epidemic or even as a pandemic of the late 20th and 21st century [2]. However, obesity has been described through the ages, starting with prehistoric statuettes from 30,000 years ago such as the Venus of Willendorf [3] and including members of the Ptolemy dynasty who ruled Egypt from 305-30BC [4]. Physicians from the Greco-Roman and Byzantine world described the aetiology and clinical manifestations of obesity with suggestions for therapy including a Mediterranean-like diet with an active lifestyle [5]. In European art, overweight and obesity were indications of health, beauty and vitality [6]. In 1980, there were 921 million overweight and obese people; this had increased to 2.1 billion in 2013, about 29% of the world's total population in 2013 of 7.3 billion with the proportion of adult males with a body mass index (BMI) of 25 or greater increasing from 28.8% in 1980 to 36.9% in 2013 and an increase from 29.8 to 38.0% in adult females [7]. The trends in mean BMI as a measure of under- or over-nutrition from 1975 to 2014 have been reported for adults aged over 18 years in 200 countries [8]. Over these 4 decades, global prevalence of underweight has decreased from 13.8 to 8.8% in men and from 14.6 to 9.7% in women. However, the prevalence of obesity increased from 3.2 to 10.8% in men and from 6.4 to 14.9% in women. Severe obesity with BMI ≥ 35 kg/m^2 occurred in 2.3% of the world's men and in 5.0% of the world's women; the corresponding figures for morbid obesity with BMI ≥ 40 kg/m^2 were 0.64% in men and 1.6% in women. Obesity is clearly a current worldwide occurrence in adults.

Obesity is not restricted to adults, but occurs widely in children and adolescents. Analyses of data sources with measurements of height and weight on 128.9 million children and adolescents aged 5 years and older from 1975 to 2016 have shown that the prevalence of obesity increased from 0.7% in 1975 to 5.6% in 2016 in girls, and

Table 17.1 Key facts on overweight and obesity (World Health Organisation)

Key facts
Worldwide obesity nearly tripled since 1975
In 2016, more than 1.9 billion adults, 18 years and older, were overweight. Of these, over 650 million were obese
39% of adults aged 18 years and over were overweight in 2016, and 13% were obese
Most of the world's population live in countries where overweight and obesity kills more people than underweight
41 million children under the age of 5 were overweight or obese in 2016
Over 340 million children and adolescents aged 5-19 were overweight or obese in 2016
Obesity is preventable

from 0.9% in 1975 to 7.8% in 2016 in boys, with accelerated increases in east and south Asia [9]. This study also highlighted the co-existence of underweight and obesity in the world's children and adolescents with 75 million girls and 117 million boys being underweight, along with 50 million girls and 74 million boys being obese in 2016.

Despite the increasing prevalence of overweight and diabetes, life expectancy at birth has increased from 59 to 71 years over the same time-frame [10]. This apparently protective relationship is likely to be casual, rather than causal, as BMI changes in children do not appear to be protective. Also, the increases in obesity have not been evenly spread in the world, as the poor are more likely to be obese in high income countries but this group is usually underweight in low-income countries leading to increased global inequalities in the prevalence of obesity [10], making a protective effect of obesity on life expectancy less likely.

As part of the Global Burden of Disease study, patterns of death and disability-adjusted life years were determined in people with high BMI from 195 countries from 1990 to 2015 [11]. In 2015, high BMI was estimated to contribute to 4 million deaths each year and 120 million disability-adjusted life years. More than two-thirds of deaths were related to cardiovascular disease, but the improvements in survival from cardiovascular disease have reduced the rate of increase of mortality in people with high BMI. Diabetes was the second leading cause, contributing to 0.6 million deaths per year and 30.4 million disability-adjusted life years. Fewer than 10% of deaths in 2015 were associated with chronic kidney disease or cancers [11].

The World Health Organisation has summarised that an increased BMI is a risk factor for the non-communicable diseases that are listed in Table 17.2 [1].

Obesity has been defined as a disease of chronic low-grade inflammation [12] which will be further discussed in the next section. Co-morbidities with obesity include cardiovascular disease, diabetes, some cancers, non-alcoholic fatty liver disease (NAFLD) and osteoarthritis. There is a clear relationship between obesity and cardiovascular disease, with increasing incidence of cardiovascular disease with increasing obesity and length of duration of obesity [13]. However, obese individuals with high cardiorespiratory fitness, known as metabolically healthy but obese or fat-but-fit, showed reduced cardiovascular risk. This may be one of the reasons for an obesity paradox, where overweight or mildly obese individuals with cardiovascular

Table 17.2 Risk factors with increased body mass index (World Health Organisation)

Cardiovascular diseases (mainly heart disease and stroke), which were the leading cause of death in 2012
Diabetes
Musculoskeletal disorders (especially osteoarthritis—a highly disabling degenerative disease of the joints)
Some cancers (including endometrial, breast, ovarian, prostate, liver, gallbladder, kidney, and colon)

disease showed a better prognosis [13]. Obesity and type 2 diabetes are both characterised by defects in insulin action produced by increased plasma concentrations of free fatty acids and so are commonly observed in the same individuals [14]. The increased flux of free fatty acids from increased abdominal fat is likely to be the cause of the increased incidence of NAFLD in obese individuals [15]. Further, both diabetes and obesity are associated with an increased risk of cancers which could be initiated by the increased pro-inflammatory environment that characterises both diabetes and obesity [16]. Increased adiposity increases the release of a wide range of adipokines from adipocytes; together with increased mechanical loading, the pro-inflammatory cytokines such as leptin and visfatin are likely to be important in the development of osteoarthritis in obese and older patients [17]. Thus, the major common mechanism in obesity and its co-morbidities is systemic inflammation.

Inflammation in Obesity

Obesity has been described as a low-grade, chronic inflammation orchestrated by metabolic cells in response to excess nutrients and energy [18]. As described by these authors, this inflammation is different from classic inflammation producing redness, swelling, heat and pain, which is essential for the repair, remodelling and renewal of tissues. The inflammatory trigger in the development of obesity is the consumption of foods causing an increased production and secretion of an array of inflammatory cytokines, known as adipokines, by adipocytes, causing increased infiltration of immune cells such as macrophages into the metabolic tissues, including adipose, liver, muscle, pancreas and brain. Obesity-induced inflammation differs from classic inflammation in that it is moderate, creates a pro-inflammatory environment and is sustained by the constant stimulus of chronic nutrient intake. This metabolic inflammation (or metaflammation) then interferes with normal metabolism and disrupts insulin signalling to produce insulin resistance and lipo-lysis, so disrupting glucose and lipid homeostasis [18]. The precise triggers of metabolic inflammation could include intestinal antigens including lipopoly-saccharides, components of foods such as free fatty acids, or signals associated with dying or stressed adipocytes including leptin and other adipokines, or hypoxia which could involve hypoxia-inducible factor (HIF) 1α [19].

The nutrient and immune systems are fundamentally related as survival relies on both the ability to store and harness energy and to sense and fight infection. Both respond to danger signals and share many of the signalling networks. We suggested that, in metabolic disease, the initiation of multiple redundant mechanisms in response to an increase in nutrients limits endogenous nutrient output and exo-genous nutrient intake with a similar set of molecules and signalling pathways as in innate immunity [20]. Studies on the convergence of these pathways show that many levels are involved, including receptors, organelles, kinase pathways and gene expression [12]. These evolutionally conserved interactions are essential for the maintenance of healthy organs, including adipose tissue, as well as overall health, especially in chronic non-communicable diseases such as obesity [12].

The Gut Microbiota

The bacteria in the gastrointestinal tract, concentrated in the colon, are essential for many functions including protection from pathogens, digesting otherwise indigestible carbohydrates, synthesising essential vitamins and modifying the immune system. The intestinal microbiota is diverse, with at least 1000 different species providing around 2×10^{13} bacteria, similar to the number of cells in the adult human [21]. The healthy human microbiota is dominated by bacteria of two phyla, *Bacteroidetes* and *Firmicutes* together with about 90% of the total number of bacteria, although there can be marked variation of the *Bacteroidetes/Firmicutes* ratio in healthy subjects which may result from differences in nutrition or geography [22]. Further, the role of individual bacterial families is far from clear as maybe 50% of these families remain functionally uncharacterised [22]. Changes to this complexity during the development and maintenance of obesity include decreased *Bacteroidetes* and increased *Firmicutes*, but these changes are highly individualised, and depend on dietary composition and a wide range of lifestyle factors including breastfeeding, exercise, stress, use of antibiotics, sleep disturbances and cold exposure [23]. Many physiological responses have been associated with obesity-associated microbiota including host energy harvesting, insulin resistance, inflammation and fat deposition, together with regulation of adiposity, energy balance, and central appetite and food reward signalling [24]. Metabolic, immune and defence systems are strongly influenced by the gut microbiota, directly influencing human health in conditions such as obesity and metabolic diseases, undernutrition and eating disorders, inflammatory bowel disease and colorectal cancer [25]. Further, changes in the intestinal microbiota and changes in intestinal permeability may provide the triggers of the persistent low-grade systemic inflammation that characterises obesity [26]. Increased intestinal permeability allows bacterial components such as lipopolysaccharides into the body where they may disrupt vagal afferent signalling to the brain and then increase body weight [27].

The intestinal microbiota is important in host energy metabolism and clearly affects the bidirectional communication between the brain and the gut, referred to as the microbiota-gut-brain axis. These changes in the microbiota are important in metabolic disorders as diverse as anorexia nervosa, cachexia, severe malnutrition such as kwashiorkor, as well as obesity, and also after interventions such as bariatric surgery [28]. Short-chain fatty acids such as acetate, propionate and butyrate produced by colonic bacteria influence host energy metabolism and appetite by multiple mechanisms [28]. The gut microbiota is important in energy harvesting by producing these short-chain fatty acids, in particular from poly-saccharides such as cellulose, xylan and pectin which cannot be metabolised by human digestive enzymes, so termed as prebiotics. Prebiotics increase the growth of the selected intestinal microbiota including *Bifidobacterium* and *Lactobacillus*; this reduces the production of liposaccharides, increases the integrity of the gastrointestinal barrier and may prevent obesity [29].

Products of the intestinal microbiota induce activation of tissue macrophages and this low-grade inflammation contributes to metabolic diseases such as obesity and other metabolic disorders [30, 31]. Thus, changes to the microbiota through treatments including faecal transplantation and prebiotics are logical approaches to the treatment of these metabolic diseases [32]. Faecal microbiota transplantation has already shown the importance of intestinal dysbiosis in disease states such as infections with *Clostridium difficile* and inflammatory bowel disease [33]. Faecal microbiota transplantation is a potential method to delineate the effects of intestinal microbiota on the metabolic syndrome [34]. This could be an innovative option to treat human obesity [35] as studies in murine models showed that faecal transplantation from an obese human twin produced obesity in contrast to the lack of weight gain with faecal transplantation from the lean human twin [36]. However, there is no current information to show that this treatment improves human obesity [37]. Further, the gut microbiota may be heavily involved in obesity-associated diseases such as the development of NAFLD [38] as well as other chronic liver diseases including chronic hepatitis B and C, and cirrhosis [39]. However, the effectiveness of faecal microbiota transplantation in these common liver diseases in humans remains to be proved.

Hypoxia

As adipose tissue mass expands, hypoxia develops and this reduction in oxygen underlies the switch from oxidative metabolism to anaerobic glycolysis [40] with increased macrophage infiltration and the development of insulin resistance leading to hyperglycaemia [40]. Gaseous oxygen is essential for all aerobic organisms [41] and deficiency of oxygen leads to widespread cellular adaptations, mostly driven by transcription factors such as the HIFs. HIFs are important in hypothalamic control of the regulation of body weight, glucose homeostasis and liver metabolism [42] and inhibition of HIFs decreases adverse diet-induced metabolic phenotypes so that HIFs may be drug targets for metabolic diseases [43]. Hypoxic adipocytes produce a wide range of protein factors, the adipokines, that alter many physiological functions such as appetite, insulin sensitivity and blood pressure, important in the development of obesity, diabetes, NAFLD and dyslipidaemia [44]. Despite this close relationship between hypoxia, adipokine secretion and chronic systemic inflammation, intermittent hypoxia during rest or exercise may lead to improved exercise tolerance, metabolism and systemic arterial pressure as a treatment strategy to increase weight loss and improve obesity-associated disease [45]. One example is the improved body composition, physical fitness, pulmonary function and heart rate variability following 12 weeks' hypoxic training compared to normoxic training in obese 65–70 year old Korean men [46].

Current and Future Treatments of Obesity

Drug treatments for obesity are characterised by their relatively small changes in body weight, usually less than 10%, which are often not sustained [47]. More effective compounds may be unimolecular dual agonists, for example at receptors for glucagon-like peptide 1 and glucagon or glucose-dependent insulinotropic polypeptide, or triple agonists for all three peptide receptors [47]. Peptide-mediated delivery of nuclear hormones, for example covalently bound glucagon-like peptide 1-dexamethasone or -oestradiol, and glucagon-triiodothyronine, may be more acceptable as the nuclear hormone is only released in cells with the peptide receptors, so reducing the adverse effects in other tissues [47]. An alternative treatment protocol may involve functional foods, defined as foods that provide nutrition as well as being able to prevent or reverse disease states [48], or nutraceuticals, defined as pharmaceutical-grade and standardised nutrients derived from foods such as prebiotics and probiotics [49]. In obesity, these functional foods and nutraceuticals are likely to be anti-inflammatory so as to be effective in a disease state of chronic low-grade inflammation. This concept is supported by the effectiveness of selective semi-synthetic anti-inflammatory compounds including protease-activated receptor 2 antagonists [50], complement 3a and 5a receptor antagonists [51] and phospholipase A_2 group IIA inhibitors [52] to reduce abdominal obesity. While using foods to treat an inflammatory state produced by increased food intake sounds counter-intuitive, plants produce a wide range of secondary metabolites that may have developed to protect the plants. The hypothesis is that this protective role could potentially be translated to decreasing chronic inflammatory states in humans such as obesity. There are also practical reasons for testing functional foods since the production of these foods is likely to be sustainable and adherence to dietary treatment will probably be higher than with pharmaceutical agents.

Food as the Source of Anti-obesity Agents

Functional foods are foods that provide basic nutritional requirements along with health benefits [53]. Obesity is a chronic human disease that functional foods may have a role in treating or preventing [48, 54]. Over the past few decades, many animal models have been developed for the study of anti-obesity effects of many foods and their components [55, 56]. However, the successes in animal trials have not been well translated to overweight or obese humans [48, 57]. Nonetheless, there is widespread interest in functional foods as a viable therapeutic option for obesity and both animal and human trials are continuing to search for a solution to this worldwide problem. Here, we will summarise the findings for some key functional foods and update the literature from our previous review [48] to demonstrate the possibilities with functional foods to treat human overweight and obesity.

Anthocyanins

Anthocyanins are dark-coloured pigments from fruits and vegetables [48] that are produced as secondary metabolites by the plants [58]. These secondary metabolites are produced as a defence mechanism against stress situations including pathogen infection, low nitrogen condition and photo-oxidative damage [58–61]. Sources of anthocyanins include berries, cherries, grapes, plums, dark-coloured vegetables and pigmented grains [48]. Anthocyanins have been proven to be effective in reducing obesity and metabolic syndrome in animal models as well as in humans [48, 62–65]. Some of the proposed mechanisms of actions of anthocyanins as anti-obesity agents are inhibition of lipid absorption, increase in energy expenditure, regulation of lipid metabolism, suppression of food intake, regulation of gut microbiota, amelioration of oxidative stress and resolution of inflammation [66].

Recent studies have extended these observations. Intervention with blackberry and blueberry anthocyanins in high-fat diet-fed mice for 12 weeks resulted in inhibition of body weight gain, reduction in serum and hepatic lipid concentrations, increased faecal acetate and butyrate concentrations and reduced expression of tumour necrosis factor α, interleukin-6 and nuclear factor-κB genes in the pathways to inflammation. These anthocyanins also modulated hepatic lipid and glucose metabolism, and the insulin signalling pathway [67]. In obese rats fed a high-carbohydrate, high-fat diet, chokeberry and purple maize, both containing cyanidin 3-glucoside, reduced visceral adiposity index, total body fat mass and systolic blood pressure while improving glucose tolerance, liver and cardiovascular structure and function with decreases in plasma triglycerides and total cholesterol [68]. Cyanidin glycosides from fermented chokeberries with reduced bitter taste attenuated increases in weight and serum triglyceride concentrations along with improved glucose tolerance and insulin sensitivity, when treatment was given to high-fat diet-fed mice for 8 weeks [69]. Cyanidin 3-glucoside from Queen Garnet plums and in purified form reduced obesity and metabolic syndrome symptoms in high-carbohydrate, high-fat diet-fed rats [70]. Cyanidin glucoside from Davidson's plum, a native Australian fruit, reduced visceral fat accumulation, total abdominal fat weight, size of retroperitoneal adipocytes, and plasma triglycerides and non-esterified fatty acids, normalised blood pressure, reduced left ventricular stiffness, decreased infiltration of inflammatory cells in both left ventricle and liver, decreased collagen deposition in heart, and reduced both fat vacuoles in liver and obesity-induced degeneration of knee cartilage [71]. Cyanidin and delphinidin improved insulin sensitivity and inhibited oxidative stress, NF-κB and JNK activation and PTP1B overexpression in high-fat diet-fed mice [72]. In high-fat diet-fed mice, boysenberry anthocyanins caused no changes in body weight, systolic or diastolic blood pressure, heart rate and systemic glucose intolerance while increasing nitric oxide in the aorta and improving endothelium-dependent vasodilatation in the iliac artery [73]. Delphinidin did not affect body weight, hyperglycaemia, insulin resistance or histological liver abnormalities induced by high-fat, high-carbohydrate diet in mice [74]. Delphinidin reduced triglyceride accumulation in vitro through the modulation of lipid metabolic gene expression but it failed to induce this change in high-fat, high-carbohydrate diet-fed mice [74]. Blueberry supplementation in high-fat diet-fed rats improved gut microbiota through an increase in *Gammaproteobacteria* abundance and increases in ileal villus height and ileal expression of Muc2. Tumour necrosis factor α and interleukin 1β expression in visceral fat were normalised by blueberry supplementation along with improved insulin sensitivity and hepatic insulin receptor substrate 1 phosphorylation [75]. The positive responses of anthocyanins on adipocytes, endothelial cells, inflammatory cells, hepatocytes, intestinal cells and gut microbiota have been recently reviewed [76]. This review also highlighted the lack of evidence for effects of anthocyanins on other cells, including platelets, skeletal muscle cells, hepatic stellate cells and pancreatic β-cells [76].

Fewer studies have investigated the responses to anthocyanins in humans. In a pilot crossover study, anthocyanin-rich Queen Garnet plum juice reduced ambulatory blood pressure without improving acute cognitive function in younger and older adults [77]. In mildly hypertensive overweight or obese subjects, Queen Garnet plum juice decreased systolic and diastolic blood pressures, and plasma concentrations of insulin, glucose and leptin while increasing plasma concentrations of adiponectin [78]. In an open-label study, the blend of 215 mg anthocyanins and 2.7 g prebiotic fibre daily in obese subjects reduced proportion of *Firmicutes* and increased *Bacteroidetes* with reductions in HbA1c without any changes in body weight [79]. In a randomised, placebo-controlled, crossover study in overweight or obese men, 600 g/day blackberries for seven days reduced average 24-h respiratory quotient, possibly through increased fat oxidation with no change in glucose area under the curve and reduced insulin area under the curve [80]. In a double blinded, randomised, placebo-controlled clinical study, blueberries improved insulin sensitivity without changing adiposity, energy intake or inflammatory biomarkers in obese, insulin-resistant men and women [81]. Red orange juice containing anthocyanins, flavone glycosides and hydroxycinnamic acids for 12 weeks reduced BMI, body weight, waist and hip circumference in overweight healthy human volunteers suggesting its usefulness in obesity management [82].

Coffee Components

Coffee, a complex mixture of more than thousand phytochemicals, is a functional food containing alkaloids, phenolic compounds, vitamins, carbohydrates, lipids, minerals and nitrogenous compounds [83]. Coffee has been linked with health benefits [84–86], in some studies through its action on controlling serum concentrations of leptin and adiponectin [87].

Coffee extract in high-carbohydrate, high-fat diet-fed rats improved cardiovascular and hepatic structure and function without reducing obesity [88]. A similar dose of caffeine from this study in obese rats reduced body weight and body fat along with improved cardiovascular and hepatic structure and function [89]. Coffee extract in high-fat diet-fed mice reduced the increase in body weight, prevented the decrease in the concentrations of glutathione and ascorbic acid in lens and prevented the increase in plasma cholesterol and triglycerides. These results were greater in roasted coffee-treated mice than in green coffee-treated mice [90]. Coffee and its components, caffeine and chlorogenic acid, improved liver inflammation without changing body weight, visceral fat, blood glucose and liver steatosis in Tsumura Suzuki obese diabetic mice, a spontaneous model of metabolic syndrome. This effect was seen through the effects on gut microbiota and increase in short-chain fatty acid production [91]. Green coffee bean extract in high-fat diet-fed mice reduced body weight gain, liver steatosis, white adipose tissue weights, fat mass, adipocyte size, plasma lipids and leptin [92]. Effects against metabolic syndrome were also observed with decaffeinated green coffee in diet-induced metabolic syndrome [93]. Chlorogenic acid reduced visceral fat, abdominal circumference, systolic blood pressure, left ventricular diastolic stiffness, ventricular infiltration of inflammatory cells and collagen deposition, inflammation and fat deposition in the liver, and plasma liver enzyme activities without changing plasma lipid profile along with increased diversity of gut microbiota in high-carbohydrate, high-fat diet-fed rats [94]. Chlorogenic acid treatment reduced hepatic steatosis, inflammation and insulin resistance while suppressing hepatic gene expression of Pparγ, Cd36, Fabp4 and Mgat1 in high-fat diet-fed mice [95]. Other effects of coffee and its components in animal models have been reviewed recently [96].

Moderate coffee consumption (3–4 times daily) was associated with reduced incidence of metabolic syndrome in Korean adults [97]. In a comprehensive study of longitudinal associations in a Danish cohort, an association was found between increased coffee consumption over a 6-year period and decreased concurrent gain in BMI, fat mass index, body fat percentage and waist circumference [98]. In a randomised clinical trial in individuals with metabolic syndrome, green coffee extract reduced systolic blood pressure, fasting blood glucose, HOMA-IR, waist circumference or appetite score with no impact on lipid profile and glycated haemoglobin [99]. In a randomised clinical trial in obese women, green coffee bean extract with energy restriction reduced body weight, BMI, fat mass index and waist-to-hip circumference ratio. These changes were also accompanied by reductions in serum total cholesterol, low-density lipoprotein, leptin, and plasma free fatty acids and increases in serum adiponectin [100]. The effects of coffee and chlorogenic acid in metabolic syndrome through epidemiological studies have been reviewed recently [101]. Prospective cohort studies have confirmed a link between habitual coffee consumption and reduced all-cause mortality rates with reduced risk of cardiovascular death, type 2 diabetes and liver disease [102]. These outcomes highlight the potential health benefits of coffee and its components.

Tropical Fruits

Fruits grown in tropical regions are high in their nutritive values. However, their medicinal properties are unknown for the majority of the tropical fruits [103] as these fruits are not studied to the same extent as the fruits grown in temperate regions of the world, although tropical fruits contain many bioactive compounds [104]. Although some tropical fruits have been commercialised, many of these fruits are still underutilised [103].

As an example of tropical fruits, the genus *Garcinia* from the Clusiaceae or Guttiferae family comprises evergreen, dioecious trees and shrubs that flourish in lowland tropical forests [105]. Most *Garcinia* species are sources of secondary metabolites, including simple organic acids, xanthones, flavonoids, benzophenones, lactones and phenolic acids [106]. The beneficial effects of some of the *Garcinia* species have recently been reviewed [54]. *Garcinia mangostana* xanthones are promising compounds for the development of new drugs that may interact with multiple biological targets for improving cancer, pain, insulin resistance and neurological impairment [107, 108]. In high-fat diet-fed mice, garcinol reduced body weight gain, visceral adipose tissue weight, plasma activity of alanine transaminase, plasma concentrations of total cholesterol and triglycerides, *Firmicutes* to *Bacteroidetes* ratio in gut microbiota and gut inflammation by increasing *Akkermansia* [109]. Mangosteen pericarp extract decreased concentrations of plasma free fatty acids, hepatic triglycerides and hepatic thiobarbituric acid reactive substances while increasing liver activities of antioxidant enzymes, NADH-cytochrome c reductase and succinate-cytochrome c reductase [110]. In high-fat diet-fed mice, mangosteen exerted anti-obesity effects by activating AMPK and sirtuin 1 and by suppressing PPARγ expression in the liver along with reduced body weight gain, adipose mass, and serum concentrations of triglycerides, total cholesterol and LDL [111]. In high-fat diet-fed rats, *Garcinia cambogia* extract decreased body weight gain, glucose intolerance, and plasma leptin and TNF-α concentrations [112]. In a randomised, controlled pilot trial in obese females, mangosteen improved insulin sensitivity [113]. *Garcinia cambogia*, a source of hydroxycitric acid, has been shown to be effective in weight loss [54, 114]. However, there are other tropical fruits such as durian (*Durio zibethinus* Murr.) [115] and jamun (*Syzygium cumini*) [116, 117] with only very limited evaluation for health benefits and many tropical fruits that have not yet been tested.

Food Waste as Source of Nutraceuticals

Food waste is an increasing problem in the modern world. One third of all food produced in the world is lost or wasted ($\sim$1.3 billion tonnes of food each year) costing the global economy close to $990 billion each year [118]. Much of the food waste, especially from the agriculture and food processing industries, can be a good source of nutraceuticals [119, 120]. As one example, achacha (*Garcinia humilis*) rind as the food waste reduced systolic blood pressure, diastolic stiffness, left ventricular inflammatory cell infiltration and collagen deposition in high-carbohydrate, high-fat diet-fed rats [121]. The responses of achacha rind were greater than the pulp which is the edible part of the fruit, suggesting the increased value of the rind that is generally discarded [121].

Parts of many tropical fruits such as rind, peel, seed, flower and leaf are sources of polyphenols including flavonoids that are also present in many other functional foods, thus suggesting potential health benefits in these waste components of foods [122, 123]. Temperate fruits such as citrus are rich sources of flavonoids and these phytochemicals have shown promising health benefits [124]. Similarly, citrus waste is a readily-available source of similar phytochemicals and can serve as the source of extracting these compounds for the development of nutraceuticals and pharmaceuticals [125–127]. Grape pomace is a waste from wine production and is a rich source of polyphenols including anthocyanins. In high-fat diet-fed rats, grape pomace improved glucose tolerance and insulin sensitivity suggesting the potential for further studies and nutraceutical development [128]. In a prospective, randomised, controlled, parallel-group trial, red wine grape pomace flour decreased systolic and diastolic blood pressure as well as fasting glucose concentrations [129]. Although there are suggestions of these potential benefits, there are limited high-quality studies confirming the clinical benefits of grape polyphenols [130]. Similarly, many other food wastes are sources of valuable compounds and can serve as alternatives for nutraceutical development [131–135]. These food wastes can provide opportunity for improving public health through the development of sustainable products that can be used for human consumption.

Human Trials with Functional Foods

In our previous review, we summarised the potential health benefits of many functional foods and their components [48]. In that review, we highlighted that there were remarkably few natural products with unambiguous evidence for efficacy in individuals with metabolic syndrome, especially obesity [48]. We also highlighted duration and the dose of interventions as the contributing factors in creating difficulties in translation of studies into suitable products. Recently, we reviewed other confounding factors that can create difficulties in suitable translation of functional foods in human studies [136]. These include obtaining adequate funding support for the trial, technical knowledge to initiate the trial, identifying an appropriate placebo, food delivery to participants, food quality, food acceptability, unwanted access to functional foods, compliance, appropriate biomarkers, intake of functional food by the control participants, statistical analysis, the response of the public and the response of the medical community [136].

Conclusions

Overweight and obesity are common disorders throughout the world, increasing the risk of chronic cardiovascular, metabolic and musculoskeletal diseases as well as some cancers. Obesity is described as a chronic, low-grade inflammation. Thus, functional foods delivering adequate amounts of anti-inflammatory compounds would be expected to improve the range of pathophysiological changes in overweight and obese humans. There is much evidence to support this concept, mostly in animal models but increasingly in clinical trials in humans. The studies in animal models have investigated an interesting range of compounds in functional foods that could provide relevant advances in the currently inadequate treatment of obesity in humans. Further, food waste is an underutilised source of sustainable interventions that could prevent or treat human obesity.

References

1. World Health Organisation (2018) Obesity and overweight. https://www.who.int/news-room/fact-sheets/detail/obesity-and-overweight
2. Meldrum DR, Morris MA, Gambone JC (2017) Obesity pandemic: causes, consequences, and solutions-but do we have the will? Fertil Steril 107:833–839
3. Haslam D (2007) Obesity: a medical history. Obes Rev 8(Suppl 1):31–36
4. Michalopoulos A, Tzelepis G, Geroulanos S (2003) Morbid obesity and hypersomnolence in several members of an ancient royal family. Thorax 58:281–282
5. Papavramidou N, Christopoulou-Aletra H (2007) Greco-Roman and Byzantine views on obesity. Obes Surg 17:112–116
6. Ferrucci L, Studenski SA, Alley DE et al (2010) Obesity in aging and art. J Gerontol A Biol Sci Med Sci 65:53–56
7. Ng M, Fleming T, Robinson M et al (2014) Global, regional, and national prevalence of overweight and obesity in children and adults during 1980–2013: a systematic analysis for the Global Burden of Disease Study 2013. Lancet 384:766–781
8. NCD Risk Factor Collaboration (NCD-RisC) (2016) Trends in adult body-mass index in 200 countries from 1975 to 2014: a pooled analysis of 1698 population-based measurement studies with 19.2 million participants. Lancet 387:1377–1396
9. NCD Risk Factor Collaboration (NCD-RisC) (2017) Worldwide trends in body-mass index, underweight, overweight, and obesity from 1975 to 2016: a pooled analysis of 2416 population-based measurement studies in 128.9 million children, adolescents, and adults. Lancet 390:2627–2642
10. Smith GD (2016) A fatter, healthier but more unequal world. Lancet 387:1349–1350
11. The GBD 2015 Obesity Collaborators (2017) Health effects of overweight and obesity in 195 countries over 25 years. N Engl J Med 377:13–27
12. Hotamisligil GS (2017) Inflammation, metaflammation and immunometabolic disorders. Nature 542:177–185
13. Ortega FB, Lavie CJ, Blair SN (2016) Obesity and cardiovascular disease. Circ Res 118:1752–1770
14. Verma S, Hussain ME (2017) Obesity and diabetes: an update. Diabetes Metab Syndr 11:73–79

15. Lim S, Taskinen MR, Boren J (2019) Crosstalk between nonalcoholic fatty liver disease and cardiometabolic syndrome. Obes Rev 20:599–611
16. Garg SK, Maurer H, Reed K, Selagamsetty R (2014) Diabetes and cancer: two diseases with obesity as a common risk factor. Diabetes Obes Metab 16:97–110
17. Tu C, He J, Wu B et al (2019) An extensive review regarding the adipokines in the pathogenesis and progression of osteoarthritis. Cytokine 113:1–12
18. Gregor MF, Hotamisligil GS (2011) Inflammatory mechanisms in obesity. Annu Rev Immunol 29:415–445
19. Karczewski J, Sledzinska E, Baturo A et al (2018) Obesity and inflammation. Eur Cytokine Netw 29:83–94
20. Iyer A, Brown L, Whitehead JP et al (2015) Nutrient and immune sensing are obligate pathways in metabolism, immunity, and disease. FASEB J 29:3612–3625
21. Sender R, Fuchs S, Milo R (2016) Revised estimates for the number of human and bacteria cells in the body. PLoS Biol 14:e1002533
22. Lloyd-Price J, Abu-Ali G, Huttenhower C (2016) The healthy human microbiome. Genome Med 8:51
23. Cuevas-Sierra A, Ramos-Lopez O, Riezu-Boj JI et al (2019) Diet, gut microbiota, and obesity: Links with host genetics and epigenetics and potential applications. Adv Nutr 10: S17–S30
24. Torres-Fuentes C, Schellekens H, Dinan TG, Cryan JF (2017) The microbiota-gut-brain axis in obesity. Lancet Gastroenterol Hepatol 2:747–756
25. Requena T, Martinez-Cuesta MC, Pelaez C (2018) Diet and microbiota linked in health and disease. Food Funct 9:688–704
26. Cox AJ, West NP, Cripps AW (2015) Obesity, inflammation, and the gut microbiota. Lancet Diabetes Endocrinol 3:207–215
27. Hamilton MK, Raybould HE (2016) Bugs, guts and brains, and the regulation of food intake and body weight. Int J Obes Suppl 6(Suppl 1):S8–S14
28. van de Wouw M, Schellekens H, Dinan TG, Cryan JF (2017) Microbiota-gut-brain axis: modulator of host metabolism and appetite. J Nutr 147:727–745
29. Choque Delgado GT, Tamashiro W (2018) Role of prebiotics in regulation of microbiota and prevention of obesity. Food Res Int 113:183–188
30. Hersoug LG, Moller P, Loft S (2016) Gut microbiota-derived lipopolysaccharide uptake and trafficking to adipose tissue: implications for inflammation and obesity. Obes Rev 17: 297–312
31. Hersoug LG, Moller P, Loft S (2018) Role of microbiota-derived lipopolysaccharide in adipose tissue inflammation, adipocyte size and pyroptosis during obesity. Nutr Res Rev 31:153–163
32. Belizario JE, Faintuch J, Garay-Malpartida M (2018) Gut microbiome dysbiosis and immunometabolism: new frontiers for treatment of metabolic diseases. Mediators Inflamm 2018:2037838
33. Sunkara T, Rawla P, Ofosu A, Gaduputi V (2018) Fecal microbiota transplant: a new frontier in inflammatory bowel disease. J Inflamm Res 11:321–328
34. de Groot PF, Frissen MN, de Clercq NC, Nieuwdorp M (2017) Fecal microbiota transplantation in metabolic syndrome: history, present and future. Gut Microbes 8:253–267
35. Bouter KE, van Raalte DH, Groen AK, Nieuwdorp M (2017) Role of the gut microbiome in the pathogenesis of obesity and obesity-related metabolic dysfunction. Gastroenterology 152:1671–1678
36. Ridaura VK, Faith JJ, Rey FE et al (2013) Gut microbiota from twins discordant for obesity modulate metabolism in mice. Science 341:1241214
37. Lee P, Yacyshyn BR, Yacyshyn MB (2019) Gut microbiota and obesity: an opportunity to alter obesity through faecal microbiota transplant (FMT). Diabetes Obes Metab 21:479–490
38. Mokhtari Z, Gibson DL, Hekmatdoost A (2017) Nonalcoholic fatty liver disease, the gut microbiome, and diet. Adv Nutr 8:240–252

39. Milosevic I, Vujovic A, Barac A et al (2019) Gut-liver axis, gut microbiota, and its modulation in the management of liver diseases: a review of the literature. Int J Mol Sci 20:395
40. Trayhurn P (2013) Hypoxia and adipose tissue function and dysfunction in obesity. Physiol Rev 93:1–21
41. Trayhurn P (2019) Oxygen: a critical, but overlooked, nutrient. Front Nutr 6:10
42. Gaspar JM, Velloso LA (2018) Hypoxia inducible factor as a central regulator of metabolism - implications for the development of obesity. Front Neurosci 12:813
43. Gonzalez FJ, Xie C, Jiang C (2018) The role of hypoxia-inducible factors in metabolic diseases. Nat Rev Endocrinol 15:21–32
44. Jung UJ, Choi MS (2014) Obesity and its metabolic complications: the role of adipokines and the relationship between obesity, inflammation, insulin resistance, dyslipidemia and nonalcoholic fatty liver disease. Int J Mol Sci 15:6184–6223
45. Kayser B, Verges S (2013) Hypoxia, energy balance and obesity: from pathophysiological mechanisms to new treatment strategies. Obes Rev 14:579–592
46. Park HY, Jung WS, Kim J, Lim K (2019) Twelve weeks of exercise modality in hypoxia enhances health-related function in obese older Korean men: a randomized controlled trial. Geriatr Gerontol Int 19:311–316
47. Muller TD, Clemmensen C, Finan B et al (2018) Anti-obesity therapy: from rainbow pills to polyagonists. Pharmacol Rev 70:712–746
48. Brown L, Poudyal H, Panchal SK (2015) Functional foods as potential therapeutic options for metabolic syndrome. Obes Rev 16:914–941
49. Tsai YL, Lin TL, Chang CJ et al (2019) Probiotics, prebiotics and amelioration of diseases. J Biomed Sci 26:3
50. Lim J, Iyer A, Liu L et al (2013) Diet-induced obesity, adipose inflammation, and metabolic dysfunction correlating with PAR2 expression are attenuated by PAR2 antagonism. FASEB J 27:4757–4767
51. Lim J, Iyer A, Suen JY et al (2013) C5aR and C3aR antagonists each inhibit diet-induced obesity, metabolic dysfunction, and adipocyte and macrophage signaling. FASEB J 27: 822–831
52. Iyer A, Lim J, Poudyal H et al (2012) An inhibitor of phospholipase A_2 group IIA modulates adipocyte signaling and protects against diet-induced metabolic syndrome in rats. Diabetes 61:2320–2329
53. Ozen AE, Pons A, Tur JA (2012) Worldwide consumption of functional foods: a systematic review. Nutr Rev 70:472–481
54. John OD, Brown L, Panchal SK (2019) Chapter 3. *Garcinia* fruits: their potential to combat metabolic syndrome. In: Ullah M, Ahmad A (eds) Nutraceuticals and natural product derivatives: disease prevention and drug discovery. Wiley-Blackwell, Hoboken, NJ, USA, pp 39–80
55. Panchal SK, Brown L (2011) Rodent models for metabolic syndrome research. J Biomed Biotechnol 2011:351982
56. Kleinert M, Clemmensen C, Hofmann SM et al (2018) Animal models of obesity and diabetes mellitus. Nat Rev Endocrinol 14:140–162
57. Rios-Hoyo A, Gutierrez-Salmean G (2016) New dietary supplements for obesity: what we currently know. Curr Obes Rep 5:262–270
58. Zaynab M, Fatima M, Abbas S et al (2018) Role of secondary metabolites in plant defense against pathogens. Microb Pathog 124:198–202
59. Liang J, He J (2018) Protective role of anthocyanins in plants under low nitrogen stress. Biochem Biophys Res Commun 498:946–953
60. Steyn WJ, Wand SJE, Holcroft DM, Jacobs G (2002) Anthocyanins in vegetative tissues: a proposed unified function in photoprotection. New Phytol 155:349–361
61. Gould KS (2004) Nature's swiss army knife: the diverse protective roles of anthocyanins in leaves. J Biomed Biotechnol 2004:314–320

62. Riaz M, Zia-Ul-Haq M, Saad B (2016) The role of anthocyanins in obesity and diabetes. In: Riaz M, Saad B (eds) Anthocyanins and human health: biomolecular and therapeutic aspects, Zia Ul Haq, M. Springer International Publishing, Cham, pp 109–123
63. Azzini E, Giacometti J, Russo GL (2017) Antiobesity effects of anthocyanins in preclinical and clinical studies. Oxid Med Cell Longev 2017:2740364
64. Lee Y-M, Yoon Y, Yoon H et al (2017) Dietary anthocyanins against obesity and inflammation. Nutrients 9:1089
65. Naseri R, Farzaei F, Haratipour P et al (2018) Anthocyanins in the management of metabolic syndrome: a pharmacological and biopharmaceutical review. Front Pharmacol 9:1310
66. Xie L, Su H, Sun C et al (2018) Recent advances in understanding the anti-obesity activity of anthocyanins and their biosynthesis in microorganisms. Trends Food Sci Technol 72:13–24
67. Wu T, Gao Y, Guo X et al (2018) Blackberry and blueberry anthocyanin supplementation counteract high-fat-diet-induced obesity by alleviating oxidative stress and inflammation and accelerating energy expenditure. Oxid Med Cell Longev 2018:4051232
68. Bhaswant M, Shafie SR, Mathai ML et al (2017) Anthocyanins in chokeberry and purple maize attenuate diet-induced metabolic syndrome in rats. Nutrition 41:24–31
69. Kim N-H, Jegal J, Kim YN et al (2018) Antiobesity effect of fermented chokeberry extract in high-fat diet-induced obese mice. J Med Food 21:1113–1119
70. Bhaswant M, Fanning K, Netzel M et al (2015) Cyanidin 3-glucoside improves diet-induced metabolic syndrome in rats. Pharmacol Res 102:208–217
71. John OD, Mouatt P, Prasadam I et al (2019) The edible native Australian fruit, Davidson's plum (*Davidsonia pruriens*), reduces symptoms in rats with diet-induced metabolic syndrome. J Funct Foods 56:204–215
72. Daveri E, Cremonini E, Mastaloudis A et al (2018) Cyanidin and delphinidin modulate inflammation and altered redox signaling improving insulin resistance in high fat-fed mice. Redox Biol 18:16–24
73. Furuuchi R, Shimizu I, Yoshida Y et al (2018) Boysenberry polyphenol inhibits endothelial dysfunction and improves vascular health. PLoS ONE 13:e0202051
74. Parra-Vargas M, Sandoval-Rodriguez A, Rodriguez-Echevarria R et al (2018) Delphinidin ameliorates hepatic triglyceride accumulation in human HepG2 cells, but not in diet-induced obese mice. Nutrients 10:1060
75. Lee S, Keirsey KI, Kirkland R et al (2018) Blueberry supplementation influences the gut microbiota, inflammation, and insulin resistance in high-fat-diet-fed rats. J Nutr 148: 209–219
76. Jiang X, Li X, Zhu C et al (2018) The target cells of anthocyanins in metabolic syndrome. Crit Rev Food Sci Nutr 59:921-946
77. Igwe EO, Charlton KE, Roodenrys S et al (2017) Anthocyanin-rich plum juice reduces ambulatory blood pressure but not acute cognitive function in younger and older adults: a pilot crossover dose-timing study. Nutr Res 47:28–43
78. Bhaswant M, Brown L, Mathai ML (2019) Queen Garnet plum juice and raspberry cordial in mildly hypertensive obese or overweight subjects: a randomized, double-blind study. J Funct Foods 56:119–126
79. Hester SN, Mastaloudis A, Gray R et al (2018) Efficacy of an anthocyanin and prebiotic blend on intestinal environment in obese male and female subjects. J Nutr Metab 2018:7497260
80. Solverson PM, Rumpler WV, Leger JL et al (2018) Blackberry feeding increases fat oxidation and improves insulin sensitivity in overweight and obese males. Nutrients 10:1048
81. Stull AJ, Cash KC, Johnson WD et al (2010) Bioactives in blueberries improve insulin sensitivity in obese, insulin-resistant men and women. J Nutr 140:1764–1768
82. Cardile V, Graziano AC, Venditti A (2015) Clinical evaluation of Moro (*Citrus sinensis* (L.) Osbeck) orange juice supplementation for the weight management. Nat Prod Res 29: 2256–2260

83. Saeed M, Naveed M, BiBi J et al (2019) Potential nutraceutical and food additive properties and risks of coffee: a comprehensive overview. Crit Rev Food Sci Nutr. https://doi.org/10.1080/10408398.2018.1489368
84. Nieber K (2017) The impact of coffee on health. Planta Med 83:1256–1263
85. Grosso G, Godos J, Galvano F, Giovannucci EL (2017) Coffee, caffeine, and health outcomes: an umbrella review. Annu Rev Nutr 37:131–156
86. El-Sohemy A (2019) Coffee and health: what we still don't know. Am J Clin Nutr 109: 489–490
87. Izadi V, Larijani B, Azadbakht L (2018) Is coffee and green tea consumption related to serum levels of adiponectin and leptin? Int J Prev Med 9:106
88. Panchal SK, Poudyal H, Waanders J, Brown L (2012) Coffee extract attenuates changes in cardiovascular and hepatic structure and function without decreasing obesity in high-carbohydrate, high-fat diet-fed male rats. J Nutr 142:690–697
89. Panchal SK, Wong WY, Kauter K et al (2012) Caffeine attenuates metabolic syndrome in diet-induced obese rats. Nutrition 28:1055–1062
90. Nakazawa Y, Ishimori N, Oguchi J et al (2019) Coffee brew intake can prevent the reduction of lens glutathione and ascorbic acid levels in HFD-fed animals. Exp Ther Med 17:1420–1425
91. Nishitsuji K, Watanabe S, Xiao J et al (2018) Effect of coffee or coffee components on gut microbiome and short-chain fatty acids in a mouse model of metabolic syndrome. Sci Rep 8:16173
92. Choi BK, Park SB, Lee DR et al (2016) Green coffee bean extract improves obesity by decreasing body fat in high-fat diet-induced obese mice. Asian Pac J Trop Med 9:635–643
93. Song SJ, Choi S, Park T (2014) Decaffeinated green coffee bean extract attenuates diet induced obesity and insulin resistance in mice. Evid Based Complement Alternat Med 2014:718379
94. Bhandarkar NS, Brown L, Panchal SK (2019) Chlorogenic acid attenuates high-carbohydrate, high-fat diet-induced cardiovascular, liver, and metabolic changes in rats. Nutr Res 62:78–88
95. Ma Y, Gao M, Liu D (2015) Chlorogenic acid improves high fat diet induced hepatic steatosis and insulin resistance in mice. Pharm Res 32:1200–1209
96. Pan MH, Tung YC, Yang G et al (2016) Molecular mechanisms of the anti-obesity effect of bioactive compounds in tea and coffee. Food Funct 7:4481–4491
97. Kim Y, Je Y (2018) Moderate coffee consumption is inversely associated with the metabolic syndrome in the Korean adult population. Br J Nutr 120:1279–1287
98. Larsen SC, Mikkelsen ML, Frederiksen P, Heitmann BL (2018) Habitual coffee consumption and changes in measures of adiposity: a comprehensive study of longitudinal associations. Int J Obes (Lond) 42:880–886
99. Roshan H, Nikpayam O, Sedaghat M, Sohrab G (2018) Effects of green coffee extract supplementation on anthropometric indices, glycaemic control, blood pressure, lipid profile, insulin resistance and appetite in patients with the metabolic syndrome: a randomised clinical trial. Br J Nutr 119:250–258
100. Haidari F, Samadi M, Mohammadshahi M et al (2017) Energy restriction combined with green coffee bean extract affects serum adipocytokines and the body composition in obese women. Asia Pac J Clin Nutr 26:1048–1054
101. Yamagata K (2018) Do coffee polyphenols have a preventive action on metabolic syndrome associated endothelial dysfunctions? An assessment of the current evidence. Antioxidants (Basel) 7: 26
102. O'Keefe JH, DiNicolantonio JJ, Lavie CJ (2018) Coffee for cardioprotection and longevity. Prog Cardiovasc Dis 61:38–42
103. Khoo HE, Azlan A, Kong KW, Ismail A (2016) Phytochemicals and medicinal properties of indigenous tropical fruits with potential for commercial development. Evid Based Complement Alternat Med 2016:7591951

104. Pierson JT, Dietzgen RG, Shaw PN et al (2012) Major Australian tropical fruits biodiversity: bioactive compounds and their bioactivities. Mol Nutr Food Res 56:357–387
105. Sweeney PW (2008) Phylogeny and floral diversity in the genus *Garcinia* (Clusiaceae) and relatives. Int J Plant Sci 169:1288–1303
106. Deachathai S, Mahabusarakam W, Phongpaichit S, Taylor WC (2005) Phenolic compounds from the fruit of *Garcinia dulcis*. Phytochemistry 66:2368–2375
107. Ovalle-Magallanes B, Eugenio-Perez D, Pedraza-Chaverri J (2017) Medicinal properties of mangosteen (*Garcinia mangostana* L.): a comprehensive update. Food Chem Toxicol 109:102–122
108. Tousian Shandiz H, Razavi BM, Hosseinzadeh H (2017) Review of *Garcinia mangostana* and its xanthones in metabolic syndrome and related complications. Phytother Res 31:1173–1182
109. Lee PS, Teng CY, Kalyanam N et al (2019) Garcinol reduces obesity in high-fat-diet-fed mice by modulating gut microbiota composition. Mol Nutr Food Res 63:e1800390
110. Tsai SY, Chung PC, Owaga EE et al (2016) Alpha-mangostin from mangosteen (*Garcinia mangostana* Linn.) pericarp extract reduces high fat-diet induced hepatic steatosis in rats by regulating mitochondria function and apoptosis. Nutr Metab (Lond) 13:88
111. Chae HS, Kim YM, Bae JK et al (2016) Mangosteen extract attenuates the metabolic disorders of high-fat-fed mice by activating AMPK. J Med Food 19:148–154
112. Sripradha R, Magadi SG (2015) Efficacy of *Garcinia cambogia* on body weight, inflammation and glucose tolerance in high fat fed male wistar rats. J Clin Diagn Res 9: BF01–BF04
113. Watanabe M, Gangitano E, Francomano D et al (2018) Mangosteen extract shows a potent insulin sensitizing effect in obese female patients: a prospective randomized controlled pilot study. Nutrients 10:586
114. Haber SL, Awwad O, Phillips A et al (2018) *Garcinia cambogia* for weight loss. Am J Health Syst Pharm 75:17–22
115. A Aziz AN, Mhd Jalil MA (2019) Bioactive compounds, nutritional value, and potential health benefits of indigenous durian (*Durio zibethinus* Murr.): a review. Foods 8:96
116. Xu J, Liu T, Li Y et al (2019) Jamun (*Eugenia jambolana* Lam.) fruit extract prevents obesity by modulating the gut microbiome in high-fat-diet-fed mice. Mol Nutr Food Res 63: e1801307
117. Ulla A, Alam MA, Sikder B et al (2017) Supplementation of *Syzygium cumini* seed powder prevented obesity, glucose intolerance, hyperlipidemia and oxidative stress in high carbohydrate high fat diet induced obese rats. BMC Complement Altern Med 17:289
118. Food and Agriculture Organization of the United Nations (2018). http://www.fao.org/save-food/resources/keyfindings/en/
119. Lai WT, Khong NMH, Lim SS et al (2017) A review: modified agricultural by-products for the development and fortification of food products and nutraceuticals. Trends Food Sci Technol 59:148–160
120. Frenkel VS, Cummings GA, Maillacheruvu KY, Tang WZ (2017) Food-processing wastes. Water Environ Res 89:1360–1383
121. John OD, Wanyonyi S, Mouatt P et al (2018) Achacha (*Garcinia humilis*) rind improves cardiovascular function in rats with diet-induced metabolic syndrome. Nutrients 10:1425
122. Asyifah MR, Lu K, Ting HL, Zhang D (2014) Hidden potential of tropical fruit waste components as a useful source of remedy for obesity. J Agric Food Chem 62:3505–3516
123. Cheok CY, Mohd Adzahan N, Abdul Rahman R et al (2018) Current trends of tropical fruit waste utilization. Crit Rev Food Sci Nutr 58:335–361
124. Mulvihill EE, Burke AC, Huff MW (2016) Citrus flavonoids as regulators of lipoprotein metabolism and atherosclerosis. Annu Rev Nutr 36:275–299
125. Mahato N, Sharma K, Sinha M, Cho MH (2018) Citrus waste derived nutra-/pharmaceuticals for health benefits: current trends and future perspectives. J Funct Foods 40:307–316
126. Sharma K, Mahato N, Cho MH, Lee YR (2017) Converting citrus wastes into value-added products: economic and environmentally friendly approaches. Nutrition 34:29–46

127. Esparza-Martinez FJ, Miranda-Lopez R, Mata-Sanchez SM, Guzman-Maldonado SH (2016) Extractable and non-extractable phenolics and antioxidant capacity of mandarin waste dried at different temperatures. Plant Foods Hum Nutr 71:294–300
128. Khanal RC, Howard LR, Rogers TJ et al (2011) Effect of feeding grape pomace on selected metabolic parameters associated with high fructose feeding in growing Sprague-Dawley rats. J Med Food 14:1562–1569
129. Urquiaga I, D'Acuña S, Pérez D et al (2015) Wine grape pomace flour improves blood pressure, fasting glucose and protein damage in humans: a randomized controlled trial. Biol Res 48:49
130. Woerdeman J, van Poelgeest E, Ket JCF et al (2017) Do grape polyphenols improve metabolic syndrome components? A systematic review. Eur J Clin Nutr 71:1381–1392
131. Vu HT, Scarlett CJ, Vuong QV (2018) Phenolic compounds within banana peel and their potential uses: a review. J Funct Foods 40:238–248
132. Muhlack RA, Potumarthi R, Jeffery DW (2018) Sustainable wineries through waste valorisation: a review of grape marc utilisation for value-added products. Waste Manag 72:99–118
133. Elkahoui S, Bartley GE, Yokoyama WH, Friedman M (2018) Dietary supplementation of potato peel powders prepared from conventional and organic russet and non-organic gold and red potatoes reduces weight gain in mice on a high-fat diet. J Agric Food Chem 66:6064–6072
134. Edrisi F, Salehi M, Ahmadi A et al (2018) Effects of supplementation with rice husk powder and rice bran on inflammatory factors in overweight and obese adults following an energy-restricted diet: a randomized controlled trial. Eur J Nutr 57:833–843
135. Nair S, Gagnon J, Pelletier C et al (2017) Shrimp oil extracted from the shrimp processing waste reduces the development of insulin resistance and metabolic phenotypes in diet-induced obese rats. Appl Physiol Nutr Metab 42:841–849
136. Brown L, Caligiuri SPB, Brown D, Pierce GN (2018) Clinical trials using functional foods provide unique challenges. J Funct Foods 45:233–238

Chapter 18
Attenuation of Obesity-Associated Oxidative Stress by *Cucurbita maxima* Seed Oil in High Fat Diet-Induced Obese Rats

A. Kalaivani, S. Vadivukkarasi, V. V. Sathibabu Uddandrao and G. Saravanan

Abstract Obesity is a major risk factor for associated deleterious pathologies such as diabetes type 2, liver and heart disease. Many pieces of evidence of accumulated oxidative stress induced by obesity over the past two decades and play an important part in the growth of metabolic disorders. Based on this, the current study was designed to investigate the effect of *Cucurbita maxima* seed oil (CSO) on oxidative stress associated with high fat diet (HFD)-induced obesity in rats. The obese rats were supplemented with CSO (100 mg/kg body weight) for 30 days and determined the changes on hyperglycemic markers, tissue lipid profiles and oxidative stress markers such as TBARS, lipid peroxidation, protein carbonyls groups, glutathione, glutathione peroxidase, glutathione reductase, glutathione S-transferase, superoxide dismutase and catalase. Rats fed with HFD shown the significant increase of hyperglycemia and tissue lipid profiles and concomitantly decrease in the antioxidant enzymes and increased lipid peroxidation, protein carbonyl groups and TBARS. However, treatment with CSO for 30 days exhibited the noteworthy restoration of altered hyperglycaemic markers and lipid profiles. On the other hand, CSO increased the levels of antioxidant markers and concomitant decrease in the oxidative stress. In conclusion, these results suggested that CSO has the antioxidant potential and it might be used as therapeutic agent against oxidative stress related diseases such as hyperglycaemia and hyperlipidemia.

Keywords Obesity · Oxidative stress · Antioxidants · Plant extracted oils · Natural products

A. Kalaivani, S. Vadivukkarasi, V. V. Sathibabu Uddandrao are equally contributed to this work.

A. Kalaivani · S. Vadivukkarasi · V. V. Sathibabu Uddandrao · G. Saravanan (✉)
Department of Biochemistry, Centre for Biological Sciences, K. S. Rangasamy College of Arts and Science (Autonomous), Namakkal District, Tiruchengode 637215, Tamilnadu, India
e-mail: sarabioc@gmail.com

P. S. Tappia et al. (eds.), *Pathophysiology of Obesity-Induced Health Complications*, Advances in Biochemistry in Health and Disease 19,
https://doi.org/10.1007/978-3-030-35358-2_18

Introduction

The pervasiveness of obesity over the previous years has been in consistent movement driving the World Health Organization to think about it as a pandemic pathology. Obesity is characterized as an unnecessary aggregation of muscle to fat ratio mass to the degree that person's wellbeing will be contrarily influenced. To be sure, obesity is considered as a top hazard factor to create harmful related pathologies such as type 2 diabetes, liver, and coronary heart illnesses [1]. Foundational oxidative stress is a piece of the various organic changes revealed during constant obesity [2]. Evidences regarding obesity-induced oxidative stress are derived from several clinical studies, which have established correlations of biomarkers, or end-products of free radicals-mediated oxidative stress (lipid peroxidation or protein carbonylation products) with body mass index (BMI) [3]. In contrast, an inverse relationship exists between body fat, visceral obesity, and antioxidant defence markers in obese individuals [4]. The hypothesis that oxidative stress is causative in the development of metabolic disorders, especially insulin-resistant state, has been supported by different studies where treatments reducing ROS production improve insulin sensitivity, hyperlipidemia, and hepatic steatosis [1, 5]. Furthermore, oxidative stress-associated obesity has also been shown to alter the function of many cell types or tissues (including vascular endothelial cells, myocytes, or pancreatic-β-cells) leading to consider oxidative stress as a contributor in obesity-related metabolic diseases [6].

Oxidative stress is a disequilibrium condition in which the cellular redox balance is shifted towards a more oxidizing status that may trigger adaptation of cellular functions [7]. Depending on the antioxidant level of different cell types, the concentration of reactive oxygen species (ROS) achieved, and the duration of the exposure, oxidative stress may trigger beneficial responses under mild conditions and potentially harmful ones beneath severe situations, as a typically hormetic phenomenon [8]. Oxidative stress shows up as a noteworthy contributor in the improvement of numerous metabolic intricacies related to obesity. For all intents and purposes, numerous helpful procedures utilized right now to treat obesity related metabolic issue can possibly diminish oxidative stress, which may, in any event in part, contribute in their useful impacts [9, 10]. In a previous report, we have highlighted the beneficial effects of *Cucurbita maxima* seed oil (CSO) against high fat diet (HFD)-induced obesity in rats via attenuation of lipid metabolism and its related marker genes [11, 12]. However, there was no scientific report on effect of CSO on obesity-associated oxidative stress so far. In view of this, the current study was designed to evaluate the influence of CSO against oxidative stress in HFD-induced obesity in rats.

Materials and Methods

Preparation of Cucurbita maxima *Seed Oil*

Cucurbita maxima plant was authenticated by Dr. P. Ponmurugan, Associate Professor, Department of Botany, Bharathiar University, Coimbatore, Tamilnadu, India and the specimen of the plant (Voucher No: 1632/2015) was deposited at herbarium in the Department of Biotechnology, K. S. Rangasamy College of Arts and Science, Tiruchengode, Tamilnadu. *C. maxima* seeds were washed, cleaned and dried at 50 °C for 12 h in an oven, and then crushed into powder in a grinder to a size range of 0.55–1.0 mm. The resulted powder was kept in a vacuum dryer until use. Pumpkin seed powder was mixed with water (1:6, m/V) at 100 °C and oil was extracted by using a Clevenger apparatus and extraction was repeated for 6 h and the resulting extracted oil was used for the study. All chemicals used in this study were of analytical grade.

Animals

Male Wistar rats were obtained from K.S.R. College of Technology, Tiruchengode, Namakkal District, and Tamilnadu, India. Experimental animals were kept up under standard laboratory conditions (temperature; 22 ± 2 °C: moistness; 40–60%), and permitted food and water ad libitum. Rats, at first, weighing 150–180 g were separated into four groups of six each (n = 6). All procedures involving laboratory animals were in accordance with the institutional animal ethical committee of K.S.R. College of Technology, Tiruchengode, Namakkal District, and Tamilnadu, India (Approval No: KSRCT/BT/IAEC/2015/06).

High Fat Diet Formula

High fat diet was obtained from National centre for laboratory animal sciences, National institute of nutrition (NIN-ICMR), Hyderabad. High fat diet was composed by Corn starch-15%, Sugar-27.5%, Lard oil-17.6%, Vitamin mixture-1%, Mineral mixture-3.5%, Casein-20%, Cellulose powder-5%, Corn oil-9.9%, Choline bitartrate-0.2%.

Experimental Design

In the experiment, a total of 24 animals (18 HFD, 6 normal rats) were used for this study. The animals were grouped into 4 with 6 animals in each group.

Group 1: Normal control rats
Group 2: HFD control rats
Group 3: HFD rats treated with seed oil (100 mg/kg body weight) for 30 days [11]
Group 4: HFD rats treated orally with Orlistat (10 mg/kg body weight) for 30 days (suspended in 0.5% carboxy methylcellulose).

Measurement of Body Weight

The body weight of each rat was measured. At the end of the experiment, blood was collected from overnight fasted animals under inhalation of anaesthesia by retro-orbital puncture method. Blood was collected in anticoagulant coated vials and permitted for 15 min at room temperature. Plasma was separated by centrifugation at 2500 rpm for 15 min.

Biochemical Analysis

Plasma glucose was estimated using a kit (Cat.no. 1060-500) from Stanbio Laboratory (Boerne, TX, USA). The plasma level of insulin was determined using kits from Bio-Merieux (Lyon, France). Insulin resistance (IR) was measured using the homeostasis model assessment.

Oral Glucose Tolerance Test (OGTT)

OGTT was conducted at the end of the experimentation; after overnight fasting, glucose was administered orogastrically at a dose of 200 mg/kg body weight and blood samples were collected from supra orbital sinus at 0, 30, 60, 90 and 120 min and glucose level was measured.

Estimation of Tissue Lipids

Tissue lipids were collected from experimental animals by the method of Folch et al. [13] using a chloroform-methanol mixture (2:1, v/v). The liver tissues were rinsed with ice cold physiological saline, dried, homogenized in cold chloroform-methanol (2:1, v/v) and contents were extracted after 24 h. The extraction was repeated four times. The combined filtrate was washed with 0.7% KCl and the aqueous layer was discarded. The organic layer was made up to a known volume with chloroform and used for tissue lipid analysis. Total cholesterol (TC), free fatty acids (FFA), phospholipids (PL) and triacylglyceride (TG) were analyzed according to enzymatic colorimetric methods using kits (Nicholas Piramal India Ltd., Mumbai, India).

Liver Antioxidants Analysis

Blood samples were collected after sacrificing the rats. Then the liver was excised, rinsed in ice cold normal saline followed by ice-cold 10% KCl solution, blotted, dried and weighed. A 10% w/v homogenate was prepared in ice-cold KCl solution and centrifuged at 2000 rpm for 10 min at 4 °C. The supernatants thus obtained were used for the estimation of thiobarbituric acid substances (TBARS) [14], assay of GSH [15], SOD [16], CAT [17] and GpX by Paglia and valentine [18].

Results

Body Weights

Figure 18.1 showed the effects of CSO on body weight in control and HFD-induced obese rats. Supplementation of HFD induced a momentous ($p < 0.05$) rise in body weight, compared to the normal control group. Oral administration with CSO (100 mg/kg body weight) or Orlistat for 30 days considerably ($p < 0.05$) compacted the increase in body weight when compared to the HFD control group.

Plasma Glucose, Plasma Insulin and IR

Figure 18.2 depicted the level of plasma glucose, plasma insulin and IR in control and experimental obese rats. The levels of blood glucose (Fig. 18.2a) were increased in obese rats and significant decline was observed in the plasma insulin levels (Fig. 18.2b), and there were momentous increases in IR (Fig. 18.2c) in the

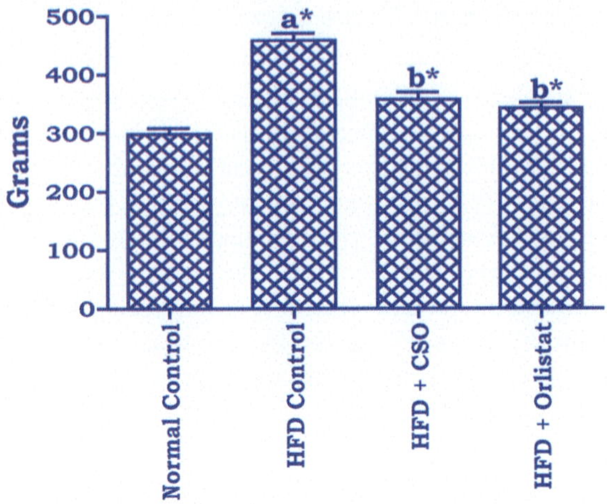

Fig. 18.1 Effect of CSO on body weights of the control and HFD-induced obese rats. Values are expressed as mean ± SD, n = 6; values are statistically significant at $p < 0.05$; [a]Significantly different from control; [b]Significantly different from HFD control

HFD induced obese rats compared with the control rats. Administration of CSO or Orlistat tended to normalize the levels of plasma insulin and IR.

OGTT

The results of OGTT of the control and treated animals were presented in Fig. 18.3. The blood glucose level in the control rats was eminent to a highest value at 60 min after glucose load and declined to near basal levels at 120 min, whereas, in HFD-induced obese rats, the peak increase in blood glucose level was noticed even after 60 min and remained high over the next 60 min. Oral administration with CSO or orlistat to obese rats demonstrated a noteworthy diminution in blood glucose level at 60 min when compared with HFD control rats.

Tissue Lipid Profiles

Table 18.1 described the concentrations of TC, FFAs, TGs and PLs in liver of control and HFD induced obese rats. The concentrations of TC, TG, FFAs and PLs were found to be increased in liver of HFD obese rats as compared to the normal rats. Supplementation with CSO or orlistat considerably ($p < 0.05$) reduced the concentrations of TC, TG, FFAs and PLs in liver in obese rats.

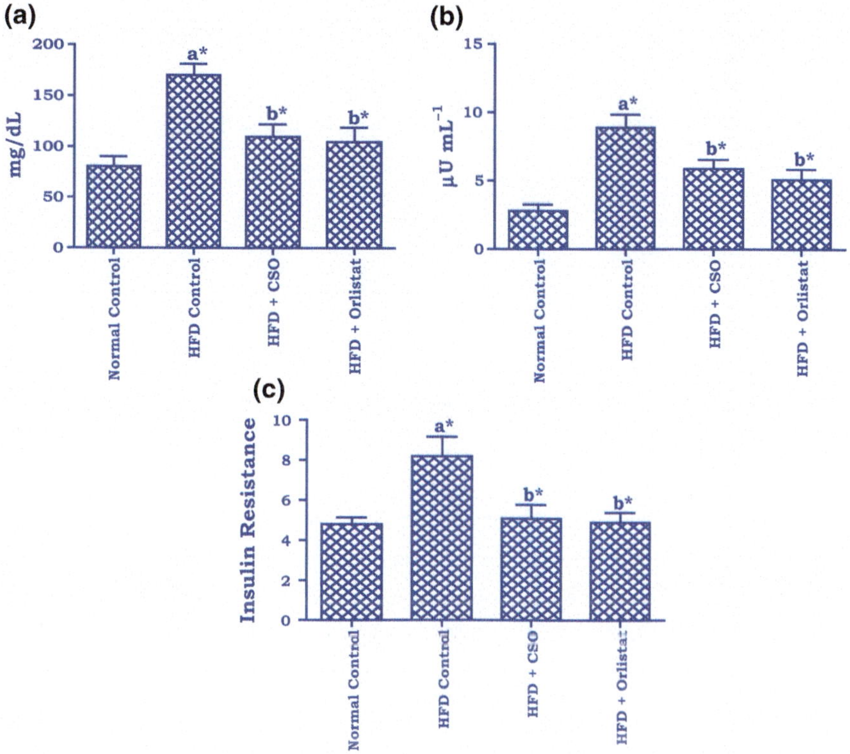

Fig. 18.2 **a** Effect of CSO on glucose levels of the control and HFD-induced obese rats. Values are expressed as mean ± SD, n = 6; values are statistically significant at $p < 0.05$; [a]Significantly different from control; [b]Significantly different from HFD control. **b** Effect of CSO on insulin levels of the control and HFD-induced obese rats. Values are expressed as mean ± SD, n = 6; values are statistically significant at $p < 0.05$; [a]Significantly different from control; [b]Significantly different from HFD control. **c** Effect of CSO on insulin resistance of the control and HFD-induced obese rats. Values are expressed as mean ± SD, n = 6; values are statistically significant at $p < 0.05$; [a]Significantly different from control; [b]Significantly different from HFD control

Effect of CSO on Obesity-Associated Oxidative Stress

The Table 18.2 expressed the hepatic oxidative stress markers in control and experimental obese rats. The obese control group showed the significant increase of the lipid peroxidation, protein carbonyl groups and TBARS. The treatment with CSO for 45 days to the obese rats concomitantly decreased the elevated oxidative stress markers by which revealed that CSO has the strong antioxidant potential.

Figure 18.4 exemplifies the effect of CSO on status of the antioxidants in obese and control rats. The study revealed that the there was significant decrease in the levels of GSH, GR, GPx, GST, SOD, and CAT in obese control group. On the other hand, administration of CSO or orlistat to the obese rats significantly increased the

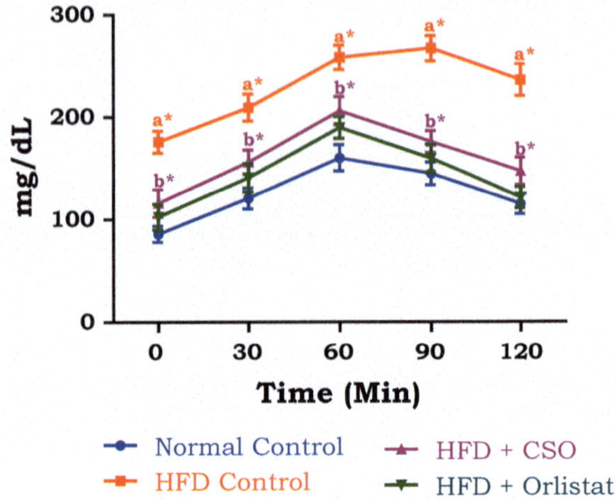

Fig. 18.3 Effect of CSO on glucose tolerance of the control and HFD-induced obese rats. Values are expressed as mean ± SD, n = 6; values are statistically significant at $p < 0.05$; [a]Significantly different from control; [b]Significantly different from HFD control

Table 18.1 Effect of CSO on tissue lipid profile in control and experimental rats

	Normal control	HFD control	HFD + CSO	HFD + orlistat
PL (mg/g tissue)	99.12 ± 6.09	175.99 ± 7.63	123.37 ± 1.36	115.99 ± 6.07
FFA (mg/g tissue)	62.21 ± 4.99[a,*]	164.26 ± 7.27[a,*]	110.16 ± 9.90[a,*]	95.80 ± 7.57[a,*]
TG (mg/g tissue)	89.43 ± 5.48[b,*]	205.05 ± 9.37[b,*]	122.58 ± 7.87[b,*]	111.12 ± 9.35[b,*]
TC (mg/g tissue)	79.20 ± 5.46[b,*]	186.29 ± 9.07[b,*]	115.21 ± 7.37[b,*]	100.38 ± 7.37[b,*]

Values are expressed as mean ± SD for 6 animals in each group
Values are statistically significant at $*p < 0.05$
[a]Significantly different from control
[b]Significantly different from HFD control

Table 18.2 Effect of CSO on lipid peroxidation, protein carbonyls groups and TBARS in control and experimental rats

	Normal control	HFD control	HFD + CSO	HFD + Orlistat
Lipid peroxidation	25.88 ± 1.02	68.89 ± 2.57[a,*]	39.55 ± 2.73[b,*]	38.10 ± 3.02[b,*]
Protein carbonyls groups	1.99 ± 0.54	3.2 ± 0.44[a,*]	2.18 ± 0.56[b,*]	2.07 ± 0.43[b,*]
TBARS	3.1 ± 1.09	6.2 ± 1.43[a,*]	4.6 ± 1.39[b,*]	4.9 ± 1.55[b,*]

Lipid peroxidation nmol of MDA formed/min/mg protein, *Protein carbonyls groups* µmol/mg protein, *TBARS* mM/100 g tissue
Values are mean ± SD, n = 6, Values are statistically significant at $p < 0.05$
[a]Significantly different from control
[b]Significantly different from HFD control

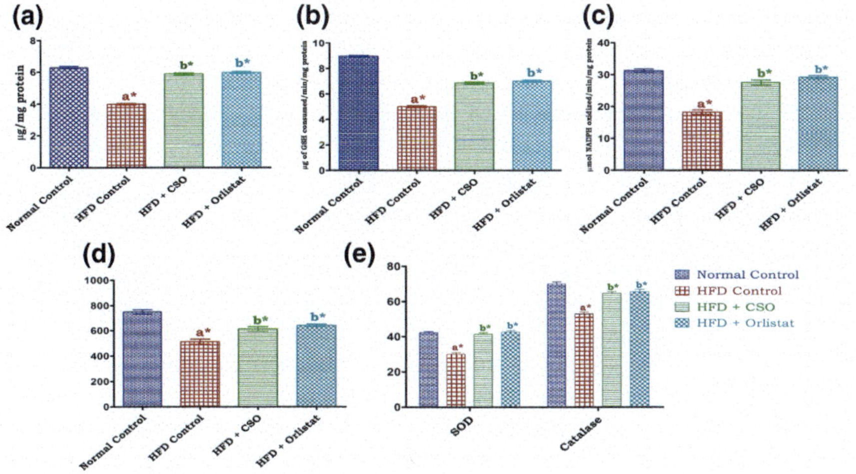

Fig. 18.4 Effect of CSO on antioxidant status **a** GSH, **b** GPx, **c** GR, **d** GST, **e** SOD and CAT in obese and control rats, values are mean ± SD, n = 6; values are statistically significant at $p < 0.05$; [a]Significantly different from control; [b]Significantly different from HFD control. GST: µmol of GSH-CDNB conjugate formed/min/mg protein; SOD: U/mg protein; CAT: mmol H_2O_2 consumed/min/mg protein

levels of GSH (Fig. 18.4a), GPx (Fig. 18.4b), GR (Fig. 18.4c), GST (Fig. 18.4d), SOD and CAT (Fig. 18.4e) in obese rats.

Discussion

Diet-induced obesity (DIO) in animals has several characteristics familiar with human obesity and signs of DIO are body weight composition, IR, dyslipidemia and impaired glucose tolerance [19]. In this study, HFD treatment to animals inveterate the nature of obesity as evidenced by escalating body weight, hyperglycemic condition and enhanced IR. This is in line with earlier reports [1, 19]. Under physiological circumstances, insulin inhibits lipolysis and enhances lipogenesis [20] principal to improved circulating levels of glucose and lipids resulting in impaired glucose enthused insulin secretion [21]. Oral administration of CSO considerably lowered plasma glucose, insulin level and IR in CSO treated rats compared with HFD control rats. CSO encourages insulin sensitivity lowering IR and reduces body weight gain and glucose level in obese animals possibly by reducing free fatty acids or regulating cell energy metabolism [22].

The obese rats displayed irregularities in tissue that mimic the instigation of a group of metabolic risks encircling dyslipidaemia. In the present study, HFD supplementation significantly increased tissue levels of TG, TC, PLs, and FFAs, in rats. Our findings are in line with Sathibabu et al. [1], Saravanan et al. [23] and Naidu et al. [19] who reported that obesity adversely affects tissue lipids, especially by increasing TC and decreasing the levels of HDL-C. During HFD-fed state, rise of chylomicrons synthesis and absorption is a familiar factor, which leads to increase in TGs, TC level and endogenous VLDL production. Undeniable substantiation from number of clinical studies on a large aggregate of patients, has established an elevated level of TG, as an independent risk factor for CVDs [24, 25].

In the present study, we assessed antioxidant enzymes activity and TBARS level in liver. Intracellular antioxidant enzymes and TBARS are the determining factors of oxidative damage at cellular level. HFD induced hyperglycemia observed in this study might be an imperative aspect to increase lipid peroxidation and causing the reduction of the antioxidant defense status and indicating the development of oxidative stress in HFD-fed rats which are interconnected with other studies [26]. During HFD induced obesity, enzymatic antioxidants level was found to be inhibited. SOD, CAT and GPx activities were reported to be lower as a result of increased production of free radicals [27]. A HFD can aggravate oxidative stress [28]. It induced IR accompanied by increased hepatic oxidative stress, which is characterized by reduction in the antioxidant enzyme activities and glutathione levels that correlate with the increase in LPO and PC levels. This may probably contribute to the progression of IR. Decreased hepatic GSH levels, decreased activities of antioxidant enzymes and increased LPO intermediates in obese group clearly indicate the development of oxidative stress in these animals. Previous studies strongly suggest that oxidative stress induced by HFD enhances lipid peroxide and diminishes antioxidant enzyme activity [29].

GSH represents the first line of defence against free radicals in the liver and is also responsible for the maintenance of protein thiols and acts as a substrate for GPx and GST [30]. GPx activity is considered to symbolize the initial protective response required for adjusting the H_2O_2 concentration under physiological condition as well as after oxidative insult. Furthermore, it has been postulated that high intake of dietary fat enhances ROS over production. ROS themselves can reduce the activity of antioxidant enzymes such as CAT and GPx [31]. The depletion in the activities of GST and GPx may result in the involvement of deleterious oxidative changes during accumulation of toxic products and the decreased activity of SOD in obese rats may be due to enhanced protein glycation by HFD. A significant decrease in the level of GSH in obese rats could be the result of decreased synthesis or increased degradation/utilization of GSH by increased oxidative stress and decreased regeneration as evident from the lower activities of GR enzyme. The antioxidant potential of CSO against HFD-induced oxidative stress are evident with lower levels of plasma glucose, insulin [11], TBARS, higher GSH levels, and increased activities of antioxidant enzymes were seen in CSO treated obese rats when compared with non treated obese rats. The in vitro antioxidant potential of CSO was well documented [32]. The reports indicate that antioxidants can modify cholesterol absorption and increase antioxidant status [33]. Several studies, Liou et al. [34] have shown that hyperlipidemia reduces the strength of the antioxidative defense system. Mehta et al. [35] indicated that HFD leads to liver injury and IR through oxidative stress.

Conclusion

In conclusion, we hypothesize that the possible explanation for reducing obesity following the consumption of CSO can be a useful add-on therapy to curtail IR, oxidative stress, and over weight in HFD-induced rats.

References

1. Uddandrao VVS, Rameshreddy P, Brahmanaidu P et al (2019) Antiobesity efficacy of asiatic acid: down-regulation of adipogenic and inflammatory processes in high fat diet induced obese rats. Arch Physiol Biochem. https://doi.org/10.1080/13813455.2018.1555668
2. Roberts CK, Sindhu KK (2009) Oxidative stress andmetabolic syndrome. Life Sci 84:705–712
3. Sankhla M, Sharma TK, Mathur K et al (2012) Relationship of oxidative stress with obesity and its role in obesity induced metabolic syndrome. Clin Lab 58:385–392
4. Chrysohoou C, Panagiotakos DB, Pitsavos C et al (2007) The implication of obesity on total antioxidant capacity in apparently healthy men and women: the ATTICA study. Nutr Metab Cardiovasc Dis 17:590–597
5. Meriga B, Parim B, Chunduri VR et al (2017) Antiobesity potential of Piperonal: promising modulation of body composition, lipid profiles and obesogenic marker expression in HFD-induced obese rats. Nutr Metab (Lond) 14:72
6. Rameshreddy P, Uddandrao VVS, Brahmanaidu P et al (2018) Obesity-alleviating potential of asiatic acid and its effects on ACC1, UCP2, and CPT1 mRNA expression in high fat diet induced obese Sprague-Dawley rats. Mol Cell Biochem 442:143–154
7. Sies H (2015) Oxidative stress: a concept in redox biology and medicine. Redox Biol 4:180–183
8. Videla LA (2010) Hormetic responses of thyroid hormone calorigenesis in the liver: association with oxidative stress. IUBMB Life 62:460–466
9. Le Lay S, Simard G, Martinez MC, Andriantsitohaina R (2014) Oxidative stress and metabolic pathologies: from an adipocentric point of view. Oxid Med Cell Longev 2014:908539
10. Sathibabu Uddandrao VV, Brahmanaidu P, Ravindarnaik R et al (2018) Restorative potentiality of *S*-allylcysteine against diabetic nephropathy through attenuation of oxidative stress and inflammation in streptozotocin-nicotinamide induced diabetic rats. Eur J Nutr. https://doi.org/10.1007/s00394-018-1795-x
11. Kalaivani A, Sathibabu Uddandrao VV, Brahmanaidu P et al (2018) Anti obese potential of *Cucurbita maxima* seeds oil: effect on lipid profile and histoarchitecture in high fat diet induced obese rats. Nat Prod Res 32:2950–2953
12. Kalaivani A, Uddandrao VVS, Parim B et al (2019) Reversal of high fat diet-induced obesity through modulating lipid metabolic enzymes and inflammatory markers expressions in rats. Arch Physiol Biochem 125:228–234
13. Folch C, Lees M, Solane SGH (1957) A simple method for isolation and purification of total lipids from animal tissues. J Biol Chem 226:497–509
14. Fraga CG, Leibovitz BE, Toppel AL (1988) Lipid peroxidation measured as TBARS in tissue slices. Characterization and comparison with homogenate and microsome. Free Radic Biol Med 4:155–161
15. Ellman GL (1959) Tissue sulphydryl. Arch Biochem Biophys 82:70–77
16. Kakkar R, Mantha SV, Radhi J et al (1998) Increased oxidative stress in rat liver and pancreas during progression of streptozotocin- induced diabetes. Clin Sci 94:623–632

17. Aebi H (1984) Catalase in vitro. Methods Enzymol 105:121–126
18. Paglia D, Valentine W (1967) Studies on the quantitative and qualitative characterization of erythrocyte glutathione peroxidase. J Lab Clin Med 70:158–169
19. Naidu PB, Sathibabu Uddandrao VV, Naik RR et al (2016) Ameliorative potential of gingerol: promising modulation of inflammatory factors and lipid marker enzymes expressions in HFD induced obesity in rats. Mol Cell Endocrinol 419:139–147
20. Saltiel AR, Kahn CR (2001) Insulin signalling and the regulation of glucose and lipid metabolism. Nature 414:799–806
21. Brahmanaidu P, Uddandrao VVS, Sasikumar V et al (2017) Reversal of endothelial dysfunction in aorta of streptozotocinnicotinamide-induced type-2 diabetic rats by S-Allylcysteine. Mol Cell Biochem 432:25–32
22. Boissonneault GA (2009) Obesity: the current treatment protocols. JAAPA 22:18–19
23. Saravanan G, Ponmurugan P, Deepa MA, Senthilkumar B (2014) Anti-obesity action of gingerol: effect on lipid profile, insulin, leptin, amylase and lipase in male obese rats induced by a high-fat diet. J Sci Food Agri 94:2972–2977
24. Sathibabu Uddandrao VV, Brahmanaidu P, Nivedha PR, Vadivukkarasi S et al (2018) Beneficial role of some natural products to attenuate the diabetic cardiomyopathy through Nrf2 pathway in cell culture and animal models. Cardiovasc Toxicol 18:199–205
25. Parim B, Sathibabu Uddandrao VV, Saravanan G (2019) Diabetic cardiomyopathy: molecular mechanisms, detrimental effects of conventional treatment, and beneficial effects of natural therapy. Heart Fail Rev 24:279–299
26. Sudhakara G, Mallaiah P, Sreenivasulu N et al (2014) Beneficial effects of hydro-alcoholic extract of *Caralluma fimbriata* against high-fat diet-induced insulin resistance and oxidative stress in Wistar male rats. J Physiol Biochem 70:311–320
27. BrahmaNaidu P, Nemani H, Meriga B et al (2014) Mitigating efficacy of piperine in the physiological derangements of high fat diet induced obesity in Sprague Dawley rats. Chem Biol Interact 221:42–51
28. Hirao K, Maruyama T, Ohno Y et al (2010) Association of increased reactive oxygen species production with abdominal obesity in type 2 diabetes. Obes Res Clin Pract 4:83–90
29. Lee SJ, Choi SK, Seo JS (2009) Grape skin improves antioxidant capacity in rats fed a high fat diet. Nutr Res Pract 3:279–385
30. Prakash J, Gupta SK, Kochupillai V et al (2001) Chemopreventive activity of *Withania somnifera* in experimentally induced fibrosarcoma tumours in Swiss albino mice. Phytother Res 15:240–244
31. Datta K, Sinha S, Chattopadhyay P (2000) Reactive oxygen species in health and diseases. Natl Med J India 13:304–310
32. Nawirska-Olszanska A, Kita A, Biesiada A et al (2013) Characteristics of antioxidant activity and composition of pumpkin seed oils in 12 cultivars. Food Chem 139:155–161
33. Nicolle C, Cardinault N, Aprikian O et al (2003) Effect of carrot intake on cholesterol metabolism and on antioxidant status in cholesterol-fed rat. Eur J Nutr 42:254–261
34. Liou W, Chang LY, Geuze HJ et al (1993) Distribution of Cu/Zn superoxide dismutase in rat liver. Free Radic Biol Med 14:201–207
35. Mehta K, Van Thiel DH, Shah N, Mobarhan S (2002) Nonalcoholic fatty liver disease: pathogenesis and the role of antioxidants. Nutr Rev 60:289–293

Chapter 19
Pathophysiology of Obesity-Related Non-communicable Chronic Diseases and Advancements in Preventive Strategies

Reena Badhwar, Ginpreet Kaur, Harvinder Popli, Deepika Yadav and Harpal S. Buttar

Abstract Obesity is characterized by an excessive accumulation of white adipose tissue (WAT) causing increases in body weight and Body Mass Index (BMI >30 kg/m^2). WAT stores energy and releases free fatty acids, but it is also a dynamic endocrine organ that triggers the production of a wide array of bioactive products such as adipokines, leptin, tumor necrosis factor (TNF-α), cytokines and interleukins(IL-1, IL-6, IL-11). Adipokines regulate appetite, insulin sensitivity, angiogenesis, blood pressure and immune response. Leptin reduces body weight. TNF-α, cytokines and interleukins are pro-inflammatory and activate atherosclerosis and oxidative stress, leading to the production of reactive oxygen species (ROS) and reactive nitrogen species (RNS). ROS and RNS adversely affects structure and function of cellular membrane and damages DNA. Obesity can be seen as the mother of numerous non-communicable chronic diseases, namely: diabetes mellitus, metabolic syndrome, dyslipidemia, several cardiovascular diseases, nephropathy, neurodegenerative disorders (dementia, Alzheimer's disease), osteoarthritis, sleep apnea, depression, and enhanced risk of breast and colon cancer, resulting in high morbidity and mortality among obese patients. The burden of healthcare for obesity-related diseases is escalating worldwide. This review highlights the incidences of obesity-related diseases and their preventive strategies through synthetic drugs and alternative therapies such as Ayurvedic remedies, herbal medicines, dietary interventions, nutraceuticals, lifestyle modifications, and

R. Badhwar · H. Popli · D. Yadav
Department of Pharmaceutics, Delhi Pharmaceutical Science and Research University, M.B Road, PushpVihar, Sec-03, New Delhi 110017, India

G. Kaur
Department of Pharmacology, SPP School of Pharmacy & Technology Management, SVKM's NMIMS, V.L Mehta Road, Vile Parle (W), Mumbai 400056, Maharashtra, India

H. S. Buttar (✉)
Department of Pathology and Laboratory Medicine, Faculty of Medicine, University of Ottawa, Ottawa, ON K1H 8M5, Canada
e-mail: hsbuttar@bell.net

P. S. Tappia et al. (eds.), *Pathophysiology of Obesity-Induced Health Complications*, Advances in Biochemistry in Health and Disease 19,
https://doi.org/10.1007/978-3-030-35358-2_19

psychotherapy. Currently nutraceuticals, probiotics, herbal remedies, and Mediterranean-type diets that are rich in fibers, fresh fruits and vegetables, olive oil, fish, poultry and dairy products are being promoted for the prevention and management of obesity-related chronic diseases and metabolic pathologies. Anti-obesity drugs assist with weight reduction by reducing appetite or decreasing the absorption of carbohydrates and saturated fats in the gut. Naturally occurring dietary polyphenols, nutraceuticals, and bioactive ingredients present in plants and spices (resveratrol, green tea catechins, quercetin, berberine, turmeric, cinnamon, red chili, black pepper, fenugreek, thymol, rosemary) are recommended for their anti-obesity actions produced through metabolism stimulation. Phytotherapies or botanical extracts attenuate adipose tissue expansion and cause adipocyte remodeling by altering inflammation-related mechanisms. Several plant-derived remedies such as *Ginko biloba, Momordica charantia, Aegle marmelos, Azadirachta indica* have anti-oxidant and anti-inflammatory potential, and are indicated for the prevention and treatment of obesity associated metabolic syndrome and chronic diseases such as type 2 diabetes, cancers, kidney pathology, as well as numerous cardiovascular and neurodegenerative disorders. The alternative therapies, healthy foods and exercise interventions are safe and effective therapeutic options to improve metabolism in body and obesity reduction. This review summarizes the mounting evidence that dietary therapies, healthy food habits, and increases in physical activity are helpful in reducing obesity and treating obesity-related complications. The purported underlying mechanisms of complementary and alternative therapies are also discussed.

Keywords Obesity · White adipose tissue · Oxidative stress · Non-communicable chronic diseases · Preventive strategies · Allopathic drugs · Herbal remedies · Nutraceuticals · Probiotics · Benefits of exercise

Introduction and Background Information

It is now well recognized that obesity results from the bio energetic imbalance between energy intake and expenditure. Generally, obesity is caused by the overconsumption of carbohydrates, saturated fats, sugar-laden drinks and a sedentary lifestyle which lead to the excessive accumulation of white adipose tissue (WAT) around the belly, hip and waist—consequently increasing body weight and Body Mass Index (BMI >30 kg/m^2) [1]. At one time, obesity was considered to be the problem of rich and developed countries. However, due to economic prosperity, abundant food supplies, and unhealthy dietary habits in developing countries, the incidence of obesity among children and adults has increased. The occurrence of obesity has amplified from 2.3 to 19.6% during the last 15 years in developing countries like India, where a BMI value >25 kg/m^2 is classified as obese among adults [2]. As per the National Family Health Survey of 2015–2016 report (NFHS-4), published by India's Union Health Ministry, a sharp increase is predicted in the prevalence of obesity by the year 2025 in adult males from 2.3 to 3.1%, and in women from 5.2 to 6.9% [3]. According to the World Health Organization (WHO) report published in the well-respected journal *The Lancet, nearly* 2.7 million adults will suffer from obesity worldwide by the year 2025 [4].

According to the 2017 Report of the World Obesity Federation, India will spend $13 million (USD) per year, for treating obesity-associated diseases by 2025. It is estimated that at this time, the cumulative cost for treating obesity-related diseases worldwide, will approach $90 billion. Considering the worldwide epidemic of obesity, the annual healthcare costs may reach $1.2 trillion by 2025 [5]. The WHO has estimated that nearly 30% of the global population is currently obese, out of which 60% live in the developing countries. The sharp increase in obesity and obesity-related diseases is not only worrisome for illness-related costs and poor quality of life, but also for its significantly adverse consequences on a country's economy [6].

Combitorial interaction of several factors including behavioral, metabolic, cellular, molecular and psychological is involved in causing corpulence due to the excessive accumulation of white adipose tissue in obese men, women and children [7]. Two important factors often linked to the occurrence of obesity are the excessive intake of carbohydrates and saturated fats, as well as sugar-loaded drinks. Overall, the excessive caloric dietary intake combined with physical inactivity are generally responsible in causing excessive accumulation of WAT, which produces several bioactive chemicals, including pro-inflammatory cytokines and mitochondrial bioenergetics reactions due to oxidative stress [8]. The white adipocytes trigger the production of pro-inflammatory cytokines, free fatty acids, adipokines, leptin, TNF-α, adiponectin, interleukin (IL-1, IL-6) and insulin resistance by phosphorylation of insulin resistance substrate (IRSs-1) [5]. Further, the obesity-related mitochondrial oxidative stress activates the production of highly reactive oxygen species (ROS) and nitrogen species (NOS) from monocytes and macrophages [9]. It is now recognized that small amounts of free radicals are needed for extra- and intra-cellular communication. However, the overproduction of ROS and NOS causes cellular membrane injury, lipid per oxidation, and DNA damage. The cellular and genetic deleterious effects caused by ROS and NOS are linked to the free-radical scavenging inability of intracellular antioxidant scavengers (glutathione, L-cysteine, superoxide and hydro peroxide enzymes, catalase), and lower levels of exogenous micronutrients (vitamin C and E, Zn and Cu), as well as lack of antioxidants like flavonoids and carotenoids present in nutraceuticals and plant-derived food. It is well known that polyphenolic compounds are potent antioxidants and free-radical scavengers. These properties of flavonoids exhibit anti-mutagenic, anti-carcinogenic, anti-inflammatory, and anti-atherogenic activities, and all these biological effects support the promising therapeutic potential of flavonoids and other phytopharmaceutical antioxidants for the treatment and prevention of malignant tumors or cancer, and different forms of CVDs. Flavonoids also suppress lipid per oxidation, platelet aggregation, and angiotensin converting enzyme (ACE) activity. Some epidemiological studies have revealed an inverse relationship between the intake of dietary flavonoids and mortality and morbidity due to CVDs. A few studies have shown that dietary intake of flavonoids reduces the incidence of atherosclerosis, hypertension and congestive heart failure [10]. In view of these observations, intakes of flavonoid rich foods or nutritional supplements containing flavonoids possess an enormous potential to promote human health and well-being through the prevention of non-communicable chronic diseases. However, randomized and placebo-controlled clinical trials are needed to ascertain the long-term safety and efficacy of flavonoids. One aspect of this review is to highlight the health benefits of flavonoids present in foods and dietary supplements in preventing and curing chronic problems, including diabetes, cancer, heart disease and stroke. Evidence will also be provided about the beneficial properties of individual flavonoids tested in the preclinical and clinical studies along with their structure-activity-relationships as well as their use in different CVDs.

It is well known that the oxidative stress generated in mitochondrial bioenergetics reactions is responsible for the pathophysiology of atherosclerosis, hypertension, coronary heart disease and stroke. Novel approaches are being taken for the diagnoses and management of CVDs that are mediated by oxidative stress: namely cellular membrane injury, lipid per oxidation, and DNA damage. This can be seen in Table 19.1.

Obesity also stimulates oxygen utilization causes the fabrication of superoxide radicals, reactive oxygen species (ROS), hydrogen peroxide and hydroxyl radicals that may originate from the electron transport chain, and increased amounts of mitochondrial respiration [12]. The over-production of ROS spoils cellular structure and activates fat deposition, over production of NO and resulting various organs dysfunction which leads to the development of others diseases like dementia,

Table 19.1 Classification of overweight and obesity by body mass index (BMI), waist circumference, and associated disease risks [adopted from Jonathan and Purnell [11]]

	BMI (kg/m^2)	Obesity class	Disease risk[a] (relative to normal weight and waist circumference†)	
			Men $\leq$ 40 in. ($\leq$ 102 cm)	>40 in. (>102 cm)
			Women $\leq$ 35 in. ($\leq$ 88 cm)	>35 in. (>88 cm)
<18.5			–	–
Normal	18.5–24.9		–	–
Overweight	25.0–29.9		Increased	High
Obesity	30.0–34.9	I	High	Very high
	35.0–39.9	II	Very high	Very high
Extreme obesity	>40	III	Extremely high	Extremely high

[a]Disease risk for type 2 diabetes, hypertension, and cardiovascular diseases

†Increased waist circumference can also be a marker for increased risk even in persons of normal weight

diabetes, various cancers, osteoarthritis and other malignancies [13]. There are three main treatment approaches that are used to manage and prevent obesity-induced chronic diseases. These are life style interventions which include daily physical exercise, dietary supplements such as the intake of fibers, nutrients, less carbohydrates, no alcohol intake etc., and pharmacotherapy which include allopathic, Ayurvedic, Fast-Moving Consumers Goods (Nutraceuticals), Homeopathic remedies and Unani methods.

Obesity is a complicated problem and is of great concern as it increases the risk of developing a number of potentially serious health problems including coronary heart disease, high blood pressure, stroke, type 2 diabetes, some cancers (breast, colon, endometrial, gall bladder, kidney and liver), sleep apnea, high LDL cholesterol, low HDL cholesterol, high levels of triglycerides, gall stones, osteoarthritis, infertility or irregular periods. Figure 19.1 shows the major non-communicable diseases caused by obesity.

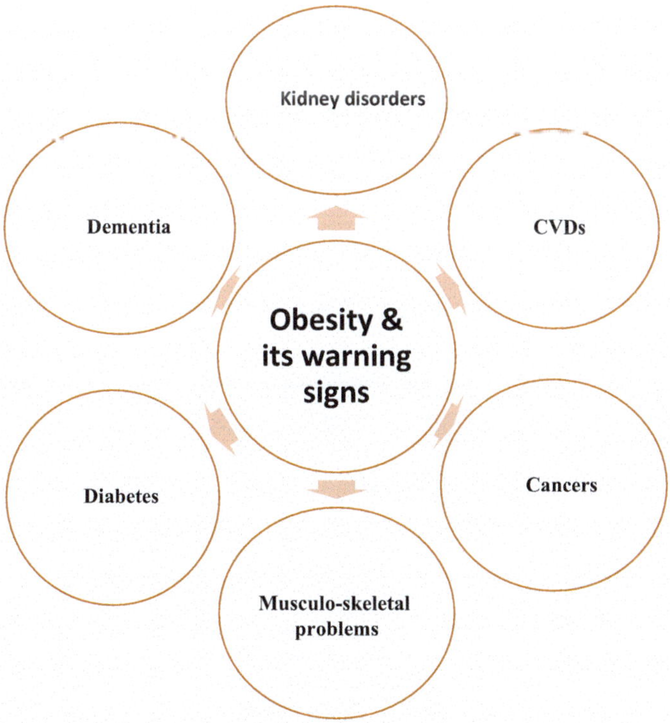

Fig. 19.1 Diagrammatic representation of obesity-induced chronic diseases

Pathophysiology of Obesity-Induced Diabetes Mellitus

Diabetes mellitus or type 2 diabetes mellitus (T2DM) is defined by high glucose and lipid levels in the systemic circulation resulting in insulin resistance (IR) [14]. IR is the key marker of (T2DM) and develops in numerous organs, as well as skeletal muscle, heart, liver and adipose tissues. As the main cause of insulin resistance is hyperglycemia and hyperlipidemia, obesity can be a chief cause of type 2 diabetes mellitus [15]. Another probable molecular mechanism i.e. inhibition of phophoinositide 3- kinase (PI_3K) via tyrosine phosphorylation leads to the activation of PDK-1 PDK-2 which stimulates Akt/PKB and PKCλ, resulting in 160KDa phosphorylation, which is insulin signaling. It was considered that persistent hyperglycemia and expanded hyper insulinaemia, and amplified nitrogen and reactive oxygen species (RNS and ROS respectively) levels can impact IR gene expression by improper functioning of transcription factors such as high mobility group AT-hook 1 (HMGA-1) [16].

Increased level of insulin and glucose in a combined pattern may lessen insulin attachment to the IR in adipocyte [17] leading to a pessimistic effect on protein kinase B/Akt actions. Increased carbohydrate metabolism and unnecessary excretion of ROS/RNS or reduced antioxidant capability cause oxidative stress and insulin resistance. In fact, pro-inflammatory cytokines like TNF-α, IL-1 and IL-6 produced insulin signaling through phosphorylation of insulin receptors substrates (IRS) reduced the interface of protein with IR. The pathophysiology of type 2 diabetes can be seen in Fig. 19.2.

Overload of glucose and fats in power house causes overproduction of ROS and hypoxia leading to oxidative stress. Obesity leads to secretion of cytokines (TNF- α,

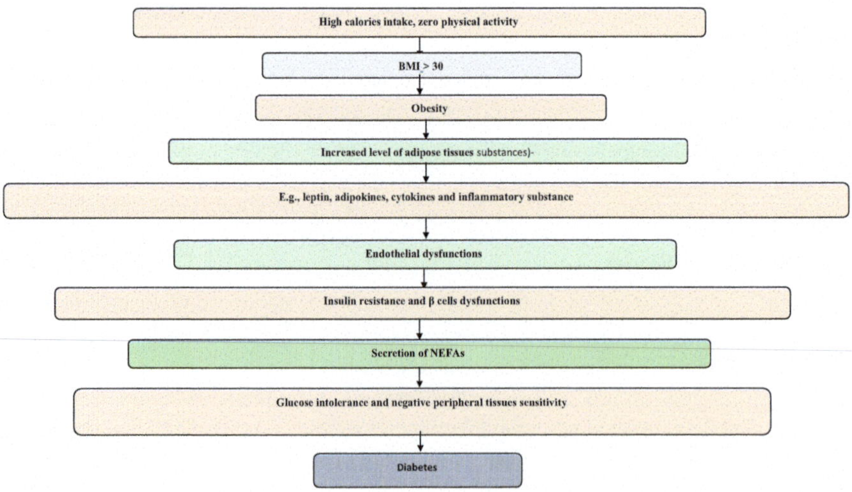

Fig. 19.2 Pathophysiology of types 2 diabetes: BMI and non-esterified fatty acids (NEFAs)

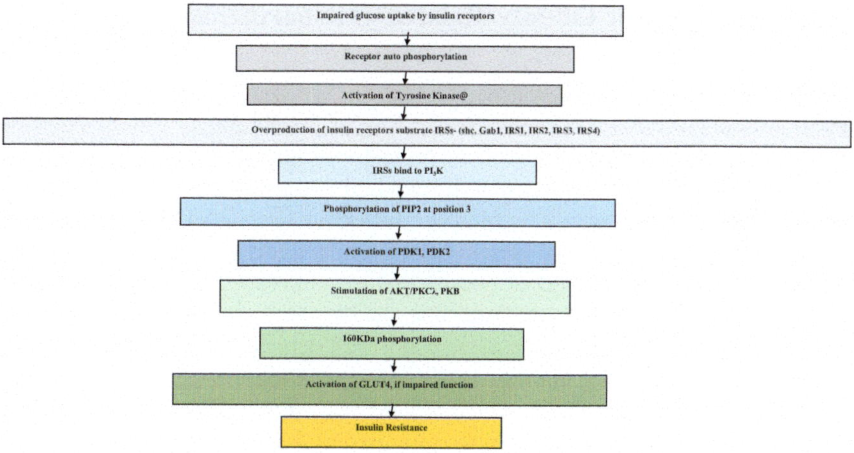

Fig. 19.3 Diagrammatic representation of insulin signaling pathways. Insulin receptor substrates (IRSs): IRS1, IRS2, IRS3 and IRS4; Shc; (PI3K) = Gab1 phosphatidylinositol 3-kinase; (PIP2) = phosphatidylinositol bisphosphate; (PDK1, PDK2); phosphatidyl protein kinase 1 and 2; (AKT/PKCλ) = atypical protein kinase; Serine/threonine kinase; (PKB) = protein kinase-B; 160KDa; (GLUT4) = glucose transport-4

ROS, NOS and free fatty acids) inhibits IRS-1 insulin signaling pathway. Proinflammatory cytokines secrete IKKβ, JNK1 inhibit PPAR and IRS-1 in insulin signaling pathway leads to insulin resistance. Further, IKKβ increase the phosphorylation of IKβa causing phosphorylation of IRS-1 (Ser_{312}, Ser_{307}, and Ser_{270}). Insulin signaling pathways can be seen in Fig. 19.3.

Preventive Strategies

In diabetes mellitus, hyperglycemia occurs due to resistance of insulin i.e., hyperinsulinemia and hyperinsulinemia. Obesity is the main risk factors for the development of this disease. It is reported that a sedentary lifestyle and improper diet are associated with diabetes mellitus. Many researchers reported that in addition of allopathic drugs, nutraceuticals, herbals, nutrients and spices play a role in the prevention and treatment of diabetes mellitus. Some herbal plants including: *Cyperus Rotundus L.* (Nut grass), *Gymnema sylvestrer. Br. Ex Schult* (Gurmar), *Nelumbo nucifera Gaerth* (Indian Lotus) and *Lagerstroemia Speciosa L.* (Pride of India), are helpful in the treatment of diabetes mellitus [18]. Non-pharmacological approaches including a low-calorie, fibrous diet, daily physical activities, controlled smoking and drinking, and weight-controlled consciousness all play important roles in prevention of obesity-induced diabetes mellitus [19].

Pathophysiology of Obesity-Related Osteoarthritis

Osteoarthritis (OA) is a joint disease that causes pain and debilitating dysfunction due to joint degradation and aging. Nearly 10% of people suffers with this disease and 37% of an aging population who are 60 years or older [20]. Further, OA affects the patients' lives due to less productivity and high healthcare maintenance. Obesity is documented as a significant problematic factor for knee OA [21, 22].

Osteoarthritis is also called degenerative joint disease with less inflammatory response. The pathophysiology of obesity linkage to knee OA specifically is multi factored [23, 24]. Although the role that obesity plays in the development of arthritis or other structural joint damage is not well defined, but the involvement of adipose tissues [25, 26], metabolic endocrine organs that secrete adipocytokines, leptin, resistin, and adiponectin, which ultimately destroys and causes remodeling of joints tissues. Adipokines impacts the joint tissues, including cartilage, synovium and bone.

In obese individuals, the overproduction of leptin creates "leptin resistance syndrome", further causing insulin resistance. Leptin and its receptor found in chondrocytes, osteophytes [27, 28], synovium, and infra-patellar fat pads [29] and may influence growth factor synthesis and anabolism. Examination of cartilage, subchrondral bone, and osteophytes shows evidence of up-synchronized leptin expression. Leptin expression openly relates to the degree of cartilage deterioration [30, 31] and synergistic associations of leptin and proinflammatory cytokines have been noticed [32]. Leptin has a direct pro-inflammatory and catabolic role in cartilage metabolism [33]. The molecular mechanism of obesity-induced osteoarthritis can be found in Fig. 19.4.

Preventive Strategies

Osteoarthritis is a chronic inflammatory disorder in which synovium and cartilage inflammation takes place and causes swelling, pain and joint complications [34]. Weight loss is the primary helping factor in the treatment of osteoarthritis, as an increased weight and BMI can affect action and pharmacokinetics of numerous drugs pharmacotherapy. Weight reduction decreases the stimulation of inflammatory cytokines (TNF-α, IL-6, IL-18 etc.) and ultimately reverses the pathophysiology of obesity. A patient suffering from osteoarthritis should reduce calories intake by a minimum of 8 kcal/Kg [35]. Weight can be reduced by FDA-approved medications such as Orlistat and Sibutramine. Surgeries are also helpful in weight and pain reduction of knee OA within 2–3 months [36]. Proper fibrous diet and disease-related physical activities are also helpful in weight and pain reduction of body [37].

The results of a randomized, double blind, placebo controlled, crossover study showed that treatment with ginger extract (250 mg of Zingiberis Rhizoma per

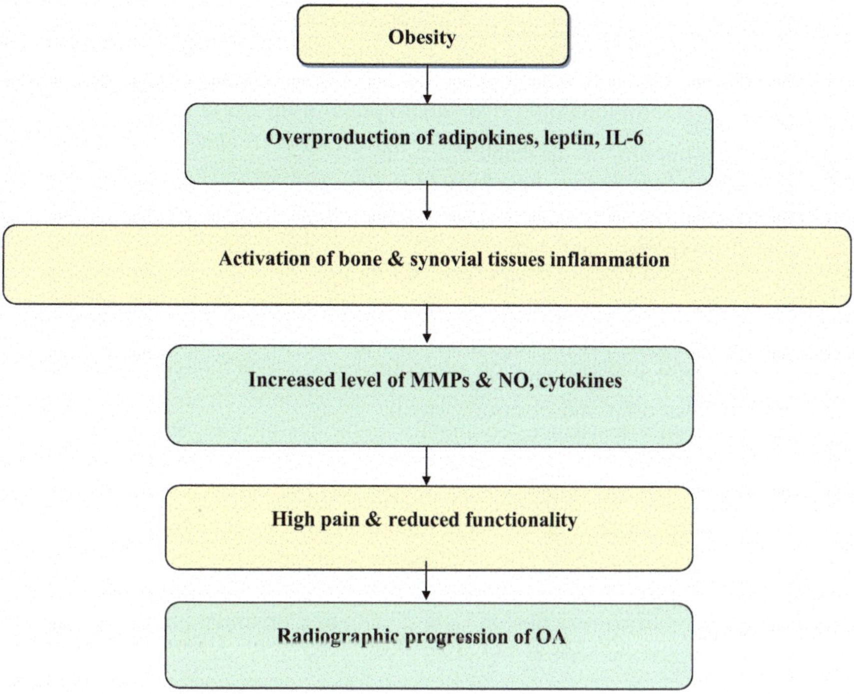

Fig. 19.4 Flow chart of molecular mechanism of obesity

capsule, qid) for six months caused significant improvement in twenty patients suffering from gonarthritis (degenerative disease of the knee joint) [38, 39]. Few antioxidants which are found in glutathione precursor (*S*-adenosylmethonine) [40] avocado, soybean [41, 42] and Oleoresin phytoconstituents from Boswellia Serrata tree are useful in the management of obesity-induced osteoarthritis [43]. Flavonoids like Harpagoside- triterpene from *Harpagophytum Procumbens* (devil's cow) [37, 44], Bromelain extract of pineapple [45] and *Shogaols* and *gingerols* (Ginger) [46] are effective against osteoarthritis disease.

Pathophysiology of Obesity-Induced Kidney Disorder

Kidney disorders occur in approximately 30% of patients leading towards cardiac surgery and significantly increase rate of morbidity and shorter life span [47, 48]. According to National Health and Nutrition in the United States, 64% of youth are obese, with a body mass index (BMI) >25. This range of BMI can be a chief reason of hypertension and diabetes, which collectively leads towards 70% of end-stage renal diseases (ESRD) [49]. Early signs of kidney failure in an obese individuals are renal vasodilation, glomerular hyper filtration, increased tubular sodium reabsorption and increased blood pressure [50]. The simultaneous elevation in glomerular filtration rate (GFR) and renal plasma flow indicates dilation of preglomerular vessels and increased glomerular hydrostatic pressure in obese patients [51]. This preglomerular vasodilation would increase transmission of elevated systemic arterial pressure to the glomeruli, resulting in higher glomerular hydrostatic pressure and wall stress. As obesity increases to a risky stage, then kidney size may also increase, but nephron numbers do not change such that each surviving nephron is exposed to increased rates of glomerular filtration.

Adipose tissues are a source of various substances including adipokines, inflammatory cytokines, leptin and TNF-α. Leptin regulates energy and intracellular lipids function and underproduction of leptin causes hyperphagia [52, 53]. The increased level of inflammatory cytokines inhibits adiponectin secretion leading to hypertension, insulin resistance, endothelial cells dysfunction, inflammation and further overproduction of Rennin angiotensin-aldosterone system (RAAS). The adipose macrophages are storage house of (RAAS) [54–56].

RAAS regulates renal structures and functionality via regulating cell proliferation and vasomotor tone [57]. The stimulation of angiotensin tensin-1 (AT-1) and RAAS results in elevation of filtration fraction (FF), cellular proliferation, and efferent arteriolar vasoconstriction and stimulation of glomerular pressure leading to renal damage, as can be seen in Fig. 19.5 [58].

Preventive Strategies

Obesity is common risk factor for developing chronic kidney disease (CKD) [59]. Obesity-induced kidney disease affects 8–16% of the population worldwide. Currently, various pharmacological treatments are used in the prevention of kidney disease, but can have various adverse effects. So, by using some non-pharmacological treatments and potential therapies kidney disease can be

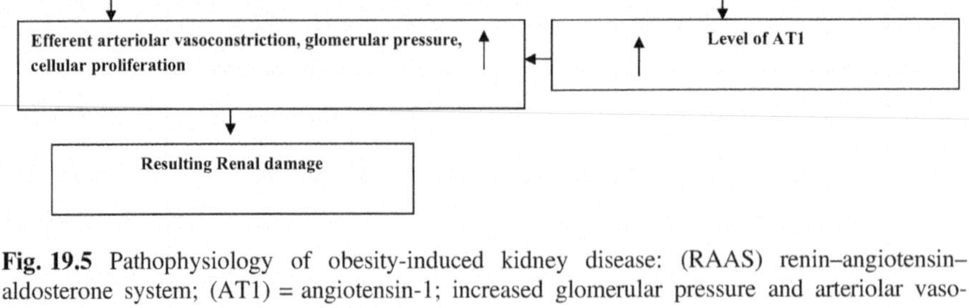

Fig. 19.5 Pathophysiology of obesity-induced kidney disease: (RAAS) renin–angiotensin–aldosterone system; (AT1) = angiotensin-1; increased glomerular pressure and arteriolar vasoconstriction; decreased adiponectin level

controlled. Some nutraceuticals approaches like the use of herbal plants and their extract and Phytochemicals can help in the prevention of certain kidney diseases. Oil seeds, nuts [60–63], *Perseaamericans* (avocado), *Juglansregia* L. (walnut), *Linumusitatissimum* L. (flaxseed), *Eruca sativa* [64]. Nutraceuticals and Xenobiotics such as curcumin (*curcuma Longa*) [65, 66], Stevia (*stevia rebaudiana*) [67], green tea [68], nitrates and nitrites [69]. Some Ayurvedic products like nettle leaf and Gokhru juice are also considered useful for the management of obesity-induced kidney diseases.

Pathophysiology of Obesity-Induced Breast Cancers

Obesity increases the risk of developing breast cancer in both pre- and post-menopausal women and also negatively impacts the recurrence of breast cancer and survival of patients [70]. Breast cancer is the most common cause of death in women and it is the second most common disease among Korean women, but is lower for women in western countries. It is rapidly rising statistic with a yearly growth rate of 6.3% [71]. In a study conducted by Eliassen et al., there was a 12% development of breast cancer among overweight, postmenopausal women, this increased risk has now risen to 25%.

The leading cause of Breast Cancer may be due to a mutation in tumor suppressor genes such as breast cancer A1 and breast cancer A2 (BRCA1 and BRCA2) [71]. Over secretion of sex hormones trigger tumor growth and production of mammary cell production [72].

Hyperinsulinaemia, or impairment of insulin, motivates the propagation of tumors cells in vitro and in vivo evaluations resulting in augmented levels of prostate, colon, bladder and breast cancers. In potential epidemiological experiments, amplified risk of postmenopausal breast, colon, pancreatic, and endometrial cancers [73] have been noticed in relation to superior levels of moving C-peptide levels. Multiple molecular pathways including hormone (estrogen), adipose tissues, insulin resistance and inflammatory cytokines (IL-6, IL-11, IL-1), leptin, and free fatty acids (FFA) are involved breast cancer growth [74]. The overproduction of estrogen in adipose tissues in postmenopausal females is due to androstenedione and testosterone aromatization. The stimulation of inflammatory cytokines like interleukin (IL-6, IL-11) and tumor necrosis factor- (TNF-α) triggers aromatization resulting in breast cancer growth as seen in Fig. 19.7. Epithelial cell lead to aromatization effecting promoter 1.3/11 region (PI.3/PII) of aromatase gene [75, 76]. Also, by stimulating TNF-α and interleukin-6, (IL-6) causes aromatase gene activation in breast cancer. In obesity, over production of free fatty acids (FFA), inflammatory cytokines (leptin, adipokines), triggers insulin resistance regulating insulin receptors substrate-1 (IRS-1). The estrogen increases E2F transcription factor via phosphorylation of retinoblastoma gene (RB) leading to aromatase gene activation. The molecular pathway includes adipose tissues, inflammatory cytokines and hormones can be seen in Fig. 19.6 [77, 78].

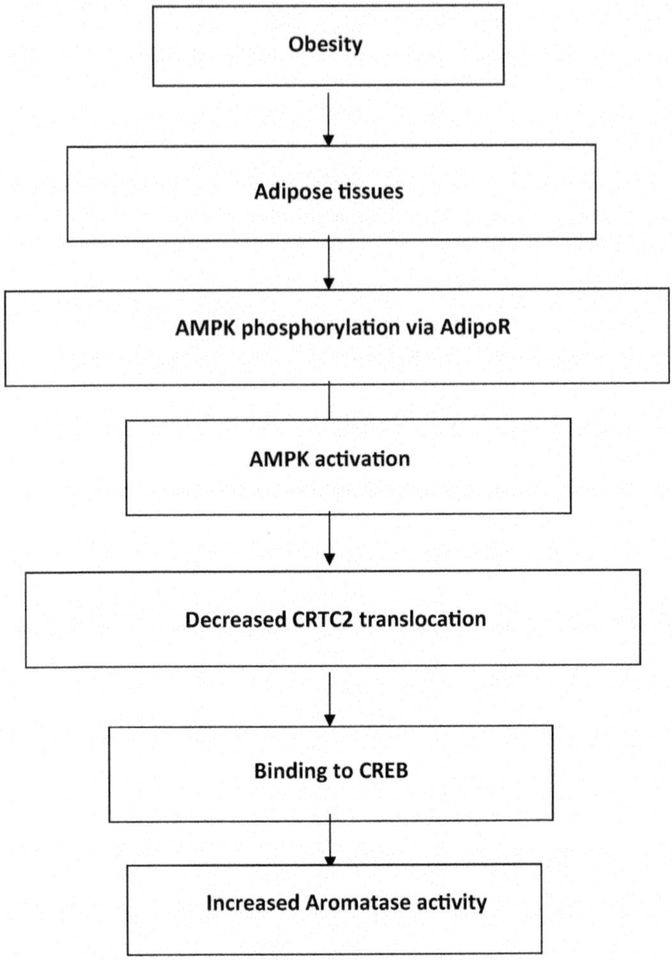

Fig. 19.6 Molecular pathways connecting obesity and breast cancer: AMPk = activated protein kinase; AdipoR = adipose receptors; CREB = Camp response element-binding protein; CRTC2 = CREB-regulated transcription co activators 2

Preventive Strategies

Breast cancer is a common cause of death in western women. Several pharmacotherapies exist including Tamoxifene, Raloxifene and aromatase inhibitors, which are available for the treatment of breast cancer, but have many adverse effects for patients [79]. So, to minimize these side effects and to make patients' lives easier, some non-pharmacological therapies show their potential role in the prevention and management of cancer. Nutraceuticals plants including *Vaccinium erythrocarpum, Commiphora mukulengl* (Indian Belelliom tree), *Nelumbo Nucifera*

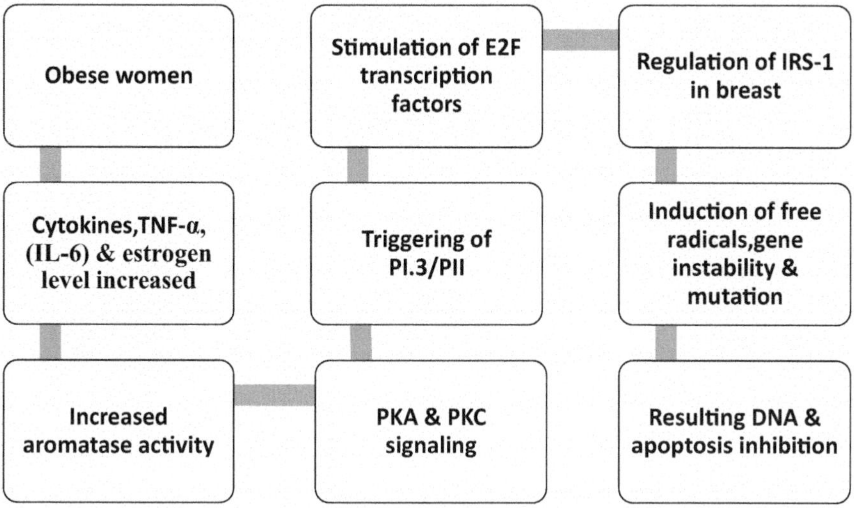

Fig. 19.7 Hormonal mechanism of breast cancer: TNF-α = tumor necrosis factor-α; (IL-6) = intrerleukin-6; E2F-prostaglandin = E2; (PI.3/PII)-promoter 1,3 region; promoter, II; (PKA and PKC)-protein kinase A,C; (IRS-1) = insulin receptors substrate

Gaerth (Indian Lotus), Phytochemicals (Ursolic acid, Scopoletin, Damnacanthal, carotenoids, stilbenes) and nutrients (Vitamins and minerals-A, E, K, C, B1, B2, B3), omega 3 fatty acids, fruits and vegetables and tree nuts show synergistic effects in the treatment of obesity-induced cancer [80]. Changes in life style, weight control, physical activity, and minimizing alcohol consumption can also reduce the risk of breast cancer. Alcohol intake can increase the risk of breast cancer by minimizing some alcohol labile nutrients [81, 82].

Pathophysiology of Obesity-Induced Cardiovascular Diseases

The rising evidence suggests that alteration in the gut microbiome, related to dietary and genetic features may contribute to metabolic disorders, which is a significant factor for obesity development, hypertension, insulin resistance and T2DM [83, 84]. In developing countries, there is a fast growth of cardiovascular disease and obesity, which negatively affects patients' lifespan [85]. The collection of multiple factors such as hip and waist circumference, increased belly fats, increased BMI >30, LDL, HDL levels, hypertension, insulin imbalance or resistance, high serum triglyceride concentrations, and high concentration of cholesterol leads towards "metabolic syndrome" [86]. This syndrome becomes a basic risk for the development of obesity among children. Currently 40–45% of youngsters are caught by this developing issue [87]. Oxidative stress is also a main disorder that leads towards obesity and can cause numerous diseases including hypertension and dyslipidemia.

Dyslipidemia is a state in which accumulation of high concentration of triglyceride (TG) and blood cholesterol takes place which enhances the danger of CVDs including stroke [88]. Obesity combined with dyslipidemia is also a leading giant of CVD. Small amounts of circulating high-density lipoprotein (HDL), amplified post-prandial TG values, augmented clearance of HDL particles, and high plasma very low-density lipoprotein (VLDL) levels cause overproduction of ROS [89]. ROS can alter the various cellular pathways including nitrogen activated protein kinase and can damage lipid or protein structures further, resulting in DNA damage [90].

The oxidative modification combined with oxidative low-density lipoproteins (Ox-LDL) and oxidation-derived products can be responsible for CVD generation [91, 92]. Additionally, Ox-LDL alters the adipokines fabrication resulting OS [66]. In obese individuals the occurrence of dyslipidemia and elevated Ox-LDL levels may be due to imbalanced antioxidant abilities that is generated by low serum movement of the antioxidant enzyme (SOD) or low HDL-associated with paraoxonase-1 (PON-1). Even though numerous mechanisms such as increased VLDL and Ox-LDL, TG level and lower levels of circulating HDL have been explained which lead towards CVDs and found that oxidative stress is also concerned in vascular blockade [93].

Hypertension connected to obesity can take place by multiple mechanisms including adipokines variations, alterations in kidney functions and structures, inappropriate sympathetic nervous system (SNS) and renin–angiotensin–aldosterone system (RAAS) activation and insulin resistance. Epigenetic factors are also responsible for obesity-related hypertension [85, 94]. Epigenetic mechanisms include histone modifications, microRNA (miRNA) regulation and changes in DNA methylation [95]. Reduced vasodilation due to imbalances between superoxide and nitrous oxide (NO) may be responsible for the growth of hypertension. The half-life of nitrous oxide is momentous because it is quickly destroyed by a superoxide anion discharged from eNOS (a vasoconstrictor). Later, eNOS develops into a peroxynitrite generator, causing a marked amplification in oxidative stress with a pleiotropic effect on vascular efficacy via oxidation of proteins and cellular lipids [96]. Improper functioning of kidneys, considering stimulation of intrarenal angiotensin II (Ang. II), is also a significant expansion of obesity-connected hypertension. Some data explains that NADPH-driven creations of ROS simultaneously motivates redox-based signaling cascades and are taking part in the working of Angiotensin-II-producing hypertension. Angiotensin-II comes in contact with a certain receptor, Angiotensin-II type 1 (AT1r), and stimulates non-phagocyte NADPH-oxidase, causing growth of hydrogen peroxides, peroxynitrite and superoxide. The stimulation of adipocyte tissues triggers the overproduction of proinflammatory cytokines (TNF-α, IL-6 and IL-1, IL-11), leptin, adipokines, FFA, and resistin. The underproduction of adiponectin activates RAAS and SNS, which results in insulin resistance via phosphorylation of IRS-1 [97–100]. This

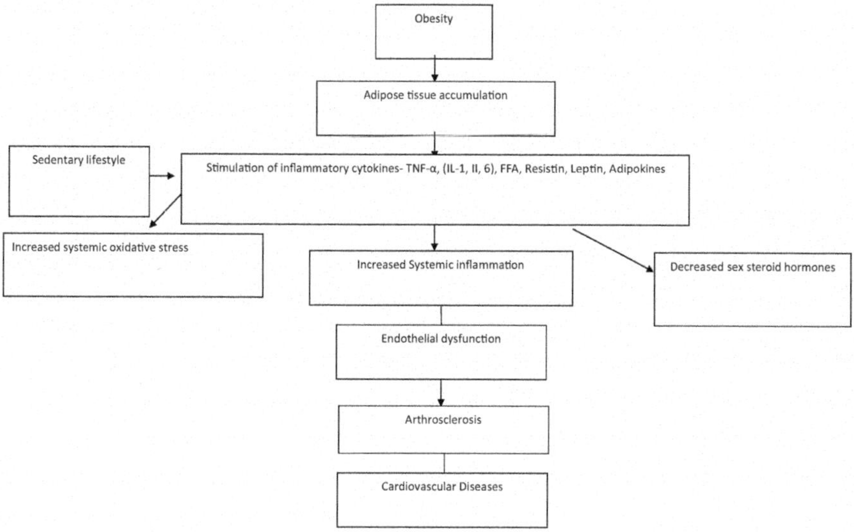

Fig. 19.8 Pathophysiology of obesity-induced cardiovascular diseases: (TNF-α) = tumor necrosis factor-α; (IL) = interleukins; (FFA) = free fatty acids

overproduction of inflammatory cytokines leads to endothelial dysfunction, further causing arthrosclerosis and hypertension development [101, 102]. The neuropeptides (α-melanocyte-stimulating hormone and neuropeptides Y) have also been considered as a further link between oxidative stress and hypertension [103, 104]. This can be seen in Fig. 19.8.

Preventive Strategies

Obesity is the leading cause of cardiovascular diseases (CVDs), in the developed and developing countries [105, 106]. This leads to nearly 17.3 million deaths globally. A cost-effective treatment and interventional strategy are needed for the prevention of overweight/obesity, metabolic syndrome and CVDs [107, 108]. The use of some herbal therapies combined with pharmacological treatment show synergistic effects in the CVD treatment. Herbal supplements like nutraceuticals (*Coleus forskohlii*, *Vitisvinifera L*), Flavonoids (*apigenin, Hesperidins*) and Phytoconstituents including Quercetin, Kaempferol and rutin are used as an alternative treatment of obesity and its complications [109–111]. Different plants contain a large variety of several components with different anti-obesity effects on body metabolism and fat oxidation, and for this reason have been investigated and reported to be useful in treatment of obesity [112, 113]. Significant amounts of polyphenols (flaxseeds, tea, coffee and cocoa) [114] and phytosterols (cereals, seeds, nuts and grains), are considered helpful in the management and treatment of cardiovascular diseases [115].

It is well known that global epidemic of obesity-related disorders are on the upswing. Weight loss appears to protect against diabetes, coronary heart disease, hypertension, kidney and neurodegenerative disorders, musculoskeletal diseases, cancer and overall mortality. Nutritional interventions, lifestyle modifications, and exercise reduce the incidence of obesity-related non-communicable chronic diseases. Abnormal deposition of white adipose tissue in the body alters the metabolism and disposition of steroidal hormones as well as components of the renin-angiotensin system and sympathetic nervous system [116, 117]. Bariatric surgery is a viable option for the treatment of severe obesity and is associated with a reduction in overall mortality in patients with mean BMI 41 kg/m^2 [118].

Pathophysiology of Obesity-Induced Neurological Disorders

Obesity is a state that occurs when the consumption of calories is greater than the body requirement. Obesity and dementia are two significant neurodegenerative problems counting Alzheimer's disease (AD). In 2015, a World Alzheimer survey reported 46.8 million cases of dementia and calculated that this figure may double every 20 years [119]. In this scenario, obesity-induced dementia has become a major concern for medical disciplines as it increases at a distressing pace [120, 121].

There is an accumulation of adipose tissues in obese individuals, which involves the high secretion of proinflammatory cytokines (FFA, TNF-α, IL-1β, IL-6) [122, 123] and chemokines ligand 2, which directly leads to AD. These mediators also cause chronic, as well as local inflammation in adipose tissues which in turn promotes cellular insulin resistance. This can cause hyperglycemia and hyperinsulinemia, which may result in AD [124]. Leptin resistance also plays an important role in AD generation, as seen in Fig. 19.9 [125].

Strategies for the Prevention of Neurodegenerative Disorders

Obesity is a risk factor in the development of neurodegenerative disorders, such as dementia and Alzheimer's disease (AD). Some forms of dementia lead to progressive brain disorders and loss of memory and cognitive skills. These may be responsible for significant changes in behavior [126]. Pharmacological and non-pharmacological interventions play important roles in the management and prevention of dementia and AD. Some basic therapies such as brain stimulation via brain-related exercise and practice are helpful in increasing neurodensity and plasticity [127]. Intake of fiber rich nutrients and vitamins (D, B6, B12, and folates) [128–133], plant extracts and Phytochemicals like terpenoid, anti-oxidants and

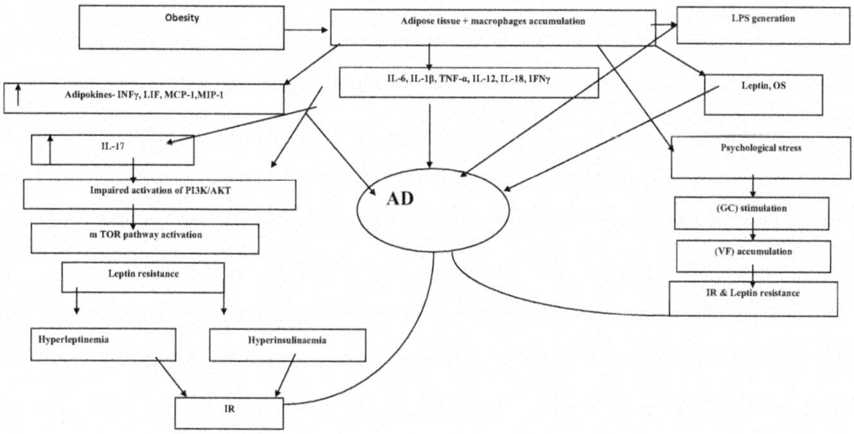

Fig. 19.9 Pathophysiology of obesity—related neurological disorders and Alzheimer's disease (AD). (INF-γ) = interferon-y; (GC) = glucocorticoids; (IR) = insulin resistance; (VF) = visceral fat; (OS) = oxidative stress; (mTOR) = mammalian target of rapamycin; (IL-17) = interleukin-17; (PI3 K/AKT) = Phosphatidylinositol 3 kinase; protein kinase B; (LIF) = leukemia inhibiting factor; (MCP-1, MIP-1) = macrophagic inflammatory proteins-1; (TNF-α) = tumor necrosis factor-α

anti-inflammatory flavonoids (luteolin) [134], polyphenols (resveratrol) found in red wine, red grapes, blueberries and peanuts [135] provides protection against AD [136]. Intake of unsaturated fatty acids (omega-3, fibrous diet, fresh fruits and vegetables are helpful in weight reduction and prevention of obesity-induced dementia and AD. The consumption of olive oil, fish oil, omega-3 fatty acid [137] green tea, curcumin, nutraceuticals (*Ginkobiloba*) [138], probiotics like lactobacillus rhamnosus [139] are also considered to play an important role in the prevention of neurodegenerative disorders.

Summary and Conclusions

Obesity is a social problem which is defined by an increase in body weight that results in excessive fat accumulation. The epidemic of obesity is increasing worldwide and threatens to continue to decrease life expectancy. This problem has been recognized as a major underlying factor for the pathogenesis of several diseases including metabolic syndrome, diabetes mellitus, cardiovascular diseases, liver disease and cancer. It can be said that this disorder not only affects the economy of a country, but also contributes to negative health care issues. For the prevention and treatment of this problem the use of pharmacological approaches is not enough. Ultimately some non- pharmacological basic therapies playa great role in the treatment and management of obesity and its associated chronic diseases.

Some basic therapies including low caloric intake, weight controlled consciousness, smoking cessation, a healthy diet plan and daily physical activities increase the antioxidant activity. Some psychotherapy (behavioral therapy) is also helpful for patients to increase the consciousness toward weigh control habits which is major risk factor for chronic disease. Although the presence of both pharmacological and non-pharmacological treatment exists, there is want for innovative policies for the prevention and control of obesity. There are also some strategies for the management and cure of obesity and its induced chronic diseases, but there is a deficiency of data for their efficacy and also have their own limitations. Lastly, for the development of novel cure and treatments, there needs to be an understanding of the basic underlying molecular mechanisms that take place in obese individuals. Prospective studies should explain the probable function of oxidative stress in obesity to control the beginning and development of inflammatory and autoimmune complications.

Conflict of Interest All authors declare no conflict of interest.

References

1. Hossain P, Kawar B, El Nahas M (2007) Obesity and diabetes in the developing world–a growing challenge. N Engl J Med 356:213–215
2. Sikaris K (2004) The clinical biochemistry of obesity. Clin Biochem Rev 25:165–181
3. National Family Health Survey (NFHS-4) 2015–2016
4. World Health Organization and the Imperial College of London (2017)
5. World Obesity Federation report (2017)
6. World Health Organization report (2017)
7. Kaufer M, Tavano L, Ávila H (2001) Obesidaden el adulto. In: Casanueva E, Kaufer M, Pérez A, Arroyo P (eds) 1st edn. Editorial Médica Panamericana, México
8. Chrysohoou C, Panagiotakos DB, Pitsavos C, Skoumas I, Papademetriou L et al (2007) The implication of obesity on total antioxidant capacity apparently healthy men and women: the ATTICA study. Nutr Metab Cardiovasc Dis 17:590–597
9. Luchsinger J, Mayeux R (2004) Cardiovascular risk factors and Alzheimer's disease. Curr Atheroscler Rep 6:261–266
10. Julia JP (2012) Do flavonoids reduce cardiovascular disease incidence or mortality in US and European populations? Nutr Rev 70(9):491–508
11. Jonathan QP (2018) Definitions, classification, and epidemiology of obesity NCBI bookshelf. A service of the National Library of Medicine, National Institutes of Health
12. Elena, Clobica, Lacramioara et al (2014) The relevance body mass index on the oxidative stress status of Alzheimer's disease pathology. AnaleleŞtiinţifice ale Universităţii, AlexandruIoanCuza, SecţiuneaGeneticăşiBiologieMoleculară, TOM XV
13. Zampetaki A, Mayr M (2012) MicroRNA in vascular and metabolic disease. Circ Res 110:508–522
14. Eckel RH, Kahn SE, Ferrannini E, Goldfine AB, Nathan DM, Schwartz MW et al (2011) Obesity and type 2 diabetes. What can be unified and what needs to be individualized? J Clin Endocrinol Metab 96:1654–1663
15. Paneni F, Costantino S, Cosentino F (2014) Insulin resistance, diabetes, and cardiovascular risk. Curr Atheroscler Rep 16:419

16. Archuleta TL, Lemieux AM, Saengsirisuwan V et al (2009) Oxidant stress-induced loss of IRS-1 and IRS-2 proteins in rat skeletal muscle: role of p38 MAPK. Free Radic Biol Med 47:1486–1493
17. Whitmer RA, Gustafson DR, Barrett-Connor E et al (2008) Central obesity and increased risk of dementia more than three decades later. Neurology 71:1057–1064
18. Heredia FP et al (2013) Functional foods and nutraceuticals as therapeutic tools for the treatment of diet related diseases. Can J Physiol Pharmacol 91:387–396
19. Khan R, Elhassan G, Qureshi K (2014) Nutraceuticals: in the treatment & prevention of diseases. Pharma Innov 3(10):47–50
20. Reijman M, Pols HA, Bergink AP, Hazes JM, Belo JN, Lievense AM et al (2007) Body mass index associated with onset and progression of osteoarthritis of the knee but not of the hip: the Rotterdam Study. Ann Rheum Dis 66:158–162
21. Xue JL, Ma JZ, Louis TA, Collins AJ (2001) Forecast of the number of patients with end-stage renal disease in the United States to the year 2010. J Am Soc Nephrol 12:2753–2758
22. Perneger TV, Brancati FL, Whelton PK, Klag MJ (1995) Studying the causes of kidney disease in humans: a review of methodologic obstacles and possible solutions. Am J Kidney Dis 25:722–731
23. Srivastava T (2006) Nondiabetic consequences of obesity on kidney. PediatrNephrol 21:463–470
24. Burdon KP, Fogarty RD, Shen W et al (2015) Genome-wide association study for sight-threatening diabetic retinopathy reveals association with genetic variation near the GRB2 gene. Diabetologia 58:2288–2297
25. Sabanayagam C, Yip W, Ting DS et al (2016) Ten emerging trends in the epidemiology of diabetic retinopathy. Ophthalmic Epidemiol 23:209–222
26. Zheng Y, Lamoureux EL, Lavanya R et al (2012) Prevalence and risk factors of diabetic retinopathy in migrant Indians in an urbanized society in Asia: the Singapore Indian eye study. Ophthalmology 119:2119–2124
27. Leong WB, Jadhakhan F, Taheri S et al (2016) Effect of obstructive sleep apnoea on diabetic retinopathy and maculopathy: a systematic review and meta analysis. Diabet Med 33:158–168
28. Azzoug S, Chentli F (2014) Diabetic retinopathy in acromegaly. Indian J Endocrinol Metab 18:407–409
29. Resmini E, Minuto F, Colao A, Ferone D (2009) Secondary diabetes associated with principal endocrinopathies: the impact of new treatment modalities. Acta Diabetol 46:85–95
30. Keenan JD, Fan AZ, Klein R (2009) Retinopathy in nondiabetic persons with the metabolic syndrome: findings from the Third National Health and Nutrition Examination Survey. Am J Ophthalmol 147:934–944
31. Van Leiden HA, Dekker JM, Moll AC et al (2003) Risk factors for incident retinopathy in a diabetic and nondiabetic population: the Hoorn study. Arch Ophthalmol 121:245–251
32. Gao L, Xin Z, Yuan MX et al (2016) High prevalence of diabetic retinopathy in diabetic patients concomitant with metabolic syndrome. PLoS One 11:e0145293
33. Blüher M (2016) Adipose tissue inflammation: a cause or consequence of obesity-related insulin resistance? Clin Sci (Lond) 130:1603–1614
34. De Block CE, De Leeuw IH, Van Gaal LF (2005) Impact of over weight on chronic micro vascular complications in type 1 diabetic patients. Diabetes Care 28:1649–1655
35. Heather K, Kendrick H, Jacob C et al (2012) Weight loss and obesity in the treatment and prevention of osteoarthritis
36. Lang P (2005) The works of archimedes. J Hell Stud 125:193–194
37. Schulze-Tanzil G, Hansen C, Shakibaei M (2004) Effect of a *Harpagophytum procumbens* DC extract on matrix metalloproteinase in human chondrocytes in vitro. Arzneimittelforschung 54:213–220
38. Wigler I, Grotto I, Caspi D, Yaron M (2003) The effects of Zintona EC (a ginger extract) on symptomatic gonarthritis. Osteoarthr Cartil 11:783–789

39. Boe C, Vangsness CT (2015) Fish oil and osteoarthritis: current evidence. Am J Orthop 44:302–305
40. Musumeci G, Trovato FM, Pichler K, Weinberg AM, Loreto C, Castrogiovanni P (2013) Extra-virgin olive oil diet and mild physical activity prevent cartilage degeneration in an osteoarthritis model: an in vivo and in vitro study on lubricin expression. J Nutr Biochem 24:2064–2075
41. Lieber CS, Packer L (2002) S-Adenosylmethionine: molecular, biological, and clinical aspects: An introduction. Am J Clin Nutr 76:1148–1150
42. Lippiello L, Nardo JV, Harlan R, Chiou T (2008) Metabolic effects of avocado/soy unsaponifiables on articular chondrocytes. Evid Based Complement Altern Med 5:191–197
43. Shakibaei M, Mobasheri A, Buhrmann C (2011) Curcumin synergizes with resveratrol to stimulate the MAPK signaling pathway in human articular chondrocytes in vitro. Genes Nutr 6:171–179
44. Siddiqui MZ (2011) Boswellia serrata, a potential anti-inflammatory agent: an overview. Indian J Pharm Sci 73:255–261
45. Fiebich BL, Muñoz E, Rose T, Weiss G, McGregor GP (2011) Molecular targets of the anti-inflammatory *Harpagophytum procumbens* (Devil0s claw): inhibition of TNFα and COX-2 gene expression by preventing activation of AP-1. Phytother Res 10:36–45
46. Brien S, Lewith G, Walker A, Hicks SM, Middleton D (2004) Bromelain as a treatment for osteoarthritis: a review of clinical studies. Evid Based Complement Altern Med 1:251–257
47. Semwal RB, Semwal DK, Combrinck S, Viljoen AM (2015) Gingerols and shogaols: important nutraceuticals principles from ginger. Phytochemistry 117:554–568
48. American Diabetes Association (2015) Classification and diagnosis of diabetes. Diabetes Care 39:S13–S22
49. Wong TY, Duncan BB, Golden SH et al (2004) Associations between the metabolic syndrome and retinal micro vascular signs: the Atherosclerosis Risk in Communities study. Invest Ophthalmol Vis Sci 45:2949–2954
50. Kaaks R, Lukanova A, Kurzer MS (2002) Obesity, endogenous hormones, and endometrial cancer risk: a synthetic review. CEBP 11(12):1531–1543
51. Key TJ, Pike MC (1998) The dose-effect relationship between 'unopposed' oestrogen and endometrial mitotic rate: its central role in explaining and predicting endometrial cancer. Br J Cancer 57(2):205–212
52. Verheus M, Peeters PH, Rinaldi S et al (2006) Serum C peptide levels and breast cancer risk: results from the European Prospective Investigation into Cancer and Nutrition (EPIC). Int J Cancer 119(3):659–667
53. Gordon S (2003) Alternative activation of macrophages. Nat Rev Immunol 3:23–35
54. Lumeng CN, Bodzin JL, Saltiel AR (2007) Obesity induces a phenotypic switch in adipose tissue macrophage polarization. J Clin Invest 117:175–184
55. Lu H, Boustany-Kari CM, Daugherty A, Cassis LA (2007) Angiotensin II increases adipose angiotensinogen expression. Am J Physiol Endocrinol Metab 292:1280–1287
56. Xu ZG, Lanting L, Vaziri ND et al (2005) Upregulation of angiotensin II type 1 receptor, inflammatory mediators, and enzymes of arachidonate metabolism in obese Zucker rat kidney: reversal by angiotensin II type 1 receptor blockade. Circulation 111:1962–1969
57. Kim S, Soltani-Bejnood M, Quignard-Boulange A et al (2006) The adipose Renin-Angiotensin system modulates systemic markers of insulin sensitivity and activates the intrarenal Renin-Angiotensin system. J Biomed Biotechnol 2006:27012
58. Yvan-Charvet L, Massiera F, Lamande N et al (2009) Deficiency of angiotensin type 2 receptor rescues obesity but not hypertension induced by over expression of angiotensinogen in adipose tissue. Endocrinology 150:1421–1428
59. Goodfriend TL, Ball DL, Egan BM et al (2004) Epoxy-keto derivative of linoleic acid stimulates aldosterone secretion. Hypertension 43:358–363
60. Maliakel DM, Kagiya TV, Nair CK (2008) Prevention of cisplatininduced nephrotoxicity by glucosides of ascorbic acid and alphatocopherol. Exp Toxicol Pathol 60(6):521–527

61. Piscopo S (2009) The Mediterranean diet as a nutrition education, health promotion and disease prevention tool. Public Health Nutr 12:1648–1655
62. Luciano RL (2014) Acute kidney injury from cherry concentrate in a patient with CKD. Am J Kidney Dis 63(3):503–505
63. Lu Y, Sun J, Petrova K et al (2013) Metabolomics evaluation of the effects of green tea extract on acetaminophen-induced hepatotoxicity in mice. Food ChemToxicol 62: 707–721
64. Peterson J, Dwyer J, Adlercreutz H et al (2010) Dietary lignans: physiology and potential for cardiovascular disease risk reduction. Nutr Rev 68:571–603
65. Dreher ML (2012) Pistachio nuts: composition and potential health benefits. Nutr Rev 70 (4):234–240
66. Deng Q, Liang L, Liu Q et al (2018) Autophagy is a major mechanism for the dual effects of curcumin on renal cell carcinoma cells. Eur J Pharmacol 826:24–30
67. Hernández-Reséndiz S, Correa F, García-Niño et al (2015) Cardio protection by curcumin post-treatment in rats with established chronic kidney disease. Cardiovasc Drugs Ther 29:111–120
68. Tada A, Takahashi K, Ishizuki K et al (2013) A commercial standard by quantitative NMR. Chem Pharm Bull 61:33–38
69. Lecumberri E, Dupertuis YM, Miralbell R et al (2013) Green tea polyphenols epigallocatechin-3-gallate (EGCG) as adjuvant in cancer therapy. Clin Nutr 32:894–903
70. McMahon GM, Preis SR, Hwang SJ et al (2014) Mid-adulthood risk factor profiles for CKD. J Am Soc Nephrol 25:2633–2641
71. Seller A, Chen MA, Brown BL, Fagundes CP (2018) Obesity, dietary factors, nutrition, and breast cancer risk. Curr Breast Cancer Rep. 10(1):14–27
72. Anderson AS, Key TJ, Norat T et al (2015) European Code against Cancer 4th edition: obesity, body fatness and cancer. Cancer Epidemiol 39:S34–S45
73. Kadouri L, Hubert A, Rotenberg Y et al (2007) Cancer risks in carriers of the BRCA1/2 Ashkenazi founder mutations. J Med Genet 44:467e471
74. PDQ Cancer Information Summary (2010) Genetics of breast and ovarian cancer 2012. National Cancer Institute, Bethesda, MD
75. TeMorenga L, Mallard S, Mann J (2012) Dietary sugars and body weight: systematic review and meta-analyses of randomized controlled trials and cohort studies. BMJ 15(346):7492
76. Chen D, Reierstad S, Fang F et al (2011) Jun D and Jun B integrate prostaglandin E2 activation of breast cancer-associated proximal aromatase promoters. Mol Endocrinol 25:767–775
77. Meng L, Zhou J, Sasano H et al (2001) Tumor necrosis factor alpha and interleukin 11 secreted by malignant breast epithelial cells inhibit adipocyte differentiation by selectively down-regulating CCAAT/ enhancer binding protein alpha and Peroxisome proliferators-activated receptor gamma: Mechanism of desmoplastic reaction. Cancer Res 61:2250–2255
78. Vona-Davis L, Rose DP (2007) Adipokines as endocrine, paracrine, and autocrine factors in breast cancer risk and progression. EndocrRelat Cancer 14:189–206
79. Subbaramaiah K, Howe LR, Bhardwaj P et al (2011) Obesity is associated with inflammation and elevated aromatase expression in the mouse mammary gland. Cancer Prev Res (Phila) 4:329–346
80. Simone V, Davenia M, Argentiero A et al (2016) Obesity and breast cancer: molecular interconnections and potential clinical applications. Oncologist 21:404–417
81. Abdelmagid SA, MacKinnon JL, Janssen SM, Ma DW (2016) Role of n-3 polyunsaturated fatty acids and exercise in breast cancer prevention: identifying common targets. Nutr Metab Insights 9:71–84
82. Tyler VE, Foster F (1996) Herbs and photochemical. In: Covington TR, Berardi RR, Young LL et al (eds) Handbook of non prescription Drugs. American Pharmaceutical Association, Washington DC
83. Yetley EA (2007) Multivitamins and multimineral dietary supplements: its bioavailabity and drug interactions. Am J Clin Nutr 85:269S–276S

84. Kaur G, Mukundan S, Wani V et al (2015) Nutraceuticals in the management and prevention of metabolic syndrome, nutraceuticals in the management and prevention of metabolic syndrome. Austin J Pharmacol Ther 3(1):1063
85. Merkel M, Heeren J, Dudeck W et al (2002) Inactive lipoprotein lipase (LPL) alone increases selective cholesterol ester uptake *in vivo*, whereas in the presence of active LPL it also increases triglyceride hydrolysis and whole particle lipoprotein uptake. J Biol Chem 277:7405–7411
86. Gaens KH, Stehouwer CD, Schalkwijk CG (2013) Advanced glycation end products and its receptor for advanced glycation end products in obesity. Curr Opin Lipidol 24:4–11
87. Krzystek-Korpacka M, Patryn E, Hotowy K et al (2013) Paraoxonase (PON)-1 activity in overweight and obese children and adolescents: Association with obesity-related inflammation and oxidative stress. Adv Clin Exp Med 22:229–236
88. Brillante DG, O'Sullivan AJ, Howes LG (2009) Arterial stiffness in insulin resistance: the role of nitric oxide and angiotensin II receptors. Vasc Health Risk Manag 5:73–78
89. DeMarco VG, Johnson MS, Whaley-Connell AT et al (2010) Cytokine abnormalities in the etiology of the cardio metabolic syndrome. Curr Hypertens Rep 12:93–98
90. Yang RL, Shi YH, Hao G (2008) Increasing oxidative stress with progressive hyperlipidemia in human: relation between malondialdehyde and atherogenicindex. J Clin Biochem Nutr 43:154–158
91. Ceriello A, Taboga C, Tonutti L et al (2002) Evidence for an independent and cumulative effect of postprandial hypertriglyceridemia and hyper-glycemia on endothelial dysfunction and oxidative stress generation: effects of short- and long-term simvastatin treatment. Circulation 106:1211–1218
92. Parthasarathy S, Raghavamenon A, Garelnabi MO et al (2010) Oxidized low-density lipoprotein. Methods Mol Biol 610:403–417
93. Nishimura I, Manabe M (2007) Adipogenesis in obesity requires close interplay between differentiating adipocytes, stromal cells, and blood vessels. Diabetes 56:1517–1526
94. Gaens KH, Stehouwer CD, Schalkwijk CG (2013) Advanced glycation end products and its receptor for advanced glycation end products in obesity. Curr Opin Lipidol 24:4–11
95. Touyz RM (2004) Reactive oxygen species, vascular oxidative stress, and redox signaling in hypertension: what is the clinical significance? Hypertension 44:248–252
96. Nagae A, Fujita M, Kawarazaki H et al (2009) Sympatho excitation by oxidative stress in the brain mediates arterial pressure elevation in obesity-induced hypertension. Circulation 119:978–986
97. Zhang X, Dong F, Ren J, Driscoll ML, Culver B (2005) High dietary fat induces NADPH oxidase-associated oxidative stress and inflammation in rat cerebral cortex. Exp Neurol 191:318–225
98. Ferrante AW (2007) Obesity-induced inflammation: A metabolic dialogue in the language of inflammation. J Intern Med 262
99. Canoy D (2008) Distribution of body fat and risk of coronary heart disease in men and women. Curr Opin Cardiol 23(6):591–598
100. Samad F, Yamamoto K, Pandey M et al (1997) Elevated expression of transforming growth factor-β in adipose tissue from obese mice. Mol Med 3(1):37–48
101. Weisberg SP, McCann D, Desai M et al (2003) Obesity is associated with macrophage accumulation in adipose tissue. J Clin Investig 112(12):1796–1808
102. Cinti S, Mitchell G, Barbatelli G et al (2005) Adipocyte death defines macrophage localization and function in adipose tissue of obese mice and humans. J Lipid Res 46 (11):2347–2355
103. Rodríguez A, Catalan V, Gómez-Ambrosi J, Fruhbeck G (2007) Visceral and subcutaneous adiposity: are both potential therapeutic targets for tackling the metabolic syndrome? Curr Pharm Des 13(21):2169–2175
104. Halaas JL, Gajiwala KS, Maffei M et al (1995) Weight-reducing effects of the plasma protein encoded by the obese gene. Science 269(5223):543–546

105. Bobbert T, Rochlitz H, Wegewitz U et al (2005) Changes of adiponectin oligomer composition by moderate weight reduction. Diabetes 54(9):2712–2719
106. Calabro P, Chang DW, Willerson JT, Yeh ETH (2005) Release of C-reactive protein in response to inflammatory cytokines by human adipocytes: linking obesity to vascular inflammation. J Am Coll Cardiol 46(6):1112–1113
107. Roth GA, Huffman MD, Moran AE et al (2015) Global and regional patterns in cardiovascular mortality from 1990 to 2013. Circulation 132:1667–1678
108. O'Keeffe C, Kabir Z, O'Flaherty M et al (2013) Modeling the impact of specific food policy options on coronary heart disease and stroke deaths in Ireland. BMJ Open 3(7):e002837
109. Ranzato E, Martinotti S, Calabrese CM, Calabrese G (2014) Role of nutraceuticals in cancer therapy. J Food Res 3(4):18
110. Khan RA, Elhassan GO, Qureshi KA (2014) Nutraceuticals: in the treatment & prevention of diseases-an overview. Pharma Innov J 3(10):47–50
111. Pena AS, Wiltshire E, MacKenzie K et al (2006) Vascular endothelial and smooth muscle function relates to body mass index and glucose in obese and nonobese children. J Clin Endocrinol Metab 91(11):4467–4471
112. Makimattila S et al (1996) Chronic hyperglycemia impairs endothelial function and insulin sensitivity via different mechanisms in insulin-dependent diabetes mellitus. Circulation 94 (6):1276–1282
113. Abdollahi M et al (2003) Obesity: risk of venous thrombosis and the interaction with coagulation factor levels and oral contraceptive use. Thromb Haemost 89(3):493–498
114. Libby P (2002) Inflammation in atherosclerosis. Nature 420(6917):868–874
115. Rott D et al (2003) IL-6 is produced by splenocytes derived from CMV-infected mice in response to CMV antigens, and induces MCP-1 production by endothelial cells: a new mechanistic paradigm for infection induced atherogenesis. Atherosclerosis 170(2):223–228
116. Davy KP (2004) The global epidemic of obesity: are we becoming more sympathetic? Curr Hypertens Rep 6(3):241–246
117. Cassis LA et al (1996) Characterization and regulation of angiotensin II receptors in rat adipose tissue. Angiotensin receptors in adipose tissue. Adv Exp Med Biol 396:39–47
118. Sjostrom L et al (2004) Lifestyle, diabetes, and cardiovascular risk factors 10 years after bariatric surgery. N Engl J Med 351(26):2683–2693
119. Mittendorfer Bettina, Peterson Linda R (2008) Cardiovascular consequences of obesity and targets for treatment. Drug Discov Today Strateg 5(1):53–61
120. Nikhra V (2018) Nutraceuticals for improving cardiovascular health and prognosis in cardiovascular disease. Res Gate
121. Whitmer RA, Gunderson EP, Barrett-Connor E et al (2005) Obesity in middle age and future risk of dementia: a 27 year longitudinal population based study. BMJ 330:1360–1362
122. Keage HAD, Gupta S, Brayne C (2011) Risk for dementia and age at measurement. Int J Geriatr Psychiatry 26:329–330
123. Kloppenborg RP, van den Berg E, Kappelle LJ et al (2008) Diabetes and other vascular risk factors for dementia: which factor matters most? A systematic review. Eur J Pharmacol 585:97–108
124. Szelenyi J, Vizi ES (2007) The catecholamine-cytokine balance: interaction between the brain and the immune system. Ann N Y Acad Sci 1113:311–324
125. Watkins LR, Maier SF, Goehler LE (1995) Cytokineto-brain communication: a review & analysis of alternative mechanisms. Life Sci 57(11):1011–1026
126. Wauman J, Tavernier J (2011) Leptin receptor signaling: pathways to leptin resistance. Front Biosci 17:2771–2793
127. Schwartz MW, Porte D (2005) Diabetes, obesity, and the brain. Science 307(5708):375–379
128. Van Cauwenberghe C, Van Broeckhoven C, Sleegers K (2015) Te genetic landscape of Alzheimer disease: clinical implications and perspectives. Genetics in Medicine 18(5):421–430

129. Jedrziewski MK, Ewbank DC, Wang H, Trojanowski JQ (2014) Te impact of exercise, cognitive activities, and socialization on cognitive function: results from the national long-term care survey. Am J Alzheimer Dis Other Dementias 29(4):372–378
130. Kim H, Kim G, Jang W, Kim SY, Chang N (2014) Association between intake of B vitamins and cognitive function in elderly Koreans with cognitive impairment. Nutr J 13(1):1–11
131. Sun Y, Lu C-J, Chien K-L, Chen S-T, Chen R-C (2007) Efficacy of multivitamin supplementation containing vitamins b6 and b12 and folic acid as adjunctive treatment with a cholinesterase inhibitor in Alzheimer's disease: a 26-week, randomized, double-blind, placebo-controlled study in Taiwanese Patients. Clin Ther 29(10):2204–2214
132. Ford AH, Flicker L, Alfonso H et al (2010) Vitamins B12, B6, and folic acid for cognition in older men. Neurology 75(17):1540–1547
133. Petersen RC, Tomas RG, Grundman M et al (2005) Vitamin E and donepezil for the treatment of mild cognitive impairment. Te N Engl J Med 352(23):2379–2388
134. Arlt S, Muller-Tomsen T, Beisiegel U, Kontush A (2012) Effect of one-year vitamin C- and E-supplementation on cerebrospinal fluid oxidation parameters and clinical course in Alzheimer's disease. Neurochem Res 37(12):2706–2714
135. Gangwar AK, Rawat A, Tiwari S et al (2015) Role of vitamin-D in the prevention and treatment of Alzheimer's disease. Indian J Physiol Pharmacol 59(1):94–99
136. Yoo DY, Choi JH, Kim W et al (2013) Effects of luteolin on spatial memory, cell proliferation, and neuroblast differentiation in the hippocampal dentate gyrus in a scopolamine-induced amnesia model. Neurol Res 35(8):813–820
137. Baur JA, Sinclair DA (2006) Therapeutic potential of resveratrol: the in vivo evidence. Nat Rev Drug Discovery 5(6):493–506
138. Casamenti F, Grossi C, Rigacci S et al (2015) Oleuropein aglycone: a possible drug against degenerative conditions. In vivo evidence of its effectiveness against Alzheimer's disease. J Alzheimer Dis 45(3):679–688
139. Hu N, Yu J-T, Tan L et al (2013) Nutrition and the risk of Alzheimer's disease. Bio Med Res Int 12

Index

P. S. Tappia et al. (eds.), *Pathophysiology of Obesity-Induced Health Complications*, Advances in Biochemistry in Health and Disease 19,
https://doi.org/10.1007/978-3-030-35358-2

The manufacturer's authorised representative in the EU is Springer Nature Customer Service Centre GmbH, Europaplatz 3, 69115 Heidelberg, Germany. If you have any concerns regarding our products, please contact ProductSafety@springernature.com

Printed and bound by CPI Group (UK) Ltd, Croydon, CR0 4YY
27/07/2026
02177370-0002